Financial Engineering
with Finite Elements

For other titles in the Wiley Finance series
please see www.wiley.com/Finance

Financial Engineering with Finite Elements

Jürgen Topper

WILEY

Other Wiley Editorial Offices

John Wiley & Sons Inc., 111 River Street, Hoboken, NJ 07030, USA

Jossey-Bass, 989 Market Street, San Francisco, CA 94103-1741, USA

Wiley-VCH Verlag GmbH, Boschstr. 12, D-69469 Weinheim, Germany

John Wiley & Sons Australia Ltd, 33 Park Road, Milton, Queensland 4064, Australia

John Wiley & Sons (Asia) Pte Ltd, 2 Clementi Loop #02-01, Jin Xing Distripark, Singapore 129809

John Wiley & Sons Canada Ltd, 22 Worcester Road, Etobicoke, Ontario, Canada M9W 1L1

Wiley also publishes its books in a variety of electronic formats. Some content that appears
in print may not be available in electronic books.

Library of Congress Cataloging-in-Publication Data

Topper, Jürgen.
    Financial engineering with finite elements / by Jürgen Topper.
      p. cm. – (Wiley finance series)
    Includes bibliographical references (p.) and index.
    ISBN 0-471-48690-6 (cloth : alk. paper)
    1. Financial engineering—Econometric models.   2. Finite element method.   I. Title.   II. Series.
    HG176.7.T66 2005
    658.15′5′015195—dc22   2004022228

British Library Cataloguing in Publication Data

A catalogue record for this book is available from the British Library

ISBN 0-471-48690-6 (HB)

Typeset in 10/12pt Times and Helvetica by TechBooks, New Delhi, India
Printed and bound in Great Britain by Antony Rowe Ltd, Chippenham, Wiltshire
This book is printed on acid-free paper responsibly manufactured from sustainable forestry
in which at least two trees are planted for each one used for paper production.

To Anne

Contents

Preface

This book delineates in great detail how to employ a numerical technique from engineering called *Finite Elements* in financial problems.[1] This method has been developed to solve problems from engineering which can be formulated as differential equations. In modern finance, and to a lesser degree in modern economics in general, many practical problems can be cast into the framework of differential equations, so that numerical techniques from engineering can be adapted.

This book would not have been possible without the many discussions with colleagues at d-fine and many of d-fine's clients. The people who have contributed to this book are too numerous to list. I am highly indebted to Hans-Peter Deutsch for setting up the *Financial and Commodity Risk Consulting* group within Andersen Germany, which ultimately resulted in founding d-fine. The terrific atmosphere in this group has shaped my professional and research interests since 1997.

This book is the result of various lectures given at universities (Bielefeld, Cologne, Frankfurt, Hanover, Michigan–Ann Arbor, Venice) and conferences: SIAM 1999 (Society for Industrial and Applied Mathematics) in Atlanta, CEF99 (Computing in Economics and Finance) in Boston, SOR99 (Society for Operations Research) in Magdeburg, Young Economists' Conference 2000 in Oxford, and RISK Mathweek 2001 in New York, Computational Challenges in Mathematical Finance 2002 at Field's Institute, Toronto, and Frankfurt MathFinance Workshop 2002.

---

[1] Throughout this book, *Finite Element(s)* and *Finite Element Method* will be abbreviated as FE and FEM, respectively. Analogously, *Finite Differences* and *Finite Difference Method* will be abbreviated as FD and FDM. *Partial differential equation* and *ordinary differential equation* will be abbreviated as PDE and ODE.

List of Symbols

The notation used in this book is kept as close as possible to Paul Wilmott's classic textbooks (Wilmott, 1998, 2000). Since both books do not have an index, a compilation of symbols can be found below. Since some topics treated here are not covered in *Derivatives*, some extra symbols are needed.

Uppercase letters

$C = C(Q)$	Total cost function
D	Dividends
E	Exercise price
$F(\cdot)$	Cumulative normal distribution
Q	Demand
P	Price
S	Value of an equity
S_{u}	Position of an up-barrier (where necessary, also $S_{\mathrm{u}}^{\mathrm{out}}$ and $S_{\mathrm{u}}^{\mathrm{in}}$ are used)
S_{d}	Position of a down-barrier (where necessary, also $S_{\mathrm{d}}^{\mathrm{out}}$ and $S_{\mathrm{d}}^{\mathrm{in}}$ are used)
V	Value of a generic option

Lowercase letters

f	Density function of normal distribution
r	Interest rate (continuously compounded)
p	Penalty function
q	Control function (in dynamic optimization problems)
q^*	Optimal control
v	Variance ($v = \sigma^2$)

Uppercase Greek letters

Δ	$\frac{\partial V}{\partial S}$
Γ	$\frac{\partial^2 V}{\partial S^2}$
Γ_Ω	Boundary of Ω
Θ	$\frac{\partial V}{\partial t}$
Π	Portfolio
Ω	Set of spatial variables on which a PDE is defined

$\bar{\Omega}$	Closure of Ω (i.e. $\bar{\Omega} = \Omega \cup \Gamma_\Omega$)
Σ	Matrix of correlation coefficients ρ_{ij}

Lowercase Greek letters

σ	Volatility
ε	Elasticity
α	Length of period (as a fraction of one year)
λ	Market price of risk
τ	Time to maturity: $\tau = T - t$
μ	Drift in an Itô process
π	Profit
ρ_{ij}	Coefficient of correlation between underlying i and underlying j

Miscellaneous

€	EURO
$[x^{\min}; x^{\max}]$	Closed interval: $\{x \mid x^{\min} \le x \le x^{\max}\}$
$(x^{\min}; x^{\max})$	Open interval: $\{x \mid x^{\min} < x < x^{\max}\}$
$a \approx b$	a approximately equals b
$\mathbb{R}^n$	Set of real numbers in n-dimensional space
$\ln(\cdot)$	*Logarithmus naturalis*
u_x	Partial derivative of u with respect to x: $\frac{\partial u}{\partial x}$
u_{xx}	Second partial derivative of u with respect to x: $\frac{\partial^2 u}{\partial x^2}$

Models and various acronyms

BC	Boundary condition
FC	Final condition
FD	Finite differences
FDM	Finite difference method
FE	Finite elements
FEM	Finite element method
IC	Initial condition
LHS	Left-hand side
MC	Monte Carlo method
MWR	Method of weighted residuals
ODE	Ordinary differential equation
PDE	Partial differential equation
PIDE	Partial integro-differential equation
RHS	Right-hand side
SDE	Stochastic differential equation
USD	US Dollar

Part I
Preliminaries

1

Introduction

The raw material of banking is not money but risk. But what is risk? Here, we follow the definition of Frank H. Knight who distinguished between decisions under uncertainty and decisions under risk (Knight, 1921). For the latter a statistical distribution of the outcome is known, for the former it is not. The art of financial engineering is to customize risk. Financial engineering is based on certain assumptions regarding the statistical behavior of equities, exchange rates and interest rates. These assumptions allow the formulation of models. Many of these models can be expressed as differential equations. Differential equations have been studied for some centuries by mathematicians, physicists and engineers, so that a great deal of knowledge from these areas is available. Techniques for finding approximate solutions for differential equations arising in finance is the topic of this book. We will concentrate on a technique called Finite Elements (FE). During the last couple of years, the Method of Finite Elements (FEM) has been put forward through a number of academic papers focusing on option pricing. Here, we want to deliver a survey of methods which belong to the rather large family of FE methods. All these papers present different versions of the *Galerkin* FEM, one of the Weighted Residual Methods. In addition to the Galerkin FEM, we will also deal with other FEMs. We will especially highlight the *collocation* FEM, which offers some advantages for the boundary conditions.

Most textbooks on FE have a mechanical origin which is based on basic physical principles. For economic applications, this approach would be quite awkward. A minority of textbooks on FE analysis (for example Bickford, 1990; Burnett, 1987; Sewell, 1985 and White, 1985) start with differential equations but put emphasis solely, or mainly, on so-called *self-adjoint* differential equations[1] and their parabolic counterparts. For this class of differential equations, error bounds can be derived. Error bounds for differential equations which are not self-adjoint are usually technically involved and often not tight. Unfortunately, most differential equations arising in finance are not self-adjoint. There are, however, approaches to see whether the numerical solution is converging to the true solution, as we will demonstrate with various examples.

This book aims to explain this method without being too rigorous in a mathematical sense. In order to be self-contained, other popular methods from the finite differences (FD) family are also discussed. The book provides many, many examples. This is due to the author's experience that examples are the best motivation. Many examples are computer-based, either as a demonstration or as an exercise. Many models are delineated in great detail. This is a rather uncommon feature in the literature of this field. Often, details necessary for implementing pricing models are not even mentioned there. For example, most papers in option pricing on some kind of PDE-based method do not give all boundary conditions. In engineering and natural sciences this would be considered a big drawback, since it makes the problem irreproducible by the reader. He can only guess what kind of boundary condition has been employed by the author.

---

[1] For a definition of *self-adjoint* or *symmetric* ODEs see Section A.3.3.

Although the two greatest inventions of post-war finance, CAPM and risk-neutral pricing, have been invented in economics departments, the techniques employed in today's banking world are beyond most students of economics and management sciences curriculae. Many of these techniques, however, are familiar to mathematicians and physicists, who have been hired by banks, brokers and consulting companies in large numbers. Since serious students of economics and management science cannot evade ordinary differential equations in the form of initial value problems, we take them as a starting point.

The book starts in Chapter 2 with a collection of prototype models from various areas of modern finance, which represent different types of differential equations. Chapter 3 briefly discusses FD. The FDM is heavily employed in economics, and especially finance. Studying FD is important from two perspectives: first, FD represent the working horse in numerical finance, and the power of FE can only be appreciated with some knowledge of FD. Secondly, FE discretizations of time-dependent PDEs usually give a system of ODEs which, in turn, is solved with FD. Part II (Chapters 4 to 9) studies FE solutions of linear differential equations. FE are most easily explained in the context of ordinary boundary value problems. Accepting this fact, we start with some simple models from finance of this type and add time as a second variable in Chapter 5. In Chapter 6, we extend the analysis to two spatial variables. Again, we first discuss static problems in Chapter 6 and later allow time in Chapter 7. In Chapter 8 and 9, the analysis is extended to three spatial variables and three spatial variables with time, respectively. For 1D problems, both the Galerkin Finite Element method and the Collocation method are delineated. Both methods can be extended to higher dimensions; however, due to space limitation, we have chosen to discuss the Galerkin method for 2D problems and the collocation method for 3D problems. While the FE discretization of a linear *static* problem reduces the problem to solving a linear system, *linear* dynamic problems lead to *linear* systems of ODEs. In Chapter 10, we discuss nonlinear differential equations. The basic message of these sections of the book is that the FE discretization of a static nonlinear problem leads to a nonlinear system. The FE discretization of a nonlinear dynamic problem leads to a system of nonlinear ODEs. Part IV is intended to make the book self-contained. The most important concepts from stochastics, analysis and linear algebra relevant to this book are introduced. Last but not least, there is a quick introduction to PDE2D, the software with which all examples in this book have been computed.

Before going into the details of the FEM, we want to outline succinctly its advantages:

- A solution for the entire domain is computed, instead of isolated nodes as in the case of FD.
- The boundary conditions involving derivatives are difficult to handle with FD. Neumann conditions, however, are often easier to obtain than Dirichlet conditions when estimating the behavior of the option as the price of the underlying goes to infinity. FE techniques can incorporate boundary conditions involving derivatives easily.
- FE can easily deal with high curvature. In most FE codes, this is achieved by adaptive remeshing, a technique well developed in theory and in practice.
- The irregular shapes of the PDE's domain can easily be handled, while in an FD setting, the placing of the gridpoints is difficult. These irregular domains arise naturally when knock-out barriers are imposed on a multiple-asset option. Irregular shapes can also arise when only parts of the PDE's domain are to be approximated numerically, because some parts can be determined by financial reasoning.

- Most academic papers are concerned with pricing only, while most practitioners are at least as much interested in measures of sensitivity to those prices. Some of these measures of sensitivity, commonly called Greeks, can be obtained more exactly with FE.
- Many FE codes (such as PDE2D, used for this book) allow local refinement. This allows precise local information without having to solve the problem with higher accuracy on the entire domain. PDE2D also employs adaptive remeshing. This feature automatically leads to local refinement in areas of high curvature, for example, near to the strike price or close to the barrier.
- FE can easily be combined with infinite or boundary elements for the treatment of (semi-) infinite domains. This is common practice in engineering while in finance usually artificial BC are introduced.

These advantages come at the cost of a more complicated method compared to FD. In the rest of the book, a detailed, but also easy-to-read, presentation of the FEM will be delivered. Many examples demonstrate the usefulness of FE for financial problems.

2
Some Prototype Models

2.1 OPTIMAL PRICE POLICY OF A MONOPOLIST

Dynamic optimization of a monopolist by means of the calculus of variations was the subject of two famous papers by G. C. Evans (1924) and G. Tintner (1937). Their models entered textbooks such as Evans (1930) and Chiang (1992), on which this section is based.

The starting point is a monopolistic company producing a single asset with a total cost function

$$C = C(Q) \tag{2.1}$$

The output Q cannot be stored, so it needs to be sold immediately after production. Demand is supposed to be a function not only of the price P, but also of the price change $\dot{P}$ in time.

$$Q = Q(P, \dot{P}) \tag{2.2}$$

By definition, profit is revenues minus costs:

$$\pi_{\text{static}} \equiv \pi = PQ - C \tag{2.3}$$

The company wants to maximize its total profit $\pi_{\text{dynamic}} = \int_0^T \pi_{\text{static}} \mathrm{d}t$ over a finite time horizon $[0, T]$. The current price of the product $P(0)$ is obviously known, while the future price $P(T)$ at the end of the fixed time horizon is set by the company. Profit maximization can be formulated as a variational problem:

$$\max \ \pi_{\text{dynamic}} = \int_0^T \pi_{\text{static}}(P, \dot{P}) \, \mathrm{d}t \tag{2.4}$$

subject to:

$$P(0) = P_0 \tag{2.5}$$
$$P(T) = P_T \tag{2.6}$$

The necessary first order condition for this optimization problem is the Euler equation:

$$\pi_{\dot{P}\dot{P}} \ddot{P} + \pi_{P\dot{P}} \dot{P} + \pi_{t\dot{P}} - \pi_P = 0 \tag{2.7}$$

Here, the Euler equation has a clearcut economic interpretation, which becomes obvious after some manipulations. First, since π is not a function of t, the term $\pi_{t\dot{P}}$ drops out. Multiplying Equation (2.7) by $\dot{P}$ leads to:

$$
\begin{aligned}
0 &= \pi_{\dot{P}\dot{P}} \ddot{P} \dot{P} + \pi_{P\dot{P}} \dot{P}\dot{P} - \pi_P \dot{P} \\
&= \pi_{\dot{P}\dot{P}} \ddot{P}\dot{P} + \pi_{P\dot{P}} \dot{P}\dot{P} - \pi_P \dot{P} + \pi_{\dot{P}} \ddot{P} - \pi_{\dot{P}} \ddot{P} \\
&= \underbrace{\pi_{\dot{P}} \ddot{P} + \dot{P}\left(\pi_{\dot{P}\dot{P}} \ddot{P} + \pi_{P\dot{P}} \dot{P}\right)}_{\frac{\mathrm{d}}{\mathrm{d}t} \dot{P}\pi_{\dot{P}}} - \underbrace{\left(\pi_P \dot{P} + \pi_{\dot{P}} \ddot{P}\right)}_{\frac{\mathrm{d}}{\mathrm{d}t} \pi} \\
&= \frac{\mathrm{d}}{\mathrm{d}t}\left(\dot{P}\pi_{\dot{P}} - \pi\right) \tag{2.8}
\end{aligned}
$$

This is equivalent to:

$$\pi - \dot{P}\pi_{\dot{P}} = \text{constant} \qquad (2.9)$$

The economic interpretation is that the (constant) dynamic monopoly profit can be separated into two parts: the static monopoly profit, π, and the adjustment for the dynamic model given by $\dot{P}\pi_{\dot{P}}$.

In order to actually compute the optimal price path over time, one has to specify the total cost function and the demand function. One possibility, as chosen by Evans (1924), is to assume a quadratic total cost function and a linear demand function:

$$C = \alpha Q^2 + \beta Q + \gamma \qquad (\alpha, \beta, \gamma > 0) \qquad (2.10)$$

$$Q = a - bP(t) + h\dot{P}(t) \qquad (a, b > 0; h \neq 0) \qquad (2.11)$$

To construct the Euler equation, one needs expressions for several derivatives of π:

$$\pi = PQ - C$$
$$= P(a - bP + h\dot{P}) - \alpha(a - bP + h\dot{P})^2 - \beta(a - bP + h\dot{P}) - \gamma \qquad (2.12)$$

$$\pi_P = -2b(1 + ab)P + (a + 2\alpha ab + \beta b) + h(1 + 2\alpha b)\dot{P} \qquad (2.13)$$

$$\pi_{\dot{P}} = -2\alpha h^2 \dot{P} - h(2\alpha a + \beta) + h(1 + 2\alpha b)P \qquad (2.14)$$

$$\pi_{\dot{P}\dot{P}} = -2\alpha h^2 \qquad (2.15)$$

$$\pi_{P\dot{P}} = h(1 + 2\alpha b) \qquad (2.16)$$

$$\pi_{t\dot{P}} = 0 \qquad (2.17)$$

After some more manipulation, one ends up with the following Euler equation:

$$\ddot{P} - \underbrace{\frac{b(1 + \alpha b)}{\alpha h^2}}_{c_1} P = -\underbrace{\frac{\alpha + 2\alpha ab + \beta b}{2\alpha h^2}}_{c_2} \qquad (2.18)$$

This is a *linear second order ordinary differential equation* which, together with the boundary conditions Equations (2.5) and (2.6), constitutes a *linear two-point boundary value problem*. Linearity results from the choice of the total cost function and demand function. More general cost and demand functions lead to *nonlinear two-point boundary value problems*.

2.2 THE BLACK–SCHOLES OPTION PRICING MODEL

Before the seminal work of Black, Scholes and Merton (Black and Scholes, 1972; Black and Scholes, 1973 and Merton, 1973) the economic literature provided two approaches which are, in principle, capable of pricing options:

1. The insurance approach: The fair value of an option V is its *expected value* plus some risk premium. The risk premium depends on the risk aversion of the writer of the option.
2. The equilibrium approach: Numerous rational investors concurrently maximize their respective utility functions. Some of them write options, others buy options. With some assumptions being met, the existence of a unique price for options can be assured. However, to actually compute this price, the investors' utility functions need to be specified.

Both approaches suffer from the fact that detailed knowledge about utility functions needs to be available. Black, Scholes and Merton showed how to price options without referring to utility functions, i.e. prices derived in this framework do not depend on the investor's attitude towards risk. They are compatible with *any* utility function. There are numerous derivations of the Black–Scholes model; here, we will concentrate on an informal derivation of the pricing PDE based on Hull (2000), Ritchken (1996) and Wilmott *et al.* (1996). First, we have to make some assumptions:

1. The underlying follows a geometric Brownian motion

$$dS = \mu S dt + \sigma S dX \tag{2.19}$$

2. The interest rate for riskfree loans, r, is constant.
3. The underlying does not pay any dividends.
4. There is continuous trading without transaction costs. Trading is, by its very nature, a discrete action: at some point in time, a certain number of stocks is bought. This number usually is an integer or a fraction of integers. However, since there are usually many market participants, this assumption seems justifiable. The absence of transaction costs just simplifies the model.
5. The main assumption is that there are no arbitrage opportunities. This is the very heart of the model.
6. Short selling is allowed.

Most of these assumptions can be dropped, or at least relaxed, with more complicated models, as will be shown in later chapters. At this point we will start to outline the seminal approach by Black, Scholes and Merton. Suppose that the price of an equity S follows Equation (2.19). Considering a call option, it is clear that its value V depends on both S and t. By Itô's Lemma compare Appendix B.3.6

$$dV = \left(\frac{\partial V}{\partial S}\mu S + \frac{\partial V}{\partial t} + \frac{1}{2}\sigma^2 S^2 \frac{\partial^2 V}{\partial S^2}\right) dt + \frac{\partial V}{\partial S}\sigma S dX \tag{2.20}$$

Both dV and dS are infinitesimal changes over the time interval dt. For a small, but not infinitesimal, time interval Δt, the expressions change to:

$$\Delta S = \mu S \Delta t + \sigma S \Delta X \tag{2.21}$$

$$\Delta V = \left(\frac{\partial V}{\partial S}\mu S + \frac{\partial V}{\partial t} + \frac{1}{2}\frac{\partial^2 V}{\partial S^2}\right) \Delta t + \frac{\partial V}{\partial S}\sigma S \Delta X \tag{2.22}$$

What combination of V and S makes ΔX disappear? In financial terms this means, how many options and shares do I have to go long or short in order to make the portfolio riskless for a short interval of time? The appropriate portfolio consists of one short position in a call and $\frac{\partial V}{\partial S}$ shares. Let us call the value of this portfolio Π:

$$\Pi = \frac{\partial V}{\partial S}S - V \tag{2.23}$$

A slight change $\Delta \Pi$ of this portfolio simplifies, by use of Equations (2.21) and (2.22), to:

$$\Delta \Pi = \frac{\partial V}{\partial S}\Delta S - \Delta V \tag{2.24}$$

$$= -\left(\frac{\partial V}{\partial t} + \frac{1}{2}\frac{\partial^2 V}{\partial S^2}\sigma^2 S^2\right) \Delta t \tag{2.25}$$

Since this portfolio does not involve any stochastic components any more, it must be riskless for Δt. To avoid arbitrage, this portfolio should earn interest for this short period of time:

$$\Delta \Pi = r \Pi \Delta t \tag{2.26}$$

$$= r \left(\frac{\partial V}{\partial S} S - V \right) \Delta t \tag{2.27}$$

Equating both expressions for $\Delta \Pi$ gives us:

$$- \left(\frac{\partial V}{\partial t} + \frac{1}{2} \frac{\partial^2 V}{\partial S^2} \sigma^2 S^2 \right) \Delta t \qquad = r \left(\frac{\partial V}{\partial S} S - V \right) \Delta t \tag{2.28}$$

$$\Longleftrightarrow \qquad \frac{\partial V}{\partial t} + \frac{1}{2} \frac{\partial^2 V}{\partial S^2} \sigma^2 S^2 \qquad = r \left(V - \frac{\partial V}{\partial S} S \right) \tag{2.29}$$

$$\Longleftrightarrow \qquad \frac{\partial V}{\partial t} + \sigma^2 S^2 \frac{1}{2} \frac{\partial^2 V}{\partial S^2} + r S \frac{\partial V}{\partial S} - r V = 0 \tag{2.30}$$

This is the celebrated Black–Scholes–Merton PDE. It is a linear parabolic PDE backwards in time. The boundary conditions and the final conditions depend on the financial product at hand. For a European call, the final condition is:

$$V(T) = \max(S - E, 0) \tag{2.31}$$

Before solving this problem, it is useful to investigate whether a unique and stable solution exists. If such a solution exists, the PDE problem is *well-posed*. It turns out that the well-posedness of this problem can be assured by the theorems given in Section A.3.5. It is possible to derive an analytical solution, which can be expressed as:

$$V_{\text{call}} = S N(d_1) - E \exp\left[-r(T - t)\right] N(d_2) \tag{2.32}$$

$$d_1 = \frac{\ln(S/E) + (r + \sigma^2/2)(T - t)}{\sigma \sqrt{T - t}} \tag{2.33}$$

$$d_2 = \frac{\ln(S/E) + (r - \sigma^2/2)(T - t)}{\sigma \sqrt{T - t}} = d_1 - \sigma \sqrt{T - t} \tag{2.34}$$

with $N(\cdot)$ representing the standardized cumulative normal distribution, which has to be approximated either by analytical approximations or numerically. At this point, we can observe that the analytical solution of a plain vanilla European call still requires approximations. This is the usual case: solutions of parabolic PDEs involve the computation of infinite series and/or indefinite integrals. In this case, this is not a big problem since the normal distribution has been studied for centuries and several simple approximations for its cumulative function are available. Besides, analytical solutions are often difficult to find. This has led some authors to suggest *not* looking for analytical solutions, but solving the pricing PDE numerically instead of approximating some hard-to-find analytical expression.

2.3   PRICING AMERICAN OPTIONS

In contrast to the European options in Section 2.2, American options can be exercised by the holder at any time. In practice, this is sometimes somewhat restricted to certain days or certain time periods. This type is usually called a *Bermudan* option, denoting that it is somewhere between European and American exercise, and alluding to the fact that the point of optimal

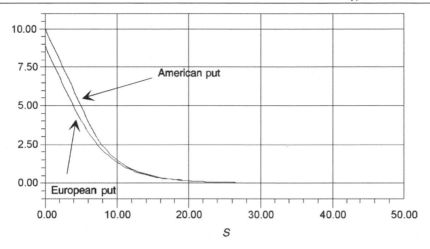

Figure 2.1 Premium of an American and a European put

exercise is often difficult to determine. Bermudan options are discussed further in Section 5.2.1. This type is the most typical case in the equity market. Banks usually allow the exercise of American options only during their business hours, which constitute the time window. Options exercised later will be processed on the next business day. In the OTC market, time windows like the first three business days each month are common. Here, we will ignore such market practice and assume that these options are literally exercisable anytime. Early exercise makes the option premium more experisive, as shown in Figure 2.1 for a put.

Within the FE setting, several methods have been suggested to incorporate early exercise. Here, we will turn our attention to a *penalty approach* introduced by Zvan *et al.* (1998a, 1998b). This method will be explained with the help of an American put; a more general outline can be found in Zvan *et al.* (1998b). It has been explained previously that for American options it can be sometimes optimal to exercise the option early. So we have two regions: one in which it is optimal to exercise early; and another where it is optimal not to exercise. The former region can be characterized by the following:

$$V = \max(E - S, 0) \tag{2.35}$$

$$\frac{\partial V}{\partial t} + \frac{1}{2}\sigma^2 S^2 \frac{\partial^2 V}{\partial S^2} + (r - D)S \frac{\partial V}{\partial S} - rV < 0 \tag{2.36}$$

The early exercise boundary is a function of time, so the PDE is defined on a domain which varies with time. Problems of this kind are called *moving boundary problems*.[1] When solving problems of this kind, one needs to determine both the moving boundary and the solution of the PDE. Equation (2.35) denotes the amount of money that goes to the holder of the option when he exercises early. Since we have presupposed that this amount is higher than without early exercise, it must have a higher value than denoted by the Black–Scholes PDE. This fact is expressed in Inequality (2.36). When early exercise is not optimal, the price of the put satisfies

---

[1] Sometimes, problems of this kind are also called *free boundary problems*, although this term originally only referred to *elliptic* problems (Crank, 1984). Nowadays, both terms, *moving boundary problems* and *free boundary problems*, are used interchangeably, probably based on the fact that techniques for solving free boundary problems have been successfully employed to solve moving boundary problems.

the Black–Scholes PDE and it is higher than its intrinsic value. This is expressed by:

$$V > \max(E - S, 0) \tag{2.37}$$

$$\frac{\partial V}{\partial t} + \frac{1}{2}\sigma^2 S^2 \frac{\partial^2 V}{\partial S^2} + (r - D)S\frac{\partial V}{\partial S} - rV = 0 \tag{2.38}$$

These four equations can be combined into one equation which is valid everywhere

$$\frac{\partial V}{\partial t} + \frac{1}{2}\sigma^2 S^2 \frac{\partial^2 V}{\partial S^2} + (r - D)S\frac{\partial V}{\partial S} - rV + p = 0 \tag{2.39}$$

with p having the following properties:

$$p = 0 \; \forall \; V > \max(E - S, 0) \tag{2.40}$$

$$p > 0 \; \forall \; V = \max(E - S, 0) \tag{2.41}$$

Zvan *et al.* (1998b) suggest the following penalty function

$$p = c_{\text{penalty}} \left\{ \min \left[V - \max(E - S, 0), 0 \right] \right\}^2 \tag{2.42}$$

with c_{penalty} being a large value (say 10^5), possibly depending on mesh size and size of time-step. There are other suggestions (Nielsen *et al.*, 2000, 2002):

$$p = \frac{\epsilon c}{V + \epsilon - g(S)} \tag{2.43}$$

with

$$0 < \epsilon \ll 1 \tag{2.44}$$

$$g(S) = E - S \tag{2.45}$$

$$c \leq rE \tag{2.46}$$

2.4 MULTI-ASSET OPTIONS WITH STOCHASTIC CORRELATION

A crucial input parameter to the pricing and hedging of multi-asset options is the correlation between the underlyings, which is known not to be constant, as can be seen in Figure 2.2. One way to come to a more realistic model is to assume correlation to be stochastic as well. To keep the discussion low-dimensional, we will concentrate on rainbow options, i.e. options with two underlyings, so that we have to deal with three risk factors. Correlation is supposed to follow an Ornstein–Uhlenbeck process,

$$\mathrm{d}\rho = \kappa(\mu - \rho)\mathrm{d}t + \sigma_\rho \mathrm{d}X \tag{2.47}$$

such that it is mean-reverting. The associated risk-neutral pricing PDE is:

$$\frac{\sigma_1^2 S_1^2}{2}\frac{\partial^2 V}{\partial S_1^2} + \frac{\sigma_2^2 S_2^2}{2}\frac{\partial^2 V}{\partial S_2^2} + \frac{\sigma_\rho^2}{2}\frac{\partial^2 V}{\partial \rho^2} + \sigma_1 S_1 \sigma_\rho \rho_1 \frac{\partial^2 V}{\partial \rho \partial S_1}$$

$$+ \sigma_2 S_2 \sigma_\rho \rho_2 \frac{\partial^2 V}{\partial \rho \partial S_2} + \rho \sigma_1 \sigma_2 S_1 S_2 \frac{\partial^2 V}{\partial S_1 \partial S_2} + (r - D_1)S_1 \frac{\partial V}{\partial S_1}$$

$$+ (r - D_2)S_2 \frac{\partial V}{\partial S_2} + [\kappa(\mu - \rho)]\frac{\partial V}{\partial \rho} - rV = -\frac{\partial V}{\partial t} \tag{2.48}$$

with ρ_i denoting the correlation between ρ given by Equation (2.47) and S_i.

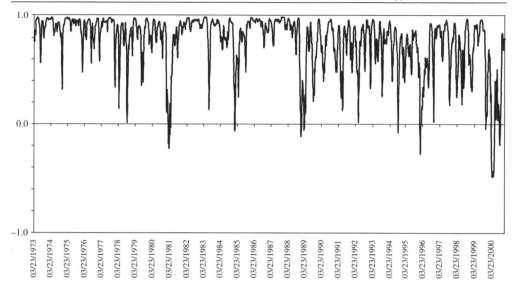

Figure 2.2   Historical 60-day correlation between Commerzbank and Dresdner Bank

Table 2.1   Popular rainbow options

| | Type of option | |
Name of product	Call	Put
Basket option	$\max[0, (Q_1 S_1 + Q_2 S_2) - E]$	$\max[0, E - (Q_1 S_1 + Q_2 S_2)]$
Option on the maximum	$\max[0, \max(Q_1 S_1, Q_2 S_2) - E]$	$\max[0, E - \max(Q_1 S_1, Q_2 S_2)]$
Option on the minimum	$\max[0, \min(Q_1 S_1, Q_2 S_2) - E]$	$\max[0, E - \min(Q_1 S_1, Q_2 S_2)]$
Dual-strike option	$\max[0, Q_1 S_1 - E_1, Q_2 S_2 - E_2]$	$\max[0, E_1 - Q_1 S_1, E_2 - Q_2 S_2]$
Reverse dual-strike option	$\max[0, Q_1 S_1 - E_1, E_2 - Q_2 S_2]$	$\max[0, E_1 - Q_1 S_1, Q_2 S_2 - E_2]$
Spread option	$\max[0, Q_1 S_1 - Q_2 S_2 - E]$	$\max[0, E - Q_1 S_1 + Q_2 S_2]$

This is a linear second order parabolic PDE in three spatial variables plus time. In order to solve this problem, boundary conditions and a final condition need to be provided. The boundary conditions depend on the product at hand. Different products also give different final conditions; the most popular rainbow options can be found in Table 2.1 (Haug, 1997; Hunziker and Koch-Medina, 1996).

The basket option is sometimes also called a portfolio option and occasionally comes with a cap, so that the payoff needs to be modified to:

$$V(S_1, S_2, T) = \min\{\text{cap}, \max[0, (Q_1 S_1 + Q_2 S_2) - E]\} \qquad (2.49)$$

The *option to exchange one asset for another* is a spread put with strike $E = 0$. Some more complicated products, such as *options on the best/worst of two assets and cash*, can be decomposed into products from Table 2.1 by techniques described in Hunziker and Koch-Medina (1996). For all these products, closed-form solutions taking stochastic correlations into account are not known.

Another interesting question comes from the model error arising from the fact that correlation is, by definition, limited to the interval $[-1, 1]$, while the density of Equation (2.47) is defined on $(-\infty, \infty)$. Obviously, one is interested in the likelihood of leaving the interval $[-1, 1]$ before t':

$$\frac{\partial C}{\partial t} = -\frac{\kappa(\mu - \rho)}{2}\frac{\partial^2 C}{\partial \rho^2} - \sigma_\rho^2 \rho \frac{\partial C}{\partial \rho} \tag{2.50}$$

$$C(t = t', \rho) = 0 \tag{2.51}$$

$$C(t, \rho = -1) = 1 \tag{2.52}$$

$$c(t, \rho = 1) = 1 \tag{2.53}$$

Another way of looking at the problem is to ask for the expected time of a current value for correlation to leave the interval $[-1, 1]$:

$$\frac{\kappa(\mu - \rho)}{2}\frac{d^2 u}{d\rho^2} + \sigma_\rho^2 \rho \frac{du}{d\rho} = -1 \tag{2.54}$$

$$u(\rho = -1) = 0 \tag{2.55}$$

$$u(\rho = 1) = 0 \tag{2.56}$$

Both differential equations are linear and of second order. Equation (2.50) is a *parabolic PDE* with incompatible data on the boundary conditions and initial condition. Equation (2.54) is a *nonhomogeneous ordinary two-point boundary value problem*.

In order to avoid the model error of correlations outside the interval $[-1, 1]$, one can make use of the following transformation:

$$d\rho = \kappa(\mu - \rho)dt + \sigma_\rho \, dX \tag{2.57}$$

$$\bar{\rho} = \frac{2}{\pi}\arctan(\rho) \tag{2.58}$$

The function $f(x) = \arctan(x)$ is a mapping from $\mathbb{R}$ to $[-\pi/2, \pi/2]$ so that the transformed variable is restricted to $[-1, 1]$.

2.5   THE STEADY-STATE DISTRIBUTION OF THE VASICEK INTEREST RATE PROCESS

Vasicek (1977) applies the idea of constructing an instantaneously riskless portfolio to bonds, in order to derive prices for fixed-income products. Although Vasicek's seminal paper derives the classical bond pricing PDE

$$\frac{dV}{dt} + \frac{w^2}{2}\frac{d^2 V}{dr^2} + u\frac{dV}{dr} - rV = 0 \tag{2.59}$$

for the more general Itô process

$$dr = u(r, t)dt + w(r, t)dX \tag{2.60}$$

the author's name is usually associated with the Ornstein–Uhlenbeck process

$$u(r, t) = (\eta - \gamma r) \tag{2.61}$$

$$w(r, t) = \beta^{\frac{1}{2}} \tag{2.62}$$

$$dr = (\eta - \gamma r)dt + \beta^{\frac{1}{2}}dX \tag{2.63}$$

which is a special case of the Itô process, for which Vasicek derived a solution to Equation (2.59) plus the associated final and boundary conditions for a zero bond. The PDE has a similar structure to the Black–Scholes PDE. Both PDEs are linear, parabolic and of second order. At this point, however, we want to turn our attention to the governing process for r in Equation (2.63). Obviously, this process is not restricted to $0 \leq r$. Since negative interest rates are rarely observed in the market, and also violate the assumption of no arbitrage in a world with the possibility to hold cash, this assumption shall be investigated in more detail. Typical questions arising during this investigation are

1. What is the distribution of r?
2. How much time will elapse before r breaches zero?

Both questions can be answered by solving ODEs, see Section B.3:

1. The steady-state distribution of $\phi_\infty(r)$ as $t \to \infty$ satisfies

$$\frac{1}{2} \frac{d^2(\beta \phi_\infty)}{dr^2} - \frac{d[(\eta - \gamma r)\phi_\infty]}{dr} = 0 \tag{2.64}$$

$$\Longleftrightarrow \qquad \frac{\beta}{2} \frac{d^2 \phi_\infty}{dr^2} - \frac{(\eta - \gamma r)d\phi_\infty}{dr} = 0 \tag{2.65}$$

$$\lim_{r \to -\infty} \phi_\infty(r) = 0 \tag{2.66}$$

$$\lim_{r \to \infty} \phi_\infty(r) = 0 \tag{2.67}$$

2. Since the amount of time it takes until r breaches some barrier is a stochastic quantity, this question can only be answered in an average sense by computing its expectation value. This is given by:

$$\frac{\beta}{2} \frac{d^2 \phi_\infty}{dr^2} + \frac{(\eta - \gamma r)d\phi_\infty}{dr} = -1 \tag{2.68}$$

$$\phi_\infty(0) = 0 \tag{2.69}$$

$$\lim_{r \to \infty} \phi_\infty(r) = 0 \tag{2.70}$$

Another question arising when studying the properties of r is: what is the probability of interest rates going beyond some unreasonable value $\bar{r}$? This question can be answered by simply changing the boundary conditions associated with Equation (2.68) to:

$$\phi_\infty(0) = 0 \tag{2.71}$$

$$\phi_\infty(\bar{r}) = 0 \tag{2.72}$$

Both ODEs are *linear two-point boundary value problems*. This makes it possible to find an analytical solution for the case of constant parameters (Wilmott, 1998, Equation (33.16)). In order to fit the current term structure, one or both parameters become dependent on time in more sophisticated models. The first ODE is *homogeneous*, while the second ODE is *nonhomogeneous* because of the term on the right-hand side. Besides this, both ODEs are *not self-adjoint*, so that we cannot expect to get tight error bounds from the numerical literature. Computing a solution with either FE or FD, however, works, as it does for self-adjoint problems. In later chapters, we will introduce tools that make this user confident that the solution computed is actually associated with the differential equation problem at hand, without being able to prove this formally.

2.6 NOTES

As can be seen from the previous section, many popular models from finance can be cast in continuous time and formulated as differential equations. Especially in the area of derivative pricing, many models based on partial differential equations have been published after the seminal work of Black, Scholes and Merton. PDEs, however, are not restricted to derivative pricing. They have not become a standard tool in economics yet, although many models from earlier decades have been formulated as PDEs. Despite some having been constructed in their own spirit, most of these models are either extensions of earlier ODE-based models enlarged by spatial variables, or applications of dynamic optimization techniques. An incomplete list of the first group comprises:

- A spatial extension of the Harrod–Domar growth model. See, for example, Puu, 1997a, Section 7.1.
- The discrete-time multiplier-accelerator model as a basis for business cycles has been formulated in continuous time by Phillips (1954). The intensity of the business cycle varies with the region. Integration of location leads to a PDE (Puu, 1997a, Section 7.2; Puu, 1997b, Chapter 7).
- Population dynamics in natural resources (Hritonenko and Yatsenko, 1999, Section 7.3).
- The seminal Lancaster (1914) model from military operations research has been extended by Chawla (1999) and Docker *et al.* (1989) to take the spatial movement of opposing forces in a battlefield into account. While the original model is a system of ODEs, the extended model is a system of parabolic PDEs (Protopopescu and Santoro, 1989).

In dynamic optimization, PDEs arise in several contexts:

- *Dynamic programming* in continuous time (Ferguson and Lim, 1998, Chapter 7 and references therein).
- *Stochastic control.* For a gentle introduction see Chapter 8 in Ferguson and Lim (1998). The pricing of passport options is based on a PDE derived from a stochastic control model, See Section 10.2.3.
- Multivariate *calculus of variations.* While applications of *uni*variate calculus of variations are numerous (Chiang, 1992; Takayama, 1985), its *multi*variate extension has so far not been widely applied to economic problems. One of the rare exceptions can be found in Chapter 3 of Puu (1997a).
- The above comment is also applicable to *optimal control.* Economic applications of optimal control mushroomed in the 1970s and 1980s. Applying Pontryagin's maximum principle to a univariate problem of optimal control leads to a two-point boundary value problem. This type of problem can be solved with techniques from Sections 3.3 and 3.4. Multivariate problems of optimal control result in PDEs.
- *Game theory.* For examples of problems from game theory leading to parabolic PDEs, see Chawla (1999) and Hofbauer and Sigmund (2003).
- *Control theory.* A modern development in this field is the extension of H^∞ control to nonlinear systems (Helton and James, 1999; Lions, 1961).

A model constructed as a PDE from the very beginning is Beckmann's flow model of interregional trade (Beckmann, 1952; Puu, 1997a, Chapter 3; Puu, 1997b, Chapter 7). In his PhD thesis from 1921 the famous economist Hotelling used PDEs to describe economies with migration (Puu, 1997a, Chapter 6). Unlike his famous later work on the extraction of natural resources, this widely neglected work is formulated as a parabolic PDE with two spatial variables describing the area to which people can migrate. A more recent application from taxation, resulting in a system of PDEs, has been given by Tarkianinen and Tuomala (1999).

A detailed delineation of the difference between actuarial and financial pricing of contingent claims can be found in Embrecht (2000).

From a more technical perspective, the derivation of the Black–Scholes PDE, Equation (2.30), has some weak points. These can be circumvented with more sophisticated methods. However, the result remains unchanged. The interested reader is encouraged to consult Bjoerk (1998), Ritchken (1996) and Rosu and Stroock (2004).

3
The Conventional Approach:
Finite Differences

3.1 GENERAL CONSIDERATIONS FOR NUMERICAL COMPUTATIONS

3.1.1 Evaluation criteria

The usefulness of a numerical algorithm can only be judged with the help of various criteria. These criteria represent the trade-off between computational cost and accuracy of the approximate solution.

- *Accuracy.* Not everybody uses the words accuracy and precision in a consistent manner at all times. However, for the most part, accuracy refers to the size of the possible error, meaning the difference between the (unknown) exact solution and the approximate one obtained. Precision, on the other hand, usually refers to the number of digits retained in the machine's internal floating point arithmetic, which is used to obtain approximate solutions. In the old days, when computers were still primitive, there used to be a lot of discussion about the relative merits of *single precision* and *double precision* arithmetic when calculating particular objects of interest. The error of an approximation stems from various sources. In addition to the error resulting from the model in use, there are three sources of error when computing a numerical solution of a differential equation:
  1. The round-off error, which will not be discussed any further since, for stable methods, this is usually not a problem.
  2. The discretization error resulting from the fact that a continuous model is mapped into its discrete counterpart.
  3. Errors resulting from the (numerical) solution of the discretized differential equation. Linear static problems are discretized into systems of linear equations; linear dynamic PDEs are discretized into systems of linear ODEs. Discretization of nonlinear static and dynamic PDEs gives systems of nonlinear equations and ODEs, respectively. Obviously, the solution of these systems of equations and differential equations is affected by errors.
- *Computational speed* is still of utmost importance for many financial applications. Front desk traders want their portfolios to be re-evaluated instantaneously, and risk managers want all positions of a financial institution to be re-evaluated at least once a day.
- *Error bounds.* The existence of (preferably tight) error bounds, or error estimates, strengthens the confidence in an approximate solution. Usually, there is a trade-off between accuracy and error, which can be more easily judged with reliable error estimates. Error bounds allow the user to find a compromise between computational cost and accuracy.
- *Programming complexity* refers to the fact that some algorithms are more complex than others. Usually, the FEM is considered to be more complex than the FDM. Increased complexity without further advantages is useless. In the context of contingent claim pricing, FE offer

some advantages in computing some of the derivatives (*Greeks*) and in the ability to cope with complex geometries.

- *Flexibility*. This criterion asks whether it is possible to use the designed algorithm for more than one task, i.e. in finance, usually various products and/or different market conditions represented by different market data.
- *Memory/storage requirements*. With the cost of memory having been decreasing for decades, this issue has lost much of its importance.

The evaluation criteria must be weighted differently, depending on the specific problem at hand. In contingent claim pricing, often not only an individual contract needs to be priced and hedged, but an entire portfolio, or even all portfolios of an entire financial institution. For front office applications, highly accurate solutions (with accuracy up to a basis point) at high speed are required. However, in the front office, usually only individual contracts (possibly including their hedges) or small portfolios are evaluated. In risk management, the solutions do not need to be as accurate as in the front office. Therefore, large numbers of deals have to be evaluated. In model and product development, computational speed usually is only a minor concern. Here, algorithmic flexibility is of high importance, since products and models change quickly. The same is true for model review at any level. A slow but well-understood method might serve well in the assessment of new numerical techniques and their application to derivative pricing.

3.1.2   Turning unbounded domains into bounded domains

Most option pricing problems have a semi-infinite domain. The usual approach is to truncate some parts of the domain to make it finite and to construct artificial boundary conditions. Although it was suggested in 1996 (McIver, 1996, p. 283) to consider numerical techniques for unbounded domains in finance, so far no efforts in this direction are known to the author. For numerical techniques for unbounded domains, see Burnett (1987), p. 431 and references therein, Givoli (1992), Groth and Müller (1997), p. 165 and Wolf and Song (1996). One possible approach with FE is to use infinite elements, as discussed by Hibbitt (1994). Although the common approach of cutting off the infinite domain at some point looks crude at first sight, it is possible to compute the error incurred by this simple device.

Artificial boundary conditions

When constructing artificial boundary conditions, two aspects of this problem have to be tackled simultaneously:

- *What* kind of boundary condition?
- *Where* to put this boundary condition?

Unfortunately, these questions can only be answered for each single problem individually. A general theory for all problems is not available and not in sight. The literature gives several approaches to the first question. These approaches will be discussed in some detail below. Coming up with appropriate boundary conditions is a crucial task in financial modeling, since the boundaries *do have an influence* on the solution in the interior of the domain. The latter question had been lying dormant in the literature until Kangro's PhD thesis was published in 1997.

We will follow the usual terminology and delineate the use of the following types of boundary conditions:

- *Dirichlet conditions.* These are the easiest boundary conditions to analyze, but it is often very hard in practice to come up with a Dirichlet condition. The Dirichlet condition states the value of the *function* which is to be computed. In mathematical terms, this can be stated as:

$$V(S_1, \ldots, S_n, t) = g(S_1, \ldots, S_n, t) \tag{3.1}$$

The Dirichlet conditions do not need to hold for the entire boundary. It is possible (and often useful) to apply different types of boundary conditions to different boundaries of the same domain.

- *Neumann conditions.* The first derivative of the function to be computed is given by:

$$\frac{\partial V(S_1, \ldots, S_n, t)}{\partial n} = g(S_1, \ldots, S_n, t) \tag{3.2}$$

with n being the normal gradient.

- *Robin conditions.* The Robin (or *mixed* or *radiation*) condition is:

$$\alpha \frac{\partial V(S_1, \ldots, S_n, t)}{\partial n} + \beta V(S_1, \ldots, S_n, t) = g(S_1, \ldots, S_n, t) \tag{3.3}$$

This type of condition is not very popular in financial applications; for an example see page 231 of Wilmott (1998).

- *Conditions based on the second derivative.* In financial applications, the function g in the Neumann condition is a constant or (rarely) a function of time only, so that the derivative with respect to the spatial variables vanishes. This fact is sometimes used in FD schemes, where a BC of the type

$$\frac{\partial^2 V(S_1, \ldots, S_n, t)}{\partial n^2} = 0 \tag{3.4}$$

can be easily integrated (see Deutsch, 2002, Section 10.2.4 and Wilmott, 1998, Section 47.3.3). It has been shown by Kangro (1997) that this nonclassical BC gives rise to a well-posed problem.

- *Differential equations located at the boundary.* The asymptotic behavior of some options can be approximated with other derivative instruments. For example, consider a basket call on two assets. When one of the assets becomes worthless, the option behaves like a call on the other asset. Deep in the money, the option behaves like a forward on the basket. This approach is especially popular with multi-asset options. A further simplification can be made by eliminating the diffusive term from the PDEs used as BCs. This, however, changes the type of the PDE. While pricing PDEs are parabolic, the equation left after eliminating the diffusive term becomes hyperbolic. This offers a further advantage of reducing the necessary BCs. To illustrate this, again consider the two-asset call on a basket. For $S_1 = 0$ we employ the pricing PDE for a plain vanilla call on S_2. While the PDE for the basket option has two spatial variables, the PDE for the plain vanilla call has only one spatial variable. BCs are needed for the spatial variable at S_2^{max} and S_2^{min}. Eliminating the diffusive term leads to a hyperbolic PDE which needs only *one* BC at either S_2^{max} or S_2^{min}.

A differential equation located at the boundary is closely connected to a Dirichlet BC, since the solution of the differential equation is the corresponding Dirichlet BC. Consider a

European call. Eliminating the diffusive term leads to:

$$-V_t + rSV_S - rV = 0 \tag{3.5}$$

Deep in the money at $S = S_{max}$ we have $V_S = 1$, so that Equation (3.5) reduces to

$$V_t + rV = rS_{max} \tag{3.6}$$

which is a linear ODE solved by $V(t) = ce^{-rt} + S_{max}$, c being a constant to be determined from the initial condition $V(0) = S_{max} - E$. It can be easily verified that $c = -E$, so that:

$$V(S_{max}, t) = S_{max} - Ee^{-rt} \tag{3.7}$$

which is the present value of a forward with strike S. Deep in the money, the price of a call approaches the price of a forward, since the probability of being in the money at maturity approaches one.

The location of the boundary depends on the contract at hand. Except in the case of a barrier, the behavior of the option at infinity is emulated at the boundary. The task is to find a point where the infinite or semi-infinite domain can be cut off without severely affecting the solution.

- *The engineering approach.* The behavior of an option as the underlying tends to infinity can usually be assessed by simple economic reasoning. The value of an out-of-the-money put approaches zero. Accordingly, the value of a call deep in the money approaches that of a forward. But, how deep in the money is 'deep enough'? A simple way to solve this question is to vary the location of the boundary and to monitor its influence on the region of interest. For a call, the boundary on the right-hand side has to be moved to the right until further moves do not influence the price of the option under inspection anymore. This approach is used in an example in Section 5.2.1.
- *The probabilistic approach.* In the following approach, we make use of the close link between parabolic PDEs and stochastic processes. Assume the pricing dynamics are governed by a geometric Brownian motion. As will be shown in later chapters, this is the most prominent stochastic process in option pricing. In the risk-neutral measure, this is $dS = \sigma SdW$. S_T is the future value of the present value reduced by discounted dividends:

$$S_T = e^{rT}(S_{t_0} - \sum_{i=1}^{m} D_i e^{-r_i(\tau_i - t_0)}) \tag{3.8}$$

Then, S_T follows a lognormal distribution at T:

$$S_T \sim f(\ln(S_T), M, \Sigma) = f\left(\frac{\ln(S_T) - M}{\Sigma}, 0, 1\right) \tag{3.9}$$

with

$$\Sigma = \sigma\sqrt{T} \tag{3.10}$$

and

$$M = -\frac{\Sigma^2}{2} \tag{3.11}$$

Next, we introduce a level of S which we call L. The probability P that S goes beyond a certain level L in either direction can be taken from:

$$P = F\left(\frac{\ln(S_T/L) - M}{\Sigma}, 0, 1\right) \tag{3.12}$$

with F denoting the cumulative normal distribution; see Section B.1.1.2.

Table 3.1 Computation and accuracy of a BC

S_T	Σ	M	$\ln(S_T/L)$	$\dfrac{\ln(S_T/L) - M}{\Sigma}$	P	Black–Scholes price	Forward	Δ
43	0.1414	−0.01	−0.0235	−0.0957	0.462	5.5623	4.9508	0.8253
44	0.1414	−0.01	−0.0465	−0.2582	0.398	6.4075	5.9508	0.8639
45	0.1414	−0.01	−0.0690	−0.4171	0.338	7.2878	6.9508	0.8956
46	0.1414	−0.01	−0.0910	−0.5726	0.283	8.1967	7.9508	0.9211
47	0.1414	−0.01	−0.1125	−0.7246	0.234	9.1282	8.9508	0.9412
48	0.1414	−0.01	−0.1335	−0.8735	0.191	10.0775	9.9508	0.9567
49	0.1414	−0.01	−0.1542	−1.0193	0.154	11.0404	10.9508	0.9685
50	0.1414	−0.01	−0.1744	−1.1622	0.123	12.0135	11.9508	0.9774
51	0.1414	−0.01	−0.1942	−1.3022	0.096	12.9943	12.9508	0.9839
52	0.1414	−0.01	−0.2136	−1.4395	0.075	13.9807	13.9508	0.9887
53	0.1414	−0.01	−0.2326	−1.5742	0.058	14.9712	14.9508	0.9921
54	0.1414	−0.01	−0.2513	−1.7064	0.044	15.9646	15.9508	0.9946
55	0.1414	−0.01	−0.2697	−1.8361	0.033	16.9601	16.9508	0.9963
56	0.1414	−0.01	−0.2877	−1.9635	0.025	17.9570	17.9508	0.9975
57	0.1414	−0.01	−0.3054	−2.0887	0.018	18.9549	18.9508	0.9983
58	0.1414	−0.01	−0.3228	−2.2116	0.013	19.9535	19.9508	0.9989
59	0.1414	−0.01	−0.3399	−2.3325	0.010	20.9526	20.9508	0.9992
60	0.1414	−0.01	−0.3567	−2.4514	0.007	21.9520	21.9508	0.9995

To visualize the usage and the effectiveness of this approach, we will turn our attention to a European call, which we will compute in Section 5.2.1 and take the data from Table 5.1. Here, we will analyze the right-hand BC. In order to use Equation (3.12), we have to compute Σ and M.

How should we interpret the numbers in Table 3.1? In the first line, we see that an equity with a spot price S_0 of 42 ending up with a price of 43 or more has a probability of 0.462. This result does not contradict intuition. Besides this, we can observe several issues. With growing S_T, the price of a forward and the Black–Scholes price of the call converge. The forward was computed via put–call parity:

$$C + E \exp(-rT) = P + S_0 \qquad (3.13)$$

The Black–Scholes values were computed using the analytical formula. The forward seems to be a good approximation of the in-the-money BC of the Dirichlet type. Also, values for Δ converge to one, since the option resembles more and more a forward the deeper it gets into the money. Consequently, taking $\Delta = 1$ of the forward is a reasonable choice for the Neumann BC. This example also shows that the domain of a PDE is often rather small. In real life, however, the user specifies L and then inverts Equation (3.12). This can only be done numerically.

- *A PDE approach.* With techniques from the theory of parabolic PDEs (such as Green's function and the comparison principle) beyond the scope of this book, we derive bounds for the error caused by the introduction of an artificial boundary (Kangro, 1997). This error estimate can be used to determine *a priori* a suitable placement for the far boundary in terms of a user's error tolerance. This approach works for multi-asset European options with variable coefficients. To give a flavor of the usefulness of this approach, the result for a

European call option with constant r and σ will be given. Suppose that $\sigma^2 \geq 2r$ and the current stock price is below strike.[1] Then an artificial BC at:

$$s_{max} = E \exp\left(\sqrt{2\sigma^2 T \ln(100)}\right) \tag{3.14}$$

gives rise to a truncated problem that provides a value for the call, as of today, that is within $E/100$ monetary units from the correct value. With $\sigma = 0.5$ and $r = 0.1$ we arrive at:

$$s_{max} = 40 \exp\left(\sqrt{2 \cdot 0.5^2 \cdot 0.5 \cdot \ln(100)}\right) \approx 116.96 \tag{3.15}$$

Transformations

In some situations, transformations can be used to find a compact domain for the PDE to be solved. In the setting of term structure models, the independent variable is a nominal interest rate r, which, by economic reasoning, is somewhere between nil and infinity. When transforming the interval $[0, \infty)$ to $[c_1, c_2]$ both the problem of finding an approximate value for infinity and the problem of putting up a BC there are circumvented (James and Webber, 2000, Section 12.3). This approach, however, does not solve the problem arising at $r = 0$. The SDE modeling the evolution of r can be inconsistent to a BC at $r = 0$ stemming from economic reasoning; for details, see d'Halluin *et al.* (2001).

For numerical purposes, transformations are generally not very popular. Transformations cannot deal with discrete dividends and time-dependent parameters. New numerical schemes and error bounds are usually derived on the basis of the respective prototype PDE. Although these results usually do not carry over to PDEs arising in finance, they might be useful for indicative purposes. In the process of finding analytical solutions, however, transformations can be of use, as first shown by Black and Scholes (1973). They transformed their famous PDE, Equation (2.30), to $u_t = u_{xx}$, which had been well-studied by mathematicians and physicists. Their approach used the fact that any second order PDE with constant coefficients can be transformed into its prototype equation. Especially for multi-asset underlyings, these transformations become rather messy; for an example, see Tomas (1996).

3.2 ORDINARY INITIAL VALUE PROBLEMS

3.2.1 Basic concepts

In this section, some basic numerical procedures for well-posed ordinary initial value problems are presented. Following the historical development, the starting point is FD techniques, since some concepts from FD carry over to FE. The differential equation under inspection needs to be well-posed because one does not attempt to solve the original problem, but only an approximation. The outline will start with techniques for a single first order ordinary initial value problem:

$$\frac{dy}{dx} = f(y, x) \tag{3.16}$$

$$x_{min} \leq x \leq x_{max} \tag{3.17}$$

$$y(x_{min}) = y_0 \tag{3.18}$$

---

[1] This implies that the following result is only applicable to options with a fairly high volatility. Some of the more general results of Kangro (1997) are applicable only when rather unrealistic conditions are met.

Note that Equation (3.16) is not required to be linear. These techniques can be adapted easily to systems of first order ordinary initial value problems, as demonstrated in some examples. Since problems of any order can be transformed to first order systems, the methods of this section apply to problems of any order. Besides, as will be shown in later chapters, FE and FD discretizations for parabolic PDEs give rise to first order systems.

As a first step, the domain $[x_{min}, x_{max}]$ is subdivided into intervals $x_{min} = x_0, x_1, \ldots, x_N = x_{max}$. The independent variable is denoted by x, although it usually represents either time or time to maturity. The collection of nonoverlapping intervals $x_0, \ldots, x_n$ is called a *grid*. If not stated differently, we assume all intervals to be of equal length:

$$x_i = x_0 + ih \tag{3.19}$$

with h being the step size. FD methods do not produce continuous approximations for the differential equation under inspection. Rather, approximations are found at certain specified points, the x_i. The basic idea is to start with the initial condition, i.e. the point where the solution starts, and to extend the solution with a reliable approximation of the derivative. When intermediate values are needed, some method of interpolation is used. For each i, a value f_i is to be found which approximates $f(y, x_i)$. To solve this task, a recurrence relation is constructed, such that:

$$f_{i+1} = G(f_{i+1}, f_i, f_{i-1}, \ldots, x_{i+1}, x_i, x_{i-1} \ldots) \tag{3.20}$$

This difference equation is solved for all f_i. The FD solution of an ordinary differential equation is a list of values for the x_i and the associated $y(x_i)$. On a computer, it is usually stored as an array of size $2 \times (N + 1)$. For a different initial value, the numerical solution has to be recomputed. Depending on the form of G, there are various classes of method:

- Linear vs. nonlinear methods. This distinction refers to the (non) linearity of Equation (3.20) but not to the ODE to be solved. In FE codes, linear methods are also used for nonlinear problems, so that here, only linear methods will be discussed.
- Singlestep vs. multistep methods. Multistep methods relate the discrete value f_{i+1} to several of its successors, while singlestep methods make use only of the previous value.
- Explicit vs. implicit methods. Explicit schemes are of the form $f_{i+1} = G(f_i, f_{i-1}, \ldots, x_i, x_{i-1} \ldots)$, i.e. f_{i+1} can be computed explicitly in terms of x_i and f_i.

In the following pages, examples will be given. A further classification of FD methods refers to their capability to cope with stiff problems. Systems resulting from a spatial discretization of parabolic PDEs with FE are usually stiff, so this is an important topic for the remainder of this book.

3.2.2 Euler's method

Derivation

Although Euler's method is not very popular in practice, we will use this simple method to pave the way for more advanced algorithms. First, we will explain how this method works. Then, we will turn our attention to the rationale and weaknesses of the method.

For any two points in the domain of the differential equation under inspection, the following relation holds:

$$y(x_{i+1}) = y(x_i) + (x_{i+1} - x_i)y'(x_i) + \frac{(x_{i+1} - x_i)^2}{2}y''(\xi_i) \tag{3.21}$$

with x_i in (x_i, x_{i+1}) because of Taylor's theorem (see Theorem 5 in Appendix A). Let $h = x_{i+1} - x_i$:

$$y(x_{i+1}) = y(x_i) + hy'(x_i) + \frac{h^2}{2}y''(\xi_i) \tag{3.22}$$

Since $y'(x) = f[x, y(x)]$:

$$y(x_{i+1}) = y(x_i) + hf[x_i, y(x_i)] + \underbrace{\frac{h^2}{2}y''(\xi_i)}_{\text{remainder}} \tag{3.23}$$

For small h

$$y(x_{i+1}) \approx y(x_i) + hf[x_i, y(x_i)] \tag{3.24}$$

holds. This way we arrive at the Euler scheme:

$$y_{i+1} = y_i + hf(x_i, y_i) \tag{3.25}$$

with y_{i+1} denoting the approximate value for $y(x_{i+1})$.

Example 1
As an example, we will take a Riccati system arising in term structure modeling (Rudolf, 2000). In general, these equations cannot be solved analytically, although for many financial applications, closed-form solutions have been found.

$$A_t(t, T) - h(t)B(t, T) + \frac{[B(t, T)]^2}{2}d(t) = 0 \tag{3.26}$$

$$1 + B_t(t, T) + g(t)B(t, T) - \frac{[B(t, T)]^2}{2}c(t) = 0 \tag{3.27}$$

$$A(T, T) = B(T, T) = 0 \tag{3.28}$$

As a first example, we take Equation (3.27) with $c = 0$ and $g = -a$, as in the Vasicek model (Vasicek, 1977). Reversing time, $\tau = T - t$ gives:

$$\dot{B}(\tau) = 1 - aB(\tau) \tag{3.29}$$

$$B(0) = 0 \tag{3.30}$$

This simple problem is solved by:

$$A(t, T) = \frac{1}{a}\left[\left(\frac{\sigma^2}{2a} - ab\right)(T - t - B(t, T)) - \frac{\sigma^2}{4}[B(t, T)]^2\right] \tag{3.31}$$

$$= \frac{1}{a}\left[\left(\frac{\sigma^2}{2a} - ab\right)(\tau - B(t, T)) - \frac{\sigma^2}{4}[B(t, T)]^2\right] \tag{3.32}$$

$$B(t, T) = \frac{1 - e^{-a(T-t)}}{a} = \frac{1 - e^{-a\tau}}{a} \tag{3.33}$$

The Euler method and all other methods discussed in this section solve problems of the form
$\dot{B} = f(x, B)$. *The variable x usually denotes time t or time to maturity τ, but in some cases,*
it can represent physical quantities as well. Assume that $h = x_i - x_{i-1}$ is constant. Then

$$B_{i+1} = B_i + hf(x_i, B_i) \tag{3.34}$$

is known as Euler's method. The initial condition is the starting value for this series: $B(x_0) = B_0$. Note that the notation has changed somewhat: in this section, $x_1, \ldots, x_N$ denotes a series of values (either in space or time), while in other, previous and later sections, the subscript indicates that there are several spatial variables. To clarify the concept, an example will be given with all details. Assume a step size h of 0.1 and $a = b = 0.1$. The first elements of this series are:

1. Iteration:

$$B_1 = B_0 + hf(x_0, B_0) \tag{3.35}$$
$$= B_0 + h(1 - 0.1B_0) \tag{3.36}$$
$$= 0.0 + 0.1\,(1 - 0.1 \cdot 0.0) \tag{3.37}$$
$$= 0.1 \tag{3.38}$$

2. Iteration:

$$B_2 = B_1 + hf(x_1, B_1) \tag{3.39}$$
$$= B_1 + h(1 - 0.1B_1) \tag{3.40}$$
$$= 0.1 + 0.1\,(1 - 0.1 \cdot 0.1) \tag{3.41}$$
$$= 0.199 \tag{3.42}$$

3. Iteration:

$$B_3 = B_2 + hf(x_2, B_2) \tag{3.43}$$
$$= B_2 + h(1 - 0.1B_2) \tag{3.44}$$
$$= 0.199 + 0.1\,(1 - 0.1 \cdot 0.199) \tag{3.45}$$
$$= 0.29701 \tag{3.46}$$

4. Iteration:

$$B_4 = B_3 + hf(x_3, B_3) \tag{3.47}$$
$$= B_3 + h(1 - 0.1B_3) \tag{3.48}$$
$$= 0.29701 + 0.1\,(1 - 0.1 \cdot 0.29701) \tag{3.49}$$
$$= 0.39404 \tag{3.50}$$

Although only five figures after the decimal point are shown, more have been used to compute the presented results. This practice will be continued for the rest of the book and should be kept in mind when reproducing the results. The above steps and some of the subsequent ones are summarized in Table 3.2.

Table 3.2 Euler's method

τ	B_{Euler}	$B_{analytical}$	$B_{Euler} - B_{analytical}$
0.0	0.0000	0.0000	0.0000
0.1	0.1000	0.0995	0.0005
0.2	0.1990	0.1980	0.0010
0.3	0.2970	0.2955	0.0015
0.4	0.3940	0.3921	0.0019
0.5	0.4901	0.4877	0.0024
0.6	0.5852	0.5824	0.0028
0.7	0.6793	0.6761	0.0033
0.8	0.7726	0.7688	0.0037
0.9	0.8648	0.8607	0.0041
1.0	0.9562	0.9516	0.0046
1.1	1.0466	1.0417	0.0050
1.2	1.1362	1.1308	0.0054
1.3	1.2248	1.2190	0.0057
1.4	1.3125	1.3064	0.0061
1.5	1.3994	1.3929	0.0065
1.6	1.4854	1.4786	0.0069
1.7	1.5706	1.5634	0.0072
1.8	1.6549	1.6473	0.0076
1.9	1.7383	1.7304	0.0079
2.0	1.8209	1.8127	0.0082

□

Error bounds

Although Euler's method is not very popular in practice, in the following it will be analyzed in some detail. Such an analysis will not be given for more sophisticated methods. In this book, the focus is on how some numerical techniques work, but not why they work. The purpose of the analysis of Euler's method is solely to give the interested reader an impression of what kind of tools exist for solving differential equations.

The numerical analyst distinguishes between *local* and *global* error. The local error is the error incurred by one iteration, assuming that all previous results are exact. The global error is the accumulated error over all iterations. The local error of Euler's method is the remainder term in Equation (3.23). It is of order $O(h^2)$, indicating that when the time-step h is halved, the local discretization error is reduced by approximately a quarter. The global error is given by the following theorem:

Theorem 1 (Burden and Faires, 1998, p. 182)
Let $f(x)$ be a well-posed solution to the problem given by Equations (3.16) to (3.18). The approximation produced by Euler's method as in Equation (3.24) is given by $y_0, y_1, \ldots, y_N$ with N being some positive integer. Suppose that f is continuous for all x in $[x^{min}, x^{max}]$ and all y in $(-\infty, \infty)$, and constants L and M exist with:

$$\left| \frac{\partial f[x, y(x)]}{\partial y} \right| \leq L \tag{3.51}$$

$$|y''(x)| \leq M \tag{3.52}$$

□

Table 3.3   Error bound on Euler's method

τ	Analytical solution B_{Euler}	Numerical solution $B_{\text{analytical}}$	Actual error $\lvert B_{\text{Euler}} - B_{\text{analytical}} \rvert$	Error bound
0.0	0.0000	0.0000	0.0000	0.0000
0.1	0.0995	0.1000	0.0005	0.0010
0.2	0.1980	0.1990	0.0010	0.0020
0.3	0.2955	0.2970	0.0015	0.0030
0.4	0.3921	0.3940	0.0019	0.0039
0.5	0.4877	0.4901	0.0024	0.0049
0.6	0.5824	0.5852	0.0028	0.0058
0.7	0.6761	0.6793	0.0033	0.0068
0.8	0.7688	0.7726	0.0037	0.0077
0.9	0.8607	0.8648	0.0041	0.0086
1.0	0.9516	0.9562	0.0046	0.0095
1.1	1.0417	1.0466	0.0050	0.0104
1.2	1.1308	1.1362	0.0054	0.0113
1.3	1.2190	1.2248	0.0057	0.0122
1.4	1.3064	1.3125	0.0061	0.0131
1.5	1.3929	1.3994	0.0065	0.0139
1.6	1.4786	1.4854	0.0069	0.0148
1.7	1.5634	1.5706	0.0072	0.0156
1.8	1.6473	1.6549	0.0076	0.0165
1.9	1.7304	1.7383	0.0079	0.0173
2.0	1.8127	1.8209	0.0082	0.0181

Then, for each $i = 0, 1, 2, \ldots, N$

$$|y(x_i) - y_i| \le \frac{hM}{2L} \left[e^{L(x_i - x_0)} - 1 \right] \tag{3.53}$$

Returning to the above example we find:

$$\left| \frac{d}{dB} [1 - aB] \right| = |-a| = 0.1 = L \tag{3.54}$$

$$\left| B''(\tau) \right| = \left| -ae^{-a\tau} \right| = 0.1e^{-0.1\tau} = M \le 0.1 \tag{3.55}$$

so that the error bound is given by $0.05e^{-0.1\tau_i} \left[e^{0.1\tau_i} - 1 \right] = 0.05 \left(1 - e^{-0.1\tau_i} \right)$. The error bound for the above example is given in Table 3.3. This table shows that the error bound is not very tight.

The global error of Euler's method is of order $O(h)$, indicating that the accumulated discretization error is approximately halved by halving the step length h. Unfortunately, it is not possible to generate arbitrarily close approximate solutions by simply decreasing the step size indefinitely. For too small h, the roundoff error starts to dominate.

Stiff problems

The Euler method does not work well with stiff problems (see Section A.3.2), resulting in a tiny step size and consequently increasing the number of iterations. As a test problem, we take the simplest problem possible:

$$y'(x) = \lambda y(x), \quad y(x_0) = y_0 \tag{3.56}$$

The exact solution is $y(x) = e^{\lambda x}$. With $\lambda > 0$, the solution explodes and so does its numerical approximation generated with Euler's method. For $\lambda < 0$, however, problems can arise. The steady-state solution equals zero, while the transient solution decays to zero. Applying the Euler scheme of Equation (3.25) to the above problem, we obtain:

$$y_{i+1} = y_i + h\lambda y_i = (1 + h\lambda)y_i \tag{3.57}$$

Inserting the initial value $y(x_0) = y_0$ leads to:

$$y_1 = (1 + h\lambda)y_0 \tag{3.58}$$
$$y_2 = (1 + h\lambda)y_1 = (1 + h\lambda)(1 + h\lambda)y_0 \tag{3.59}$$
$$\vdots$$
$$y_{i+1} = (1 + h\lambda)^{i+1}y_0 \tag{3.60}$$

The expression $(1 + h\lambda)^{i+1}y_0$ converges to $y_0 e^{\lambda x}$ as the step size h converges to zero. However, if $\lambda < 0$, then not only the error will increase if h is taken too large, but the numerical solution will grow without bounds and oscillate. For instance, choosing $\lambda = -100$ results in a rapidly decaying exact solution. Taking a step size $h = 0.1$, the Euler scheme suggests:

$$y_{i+1} = y_0(1 - 100 \cdot 0.1)^{i+1} = y_0(-9)^{i+1} \tag{3.61}$$

This recurrence formula is obviously oscillating and cannot converge to the exact solution $y(x) = e^{-100x}$. In order to guarantee a decaying solution, there is a restriction on the step size:

$$|1 + h\lambda| < 1 \quad \Leftrightarrow \quad 0 < h < -\frac{2}{\lambda} \tag{3.62}$$

To avoid oscillations in the example above (with $\lambda = -100$), the step size has to be chosen such that $h < \frac{-2}{-100} = 0.02$. This would imply more than 50 steps if the numerical solution were to be computed for the small interval $[0, 1]$.

3.2.3 Taylor methods

Derivation

Euler's method is the simplest member of a larger family of methods called *Taylor methods*. A Taylor method of order n involves the first n elements of the Taylor series. Therefore, Euler's method is a first order Taylor method. Expanding the solution $y(x)$ with a Taylor polynomial about x_i reads:

$$y(x_{i+1}) = y(x_i) + hy'(x_i) + \frac{h^2}{2}y''(x_i) + \ldots + \frac{h^n}{n!}y^{(n)}(x_i) + \frac{h^{n+1}}{(n+1)!}y^{(n+1)}(\xi_i) \tag{3.63}$$

for some $\xi_i \in [x_i, x_{i+1}]$. Differentiating the solution $y(x)$ yields:

$$y'(x) = f[x, y(x)] \tag{3.64}$$
$$y''(x) = f'[x, y(x)] \tag{3.65}$$
$$\vdots \tag{3.66}$$
$$y^{(k)}(x) = f^{(k-1)}[x, y(x)] \tag{3.67}$$

Substituting these results into the Taylor expansions leads to:

$$y(x_{i+1}) = y(x_i) + hf[x_i, y(x_i)] + \frac{h^2}{2} f'[x_i, y(x_i)] + \dots$$

$$+ \frac{h^n}{n!} f^{(n-1)}[x_i, y(x_i)] + \underbrace{\frac{h^{n+1}}{(n+1)!} f^{(n)}[\xi_i, y(\xi_i)]}_{\text{remainder}} \qquad (3.68)$$

As in the Euler method, the remainder term is ignored, so that the Taylor scheme is given by:

$$y(x_{i+1}) = y(x_i) + hf[x_i, y(x_i)] + \frac{h^2}{2} f'[x_i, y(x_i)] + \dots + \frac{h^n}{n!} f^{(n-1)}[x_i, y(x_i)] \qquad (3.69)$$

Since f is a bivariate function in both x and $y = y(x)$, the total derivative of f with respect to x needs to be computed with the chain rule:

$$f'[x, y(x)] = \frac{\partial f}{\partial x}[x, y(x)]\frac{dx}{dx} + \frac{\partial f}{\partial y}[x, y(x)]\frac{dy(x)}{dx} \qquad (3.70)$$

$$= \frac{\partial f}{\partial x}[x, y(x)] + \frac{\partial f}{\partial y}[x, y(x)]y'(x) \qquad (3.71)$$

$$= \frac{\partial f}{\partial x}[x, y(x)] + f[x, y(x)]\frac{\partial f}{\partial y}[x, y(x)] \qquad (3.72)$$

For the computation of f'' the derivative of the right-hand side of the above expression needs to be taken. This process becomes increasingly complex for higher derivatives. Higher order Taylor methods are not very popular because of the derivatives that need to be computed analytically. The scheme for a second order Taylor method is given by:

$$y(x_{i+1}) = y(x_i) + hf[x_i, y(x_i)] + \frac{h^2}{2} \left(\frac{\partial f}{\partial x}[x_i, y(x_i)] + f[x_i, y(x_i)]\frac{\partial f}{\partial y}[x_i, y(x_i)] \right)$$

$$(3.73)$$

Example 2
Returning to the example $\frac{dB(\tau)}{d\tau} = 1 - 0.1B(\tau)$ the following derivatives are needed:

$$\frac{df}{dy} \doteq \frac{d}{dB}[1 - 0.1B(\tau)] = -0.1 \qquad (3.74)$$

$$\frac{df}{dx} \doteq \frac{d}{d\tau}[1 - 0.1B(\tau)] = 0 \qquad (3.75)$$

such that the difference equation results in:

$$B_{i+1} = B_i + hf(x_i, B_i) + \frac{h^2}{2} \left(f[x_i, y(x_i)] \cdot (-0.1) \right) \qquad (3.76)$$

$$= B_i + hf(x_i, B_i) - \frac{h^2}{20} f[x_i, y(x_i)] \qquad (3.77)$$

$$= B_i + \left(h - \frac{h^2}{20} \right)(1 - 0.1B_i) \qquad (3.78)$$

□

3.2.4 Runge–Kutta methods

Derivation

Consider a second order Taylor method:

$$y(x_{i+1}) = y(x_i) + hf[x_i, y(x_i)] + \frac{h^2}{2} f'[x_i, y(x_i)] + \frac{h^3}{3!} y'''(\xi) \tag{3.79}$$

Because of Equation (3.72), this expression can be manipulated to:

$$y(x_{i+1}) = y(x_i) + h \left[f[x_i, y(x_i)] + \frac{h}{2} \frac{\partial f}{\partial x}[x_i, y(x_i)] \right.$$
$$\left. + \frac{h}{2} \frac{\partial f}{\partial y}[x_i, y(x_i)] f[x_i, y(x_i)] \right] + \frac{h^3}{3!} y'''(\xi) \tag{3.80}$$

The expression in brackets is to be replaced. Consider a bivariate Taylor expansion of f:

$$f[x_i + \alpha, y(x_i) + \beta] \approx f[x_i, y(x_i)] + \alpha \frac{\partial f}{\partial x}[x_i, y(x_i)] + \beta \frac{\partial f}{\partial y}[x_i, y(x_i)] \tag{3.81}$$

Equating the expression on the right-hand side with the expression in brackets from Equation (3.80), and rearranging, results in the following coefficients:

$$\alpha = \frac{h}{2} \tag{3.82}$$

$$\beta = \frac{h}{2} f[x_i, y(x_i)] \tag{3.83}$$

Now, the scheme for a second order Runge–Kutta method is given by:

$$y_{i+1} = y_i + h \left(f \left[x_i + \frac{h}{2}, y_i + \frac{h}{2} f(x_i, w_i) \right] \right) \tag{3.84}$$

This scheme is also called the *midpoint method*.

Example 3
First, we compare a second order Taylor method with a second order Runge–Kutta method with the help of the above example. Since $x \doteq \tau$, we obtain $y(x) \doteq B(\tau)$ and

Table 3.4   Taylor and Runge–Kutta methods of order 2

τ	$B(\tau)$	Runge–Kutta	Absolute error
0.0	0.0000000	0.0000000	0.0000000
0.1	0.0995017	0.0995000	0.0000017
0.2	0.1980133	0.1980100	0.0000033
0.3	0.2955447	0.2955398	0.0000049
0.4	0.3921056	0.3920992	0.0000065
0.5	0.4877058	0.4876978	0.0000080
0.6	0.5823547	0.5823452	0.0000095
0.7	0.6760618	0.6760508	0.0000110
0.8	0.7688365	0.7688241	0.0000124
0.9	0.8606881	0.8606743	0.0000138
1.0	0.9516258	0.9516106	0.0000152
1.1	1.0416586	1.0416421	0.0000165
1.2	1.1307956	1.1307778	0.0000179
1.3	1.2190457	1.2190265	0.0000192
1.4	1.3064176	1.3063972	0.0000204
1.5	1.3929202	1.3928986	0.0000217
1.6	1.4785621	1.4785392	0.0000229
1.7	1.5633518	1.5633278	0.0000241
1.8	1.6472979	1.6472726	0.0000252
1.9	1.7304087	1.7303823	0.0000264
2.0	1.8126925	1.8126650	0.0000275

$f[x, y(x)] \doteq f[\tau, B(\tau)] = 1 - 0.1B(\tau)$, with which the midpoint scheme reads:

$$B_{i+1} = B_i + h \left\{ 1 - 0.1 \left[B_i + \frac{h}{2}(1 - 0.1B_i) \right] \right\} \tag{3.85}$$

$$= B_i + h \left(1 - 0.1B_i - \frac{h}{20} + \frac{B_i h}{200} \right) \tag{3.86}$$

$$= B_i + \left(h - \frac{h^2}{20} \right)(1 - 0.1B_i) \tag{3.87}$$

□

This is the same difference equation as the one resulting from the second order Taylor method. It has to be emphasized that this is not generally true. For the numerical example given in Table 3.4, the step size is kept constant with $h = 0.1$.

The algebra for Runge–Kutta methods becomes increasingly messy for growing orders of the method, so they will not be followed up here. The formulae for a Runge–Kutta method of order 4 are given by:

$$B_{i+1} = B_i + \frac{1}{6}(k_1 + 2k_2 + 2k_3 + k_4) \tag{3.88}$$

$$k_1 = hf(x_i, B_i) \tag{3.89}$$

$$k_2 = hf\left(x_i + \frac{h}{2}, B_i + \frac{k_1}{2} \right) \tag{3.90}$$

$$k_3 = hf\left(x_i + \frac{h}{2}, B_i + \frac{k_2}{2} \right) \tag{3.91}$$

$$k_4 = hf(x_i + h, B_i + k_3) \tag{3.92}$$

The Runge–Kutta method of order 4 is the most popular in practice.

The computations necessary for a Runge–Kutta solution to the example from Section 3.2.2:

$$\dot{B}(\tau) = 1 - aB(\tau) \tag{3.93}$$
$$B(0) = 0 \tag{3.94}$$

will be given next. The first steps are given in detail. In the first step all significant figures are reported. In the following steps, only six decimal places will be given, although more have been used for the computations.

1. Iteration:

$$k_1 = hf(x_0, B_0) = h(1 - 0.1B_0) = 0.1(1 - 0.1 \cdot 0) = 0.1 \tag{3.95}$$

$$k_2 = hf\left(x_0 + \frac{h}{2}, B_0 + \frac{k_1}{2}\right) = h\left[1 - 0.1\left(B_0 + \frac{k_1}{2}\right)\right] \tag{3.96}$$

$$= 0.1\left[1 - 0.1\left(0 + \frac{0.1}{2}\right)\right] = 0.0995 \tag{3.97}$$

$$k_3 = hf\left(x_0 + \frac{h}{2}, B_0 + \frac{k_2}{2}\right) = h\left[1 - 0.1\left(B_0 + \frac{k_2}{2}\right)\right] \tag{3.98}$$

$$= 0.1\left[1 - 0.1\left(0 + \frac{0.0995}{2}\right)\right] = 0.0995025 \tag{3.99}$$

$$k_4 = hf(x_0 + h, B_0 + k_3) = h\left[1 - 0.1(B_0 + k_3)\right] \tag{3.100}$$

$$= 0.1\left[1 - 0.1(0 + 0.0995025)\right] = 0.099004975 \tag{3.101}$$

$$B_1 = B_0 + \frac{1}{6}(k_1 + 2k_2 + 2k_3 + k_4) \tag{3.102}$$

$$= 0 + \frac{0.1 + 2 \cdot 0.0995 + 2 \cdot 0.0995025 + 0.099004975}{6}$$

$$= 0.0995016625 \tag{3.103}$$

2. Iteration:

$$k_1 = hf(x_1, B_1) = h(1 - 0.1B_1) = 0.1(1 - 0.1 \cdot 0.0995) = 0.099005 \tag{3.104}$$

$$k_2 = hf\left(x_1 + \frac{h}{2}, B_1 + \frac{k_1}{2}\right) = h\left[1 - 0.1\left(B_1 + \frac{k_1}{2}\right)\right] \tag{3.105}$$

$$= 0.1\left[1 - 0.1\left(0.099502 + \frac{0.099005}{2}\right)\right] = 0.098510 \tag{3.106}$$

$$k_3 = hf\left(x_1 + \frac{h}{2}, B_1 + \frac{k_2}{2}\right) = h\left[1 - 0.1\left(B_1 + \frac{k_2}{2}\right)\right] \tag{3.107}$$

$$= 0.1\left[1 - 0.1\left(0.099502 + \frac{0.098510}{2}\right)\right] = 0.098512 \tag{3.108}$$

$$k_4 = hf(x_1 + h, B_1 + k_3) = h\left[1 - 0.1(B_1 + k_3)\right] \tag{3.109}$$

$$= 0.1\left[1 - 0.1(0.099502 + 0.098512)\right] = 0.098020 \tag{3.110}$$

$$B_2 = B_1 + \frac{1}{6}(k_1 + 2k_2 + 2k_3 + k_4) \tag{3.111}$$

$$= 0.099005 + \frac{0.099005 + 2 \cdot 0.098510 + 2 \cdot 0.098512 + 0.098020}{6}$$

$$= 0.198013 \tag{3.112}$$

3. Iteration:

$$k_1 = hf(x_2, B_2) = h(1 - 0.1B_2) = 0.1(1 - 0.1 \cdot 0.198013) = 0.098020 \quad (3.113)$$

$$k_2 = hf\left(x_2 + \frac{h}{2}, B_2 + \frac{k_1}{2}\right) = h\left[1 - 0.1\left(B_2 + \frac{k_1}{2}\right)\right] \quad (3.114)$$

$$= 0.1\left[1 - 0.1\left(0.198013 + \frac{0.098020}{2}\right)\right] = 0.097530 \quad (3.115)$$

$$k_3 = hf\left(x_2 + \frac{h}{2}, B_2 + \frac{k_2}{2}\right) = h\left[1 - 0.1\left(B_2 + \frac{k_2}{2}\right)\right] \quad (3.116)$$

$$= 0.1\left[1 - 0.1\left(0.198013 + \frac{0.097530}{2}\right)\right] = 0.097532 \quad (3.117)$$

$$k_4 = hf(x_2 + h, B_2 + k_3) = h\left[1 - 0.1(B_2 + k_3)\right] \quad (3.118)$$

$$= 0.1\left[1 - 0.1(0.198013 + 0.097532)\right] = 0.097045 \quad (3.119)$$

$$B_3 = B_2 + \frac{1}{6}(k_1 + 2k_2 + 2k_3 + k_4) \quad (3.120)$$

$$= 0.198013 + \frac{0.098020 + 2 \cdot 0.097530 + 2 \cdot 0.097532 + 0.097045}{6} \quad (3.121)$$

$$= 0.295545 \quad (3.122)$$

All steps are summarized in Table 3.5. A Runge–Kutta method of order n ($n \leq 4$) possesses a local error of $O(h^{n+1})$. Methods of this family are usually avoided when solving stiff problems.

3.2.5 The backward Euler method

The (forward) Euler method from Section 3.2.2 approximates the points y_{i+1} by starting from some initial point, y_0, and moving to the right using the derivative calculated at y_i. An alternative to this is to start from y_0 and move on using the derivative calculated at y_{i+1}. This means that $f(y, x)$ is evaluated at (y_{l+1}, x_{l+1}) and results in the following recurrence relation:

$$y_{i+1} = y_i + hf(y_{i+1}, x_{i+1}) \quad (3.123)$$

Since both LHS and RHS depend on y_{i+1}, the backward Euler method is an implicit method. This method is also called a first order backward difference method. For linear ordinary initial value problems, it is always possible to rearrange the recurrence formula in such a way that all terms containing y_{i+1} are on one side of the formula, so that step $i + 1$ can be computed by simply inserting the results of the previous step into the recurrence formula. For nonlinear ODEs, however, it is usually necessary to solve a nonlinear equation in each step. Applying Equation (3.123) to the example problem from Section 3.2.2:

$$\dot{B}(\tau) = 1 - aB(\tau) \quad (3.124)$$

$$B(0) = 0 \quad (3.125)$$

leads to:

$$B_{i+1} = \frac{B_i + h}{1 + ha} \quad (3.126)$$

with h denoting the step size. The results for $h = 0.1$ are given in Table 3.6.

Table 3.5   Runge–Kutta method of order 4

Step	τ	k_1	k_2	k_3	k_4	B_{RK}	$B_{analytical}$	$B_{RK} - B_{analytical}$
0	0.0	—	—	—	—	0	0	0.000000000000000
1	0.1	0.10000	0.09950	0.09950	0.09900	0.09950166250	0.09950166251	−0.000000000000008
2	0.2	0.09900	0.09851	0.09851	0.09802	0.19801326692	0.19801326693	−0.000000000000016
3	0.3	0.09802	0.09753	0.09753	0.09704	0.29554466449	0.29554466451	−0.000000000000024
4	0.4	0.09704	0.09656	0.09656	0.09608	0.39210560844	0.39210560848	−0.000000000000032
5	0.5	0.09608	0.09560	0.09560	0.09512	0.48770575495	0.48770575499	−0.000000000000040
6	0.6	0.09512	0.09465	0.09465	0.09418	0.58235466411	0.58235466416	−0.000000000000047
7	0.7	0.09418	0.09371	0.09371	0.09324	0.67606180089	0.67606180094	−0.000000000000055
8	0.8	0.09324	0.09277	0.09278	0.09231	0.76883653607	0.76883653613	−0.000000000000062
9	0.9	0.09231	0.09185	0.09185	0.09139	0.86068814722	0.86068814729	−0.000000000000069
10	1.0	0.09139	0.09094	0.09094	0.09048	0.95162581956	0.95162581964	−0.000000000000076
11	1.1	0.09048	0.09003	0.09003	0.08958	1.04165864695	1.04165864703	−0.000000000000083
12	1.2	0.08958	0.08914	0.08914	0.08869	1.13079563274	1.13079563283	−0.000000000000089
13	1.3	0.08869	0.08825	0.08825	0.08781	1.21904569070	1.21904569079	−0.000000000000096
14	1.4	0.08781	0.08737	0.08737	0.08694	1.30641764591	1.30641764601	−0.000000000000102
15	1.5	0.08694	0.08650	0.08650	0.08607	1.39292023564	1.39292023575	−0.000000000000108
16	1.6	0.08607	0.08564	0.08564	0.08521	1.47856211022	1.47856211034	−0.000000000000115
17	1.7	0.08521	0.08479	0.08479	0.08437	1.56335183392	1.56335183404	−0.000000000000121
18	1.8	0.08437	0.08394	0.08395	0.08353	1.64729788576	1.64729788589	−0.000000000000126
19	1.9	0.08353	0.08311	0.08311	0.08270	1.73040866043	1.73040866057	−0.000000000000132
20	2.0	0.08270	0.08228	0.08228	0.08187	1.81269246908	1.81269246922	−0.000000000000138

Table 3.6   The θ-method

τ	B(τ)	Backward Euler	Absolute error	Crank–Nicolson	Absolute error
0.0	0.00000	0.00000	0.00000	0.00000000	0.000000
0.1	0.09950	0.09901	0.00049	0.09950249	0.000001
0.2	0.19801	0.19704	0.00097	0.19801490	0.000002
0.3	0.29554	0.29410	0.00145	0.29554709	0.000002
0.4	0.39211	0.39020	0.00191	0.39210881	0.000003
0.5	0.48771	0.48534	0.00236	0.48770972	0.000004
0.6	0.58235	0.57955	0.00281	0.58235937	0.000005
0.7	0.67606	0.67282	0.00324	0.67606724	0.000005
0.8	0.76884	0.76517	0.00367	0.76884269	0.000006
0.9	0.86069	0.85660	0.00409	0.86069500	0.000007
1.0	0.95163	0.94713	0.00450	0.95163336	0.000008
1.1	1.04166	1.03676	0.00490	1.04166686	0.000008
1.2	1.13080	1.12551	0.00529	1.13080450	0.000009
1.3	1.21905	1.21337	0.00567	1.21905520	0.000010
1.4	1.30642	1.30037	0.00605	1.30642779	0.000010
1.5	1.39292	1.38651	0.00641	1.39293099	0.000011
1.6	1.47856	1.47179	0.00677	1.47857347	0.000011
1.7	1.56335	1.55623	0.00713	1.56336379	0.000012
1.8	1.64730	1.63983	0.00747	1.64731042	0.000013
1.9	1.73041	1.72260	0.00781	1.73042175	0.000013
2.0	1.81269	1.80456	0.00814	1.81270612	0.000014

There are no restrictions on the step size, even for stiff problems. Applying Equation (3.123) to the test problem from Section 3.2.2

$$y'(x) = \lambda y(x), \quad y(x_0) = y_0 \tag{3.127}$$

we obtain the following recursion:

$$y_{i+1} = \frac{1}{1 - h\lambda} y_i \tag{3.128}$$

This approximation might be inaccurate but it will never blow up. Thus, the backward Euler method is unconditionally stable, unlike the forward Euler method, which bears a restriction on the step size for stiff problems.

3.2.6   The Crank–Nicolson method

In the context of solving ODEs, this method is better known as the second order Adams–Moulton method. It is an implicit method. This algorithm uses the recurrence relation:

$$y_{i+1} = y_i + \frac{h}{2} [f(x_i, y_i) + f(x_{i+1}, y_{i+1})] \tag{3.129}$$

Applying Equation (3.129) to the example problem from Section 3.2.2:

$$\dot{B}(\tau) = 1 - aB(\tau) \tag{3.130}$$

$$B(0) = 0 \tag{3.131}$$

leads to:

$$B_{i+1} = \frac{\left(1 - \frac{ah}{2}\right) B_i + h}{1 + \frac{ah}{2}} \tag{3.132}$$

with h denoting the step size. The results for $h = 0.1$ are given in Table 3.6.

A generalization of this method is the θ-*method*. This algorithm uses the recurrence relation:

$$y_{i+1} = y_i + h \left[(1 - \theta) f(x_i, y_i) + \theta f(x_{i+1}, y_{i+1}) \right] \tag{3.133}$$

It contains three previously discussed methods as special cases:

- the Euler method with $\theta = 0$;
- the Crank–Nicolson method with $\theta = 0.5$;
- the backward Euler method with $\theta = 1$.

3.2.7 Predictor–corrector methods

A predictor–corrector method is a set of two equations for y_{i+1}. The first equation, the predictor, is used to predict a first approximation for y_{i+1}. The second equation, the corrector, is used to compute a second, corrected, value for y_{i+1}. Predictor–corrector methods are a combination of an explicit method for the predictor and an implicit method for the corrector. The simplest predictor–corrector method is the *modified Euler method*, which uses the (forward) Euler method as predictor. The corrector uses the average value of y' at both endpoints of the interval $[x_i, x_{i+1}]$ as an approximation to the slope of the line element for the solution over that interval:

- predictor: $y_{i+1} = y_i + h y_i' = p(y_{i+1})$
- corrector: $y_{i+1} = y_i + \frac{h}{2} \left[p(y_{i+1}') + y_i' \right]$

Since $y'(x) = f(y, x)$ it also holds that:

$$p(y_{i+1}') = f \left[p(y_{i+1}), x_{i+1} \right] \tag{3.134}$$

This way, the intermediate result from the predictor enters the formula for the corrector.

Applying this technique to the example from Section 3.2.2 leads to the following computations:

1. Iteration:
   - Compute the predictor:

$$p(B_1) = B_0 + hf[t_0, B_0] \tag{3.135}$$
$$= B_0 + h[1 - 0.1 B_0] \tag{3.136}$$
$$= 0.0 + 0.1[1 - 0.1 \cdot 0.0] \quad = \quad 0.1 \tag{3.137}$$

   - Compute $p(B_1')$:

$$p(B_1') = f[t_1, p(B_1)] \tag{3.138}$$
$$= 1 - 0.1 B_1 \tag{3.139}$$
$$= 1 - 0.1 \cdot 0.1 \quad = \quad 0.99 \tag{3.140}$$

   - Compute the corrector:

$$B_1 = B_0 + \frac{h}{2} [p(B_1') + B_0'] \tag{3.141}$$

$$= B_0 + \frac{h}{2}[p(B'_1) + (1 - 0.1B_0)] \tag{3.142}$$

$$= 0.0 + \frac{0.1}{2}[0.99 + (1 - 0.1 \cdot 0.0)] \quad = \quad 0.0995 \tag{3.143}$$

2. Iteration:
   • Compute the predictor:

$$p(B_2) = B_1 + hf[t_1, B_1] \tag{3.144}$$

$$= B_1 + h[1 - 0.1B_1] \tag{3.145}$$

$$= 0.0995 + 0.1[1 - 0.1 \cdot 0.0995] \quad = \quad 0.198505 \tag{3.146}$$

   • Compute $p(B'_1)$:

$$p(B'_2) = f[t_2, p(B_2)] \tag{3.147}$$

$$= 1 - 0.1B_2 \tag{3.148}$$

$$= 1 - 0.1 \cdot 0.198505 \quad = \quad 0.9801495 \tag{3.149}$$

   • Compute the corrector:

$$B_2 = B_1 + \frac{h}{2}[p(B'_2) + B'_1] \tag{3.150}$$

$$= B_1 + \frac{h}{2}[p(B'_2) + (1 - 0.1B_1)] \tag{3.151}$$

$$= 0.0 + \frac{0.1}{2}[0.9801495 + (1 - 0.1 \cdot 0.0995)] \quad = \quad 0.19801 \tag{3.152}$$

All further iterations are reported in Table 3.7.

Table 3.7 Modified Euler method

τ	$B_{\text{analytical}}$	Predictor	$p(B'_2)$	Corrector
0.0	0.0000000	0.0000000	—	0.0000000
0.1	0.0995017	0.1000000	0.9900000	0.0995000
0.2	0.1980133	0.1985050	0.9801495	0.1980100
0.3	0.2955447	0.2960299	0.9703970	0.2955398
0.4	0.3921056	0.3925844	0.9607416	0.3920992
0.5	0.4877058	0.4881782	0.9511822	0.4876978
0.6	0.5823547	0.5828208	0.9417179	0.5823452
0.7	0.6760618	0.6765217	0.9323478	0.6760508
0.8	0.7688365	0.7692903	0.9230710	0.7688241
0.9	0.8606881	0.8611359	0.9138864	0.8606743
1.0	0.9516258	0.9520676	0.9047932	0.9516106
1.1	1.0416586	1.0420945	0.8957905	1.0416421
1.2	1.1307956	1.1312257	0.8868774	1.1307778
1.3	1.2190457	1.2194700	0.8780530	1.2190265
1.4	1.3064176	1.3068363	0.8693164	1.3063972
1.5	1.3929202	1.3933332	0.8606667	1.3928986
1.6	1.4785621	1.4789696	0.8521030	1.4785392
1.7	1.5633518	1.5637538	0.8436246	1.5633278
1.8	1.6472979	1.6476945	0.8352306	1.6472726
1.9	1.7304087	1.7307999	0.8269200	1.7303823
2.0	1.8126925	1.8130785	0.8186922	1.8126650

3.2.8 Adaptive techniques

Adaptive techniques vary the step size of a difference equation method of order n to keep the global error within a specified bound. The local error (and thus the global error) can be approximated by using a second difference equation method of order $n + 1$. The starting point is an nth order Taylor method

$$y(x_{i+1}) = y(x_i) + h\phi(x_i, y(x_i), h) + O(h^{n+1}) \tag{3.153}$$

producing the approximation

$$w_{i+1} = w_i + h\phi(x_i, w_i, h) \qquad \text{for} \qquad i > 0 \tag{3.154}$$

and satisfying, for some K,

$$|y(x_i) - w_i| < Kh^n \tag{3.155}$$

Resulting from an $(n + 1)$th order Taylor method, another difference equation method is given by

$$\hat{w}_{i+1} = \hat{w}_i + h\hat{\phi}(x_i, \hat{w}_i, h) \qquad \text{for} \qquad i > 0 \tag{3.156}$$

satisfying

$$|y(x_i) - \hat{w}_i| < \hat{K}h^{n+1} \tag{3.157}$$

At point x_i, we assume

$$w_i = \hat{w}_i =: z(x_i) \tag{3.158}$$

and write

$$z(x_i + h) - w_{i+1} = (\hat{w}_{i+1} - w_{i+1}) + (z(x_i + h) - \hat{w}_{i+1}) \tag{3.159}$$

Since the LHS term of this equation is $O(h^{n+1})$, whereas the second term on the right side is $O(h^{n+2})$, the first term on the right side dominates, and we can approximate

$$z(x_i + h) - w_{i+1} \cong \hat{w}_{i+1} - w_{i+1} \tag{3.160}$$

Then, K can be approximated by

$$K \cong \frac{|\hat{w}_{i+1} - w_{i+1}|}{h^{n+1}} \tag{3.161}$$

Returning to the requirements we would like to meet, namely the limitation of the global error,

$$|y(x_i + qh) - w_{i+1}| < K(qh)^n < \epsilon \tag{3.162}$$

(with w_{i+1} using a new step size qh), which of course also limits the local error, such that Equations (3.161) and (3.162) lead to

$$Kq^n h^n \approx \frac{q^n |\hat{w}_{i+1} - w_{i+1}|}{h} < \epsilon \tag{3.163}$$

or

$$q < \left(\frac{\epsilon h}{|\hat{w}_{i+1} - w_{i+1}|} \right)^{1/n} \tag{3.164}$$

We can now outline the whole procedure:

1. Calculate w_{i+1} and $\hat{w}_{i+1}$ with difference equation methods of order n and $n + 1$ respectively.
2. Determine q with these values and Equation (3.163).
3. Again, calculate w_{i+1} and $\hat{w}_{i+1}$, but now using the new step size qh.
4. Take the last step size qh and the resulting w_{i+1} and $\hat{w}_{i+1}$ and predict a new step size from it:

$$q_{new} = \left(\frac{\epsilon q h}{|\hat{w}_{i+1} - w_{i+1}|} \right)^{1/n} \tag{3.165}$$

One popular adaptive method is the *Runge–Kutta–Fehlberg (RKF) method*. It uses a Runge–Kutta method of order 5,

$$\hat{w}_{i+1} = \hat{w}_i + \frac{16}{135}k_1 + \frac{6656}{12\,825}k_3 + \frac{28\,561}{56\,430}k_4 - \frac{9}{50}k_5 + \frac{2}{55}k_6 \tag{3.166}$$

to estimate the local error in a Runge–Kutta method of order 4,

$$w_{i+1} = w_i + \frac{25}{216}k_1 + \frac{1408}{2565}k_3 + \frac{2197}{4104}k_4 - \frac{1}{5}k_5 \tag{3.167}$$

where

$$k_1 = hf(x_i, w_i) \tag{3.168}$$

$$k_2 = hf\left(x_i + \frac{h}{4}, w_i + \frac{1}{4}k_1\right) \tag{3.169}$$

$$k_3 = hf\left(x_i + \frac{3}{8}h, w_i + \frac{3}{32}k_1 + \frac{9}{32}k_2\right) \tag{3.170}$$

$$k_4 = hf\left(x_i + \frac{12}{13}h, w_i + \frac{1932}{2197}k_1 - \frac{7200}{2197}k_2 + \frac{7296}{2197}k_3\right) \tag{3.171}$$

$$k_5 = hf\left(x_i + h, w_i + \frac{439}{216}k_1 - 8k_2 + \frac{3680}{513}k_3 - \frac{845}{4104}k_4\right) \tag{3.172}$$

$$k_6 = hf\left(x_i + \frac{h}{2}, w_i - \frac{8}{27}k_1 + 2k_2 - \frac{3544}{2565}k_3 + \frac{1859}{4104}k_4 - \frac{11}{40}k_5\right) \tag{3.173}$$

This choice of Runge–Kutta methods requires only six evaluations of f per step, thus reducing the evaluation cost compared to two arbitrary Runge–Kutta methods. The usual choice of q is

$$q = \left(\frac{\epsilon h}{2|\hat{w}_{i+1} - w_{i+1}|} \right)^{1/4} \tag{3.174}$$

As a demonstration, the test problem from the previous section has been solved with the above method, as implemented in routine RKFVSM55 by Burden and Faires (1998). The results are given in Table 3.8. Note that the step size is not varied much, due to the smooth character of the solution to this problem.

3.2.9 Methods for systems of equations

Numerical methods for ordinary initial value problems can usually be extended easily to systems of ordinary initial value problems. Usually, the techniques of this section are also

Table 3.8 The Runge–Kutta–Fehlberg method

τ	Step size	RKF method	$B(\tau)$	Absolute error
0.0000000000000	0.1401172714671	0.0000000000000	0.00000000000000	0.00000000000000
0.1401172714671	0.1401954558537	0.1391401978061	0.13914019780	0.00000000000007
0.2803127273208	0.1406873087141	0.2764204196553	0.27642041964	0.00000000000014
0.4210000360349	0.1411806200003	0.4122610506838	0.41226105066	0.00000000000021
0.5621806560351	0.1416774030021	0.5466703119486	0.54667031192	0.00000000000027
0.7038580590372	0.1421776788463	0.6796583375505	0.67965833752	0.00000000000034
0.8460357378835	0.1426814910450	0.8112352107399	0.81123521070	0.00000000000041
0.9887172289286	0.1431888726625	0.9414109847443	0.94141098470	0.00000000000047
1.1319061015910	0.1436998639083	1.0701956725472	1.07019567249	0.00000000000054
1.2756059654993	0.1442145044945	1.1975992529953	1.19759925294	0.00000000000060
1.4198204699938	0.1447328315082	1.3236316697592	1.32363166969	0.00000000000066
1.5645533015020	0.1452548865505	1.4483028285470	1.44830282847	0.00000000000072
1.7098081880524	0.1457807074152	1.5716226005644	1.57162260049	0.00000000000079
1.8555888954677	0.1444111045323	1.6936008187620	1.69360081868	0.00000000000085
2.0000000000000	—	1.8126924693104	1.81269246922	0.00000000000090

applied to linear systems with constant coefficients. Problems of this kind can be solved analytically by finding the eigenvalues of the matrix of coefficients. For large matrices, this often cannot be done analytically, so a numerical algorithm has to be applied. Since it is often more tricky to find eigenvalues numerically, the system of ODEs is solved directly via numerical techniques. The extension from a single ODE to a system of ODEs is straightforward and will be explained with the help of Euler's method introduced in Section 3.2.2. As an example, we will employ the Riccati system of Equations (3.26) to (3.28). With the parameter choice of the Vasicek model (Vasicek, 1977), the market data are assumed to be $a = b = 0.1$.

$$h(t) = ab = 0.1 \cdot 0.1 = 0.01 \tag{3.175}$$

$$d(t) = \sigma^2 = 0.02^2 = 0.0004 \tag{3.176}$$

$$g(t) = -a = -0.1 \tag{3.177}$$

$$c(t) = 0 \tag{3.178}$$

and, reversing time, we end up with the following system:

$$A_\tau = \frac{0.02^2}{2}B^2 - 0.1 \cdot 0.1 \cdot B \tag{3.179}$$

$$B_\tau = 1 - 0.1B \tag{3.180}$$

$$B(0) = A(0) = 0 \tag{3.181}$$

The ODEs are not independent, since $B(\tau)$ is needed to solve for $A(\tau)$. As in the previous examples, we will denote the independent variable by x. Again, we assume that $h = x_i - x_{i-1}$ is constant, i.e. a constant step size. Then, the Euler discretization of the above system is given by:

$$A_{n+1} = A_n + hf_1(x_n, A_n, B_n) \tag{3.182}$$

$$= A_n + h(0.0002B_n^2 - 0.01B_n) \tag{3.183}$$

$$B_{n+1} = B_n + hf_2(x_n, A_n, B_n) \tag{3.184}$$

$$= B_n + h(1 - 0.1B_n) \tag{3.185}$$

Assuming a step size of $h = 0.1$, the first iterations are given by:

1. Iteration:

$$B_1 = B_0 + hf_1(x_0, B_0) \tag{3.186}$$

$$= B_0 + h(1 - 0.1B_0) \tag{3.187}$$

$$= 0.0 + 0.1\,(1 - 0.1 \cdot 0.0) \tag{3.188}$$

$$= 0.1 \tag{3.189}$$

$$A_1 = A_0 + hf_2(x_0, A_0, B_0) \tag{3.190}$$

$$= A_0 + h\left(0.0002B_0^2 - 0.01B_0\right) \tag{3.191}$$

$$= 0 + 0.1\left(0.0002 \cdot 0^2 - 0.01 \cdot 0\right) \tag{3.192}$$

$$= 0 \tag{3.193}$$

2. Iteration:

$$B_2 = B_1 + hf_1(x_1, B_1) \tag{3.194}$$
$$= B_1 + h(1 - 0.1B_1) \tag{3.195}$$
$$= 0.1 + 0.1\,(1 - 0.1 \cdot 0.1) \tag{3.196}$$
$$= 0.199 \tag{3.197}$$
$$A_2 = A_1 + hf_2(x_1, A_1, B_1) \tag{3.198}$$
$$= A_1 + h\left(0.0002B_1^2 - 0.01B_1\right) \tag{3.199}$$
$$= 0 + 0.1\left(0.0002 \cdot 0.1^2 - 0.01 \cdot 0.1\right) \tag{3.200}$$
$$= -0.0000998 \tag{3.201}$$

3. Iteration:

$$B_3 = B_2 + hf_1(x_2, B_2) \tag{3.202}$$
$$= B_2 + h(1 - 0.1B_2) \tag{3.203}$$
$$= 0.199 + 0.1\,(1 - 0.1 \cdot 0.199) \tag{3.204}$$
$$= 0.29701 \tag{3.205}$$
$$A_3 = A_2 + hf_2(x_2, A_2, B_2) \tag{3.206}$$
$$= A_2 + h\left(0.0002B_2^2 - 0.01B_2\right) \tag{3.207}$$
$$= -0.0000999 + 0.1\left(0.0002 \cdot 0.199^2 - 0.01 \cdot 0.199\right) \tag{3.208}$$
$$= -0.0002980 \tag{3.209}$$

4. Iteration:

$$B_4 = B_3 + hf_1(x_3, B_3) \tag{3.210}$$
$$= B_3 + h(1 - 0.1B_3) \tag{3.211}$$
$$= 0.29701 + 0.1\,(1 - 0.1 \cdot 0.29701) \tag{3.212}$$
$$= 0.39404 \tag{3.213}$$
$$A_4 = A_3 + hf_2(x_3, A_3, B_3) \tag{3.214}$$
$$= A_3 + h\left(0.0002B_3^2 - 0.01B_3\right) \tag{3.215}$$
$$= -0.0002985 + 0.1\left(0.0002 \cdot 0.29701^2 - 0.01 \cdot 0.29701\right) \tag{3.216}$$
$$= -0.0005933 \tag{3.217}$$

The above steps and some of the subsequent ones are summarized in Table 3.9.

Next, we solve the same problem with a Runge–Kutta scheme of order 4, as introduced in Section 3.2.4. Data for this problem are shown in Tables 3.10 and 3.11. The adaptation to systems of ODEs is the same as with Euler's method.

In later chapters of this book, we will see that FE discretizations of the spatial variable(s) of a parabolic problem lead to systems of the form:

$$\mathbf{M}\dot{\mathbf{y}} + \mathbf{K}\mathbf{y} = \mathbf{J} \tag{3.218}$$

It is possible to transform the above system to the form discussed so far

$$\dot{\mathbf{z}} = \mathbf{B}\mathbf{z} + \mathbf{g} \tag{3.219}$$

Table 3.9 Euler's method. The column $A_{semiana.}$ refers to the A_is computed with Euler's method but using *analytical* B_{i-1}s as inputs

τ	A_{Euler}	$A_{anal.}$	$A_{semianal.}$	$A_{Eul.} - A_{aral.}$	Bond (numerical)	Bond (analytical)	Difference
0.0	0.0000000	0.0000	0.0000000	0.0000	1.0000	1.0000	0.000%
0.1	0.0000000	0.0000	0.0000000	0.0000	0.9900	0.9900	0.000%
0.2	−0.0000998	−0.0002	−0.0000993	0.0001	0.9802	0.9802	0.000%
0.3	−0.0002980	−0.0004	−0.0002971	0.0001	0.9704	0.9704	0.000%
0.4	−0.0005933	−0.0008	−0.0005923	0.0002	0.9608	0.9608	0.000%
0.5	−0.0009842	−0.0012	−0.0009837	0.0002	0.9512	0.9512	0.001%
0.6	−0.0014695	−0.0018	−0.0014701	0.0003	0.9418	0.9418	0.001%
0.7	−0.0020478	−0.0024	−0.0020504	0.0003	0.9324	0.9324	0.001%
0.8	−0.0027179	−0.0031	−0.0027235	0.0004	0.9231	0.9231	0.002%
0.9	−0.0034786	−0.0039	−0.0034883	0.0004	0.9140	0.9139	0.003%
1.0	−0.0043284	−0.0048	−0.0043438	0.0004	0.9049	0.9049	0.004%
1.1	−0.0052663	−0.0058	−0.0052888	0.0005	0.8959	0.8959	0.005%
1.2	−0.0062910	−0.0068	−0.0063222	0.0005	0.8870	0.8870	0.006%
1.3	−0.0074014	−0.0080	−0.0074432	0.0005	0.8782	0.8781	0.007%
1.4	−0.0085962	−0.0092	−0.0086505	0.0005	0.8695	0.8694	0.009%
1.5	−0.0098742	−0.0105	−0.0099433	0.0005	0.8609	0.8608	0.011%
1.6	−0.0112345	−0.0119	−0.0113204	0.0006	0.8524	0.8522	0.013%
1.7	−0.0126758	−0.0134	−0.0127810	0.0006	0.8439	0.8438	0.016%
1.8	−0.0141970	−0.0149	−0.0143240	0.0006	0.8356	0.8354	0.018%
1.9	−0.0157971	−0.0166	−0.0159484	0.0006	0.8273	0.8271	0.021%
2.0	−0.0174750	−0.0183	−0.0176534	0.0006	0.8191	0.8189	0.024%

Table 3.10 Runge–Kutta method of order 4

τ	A_{RK}	$A_{anal.}$	$A_{RK} - A_{anal.}$	B_{RK}	$B_{anal.}$	$B_{RK} - B_{anal.}$
0.0	0.0000000000	0.0000000000	0.0000000000	0.0000000000	0.0000000000	0.0000000000
0.1	−0.0000497676	−0.0000497676	0.0000000000	0.0995016690	0.0995016625	0.0000000065
0.2	−0.0001981479	−0.0001981479	0.0000000000	0.1980132800	0.1980132669	0.0000000131
0.3	−0.0004437735	−0.0004437735	0.0000000000	0.2955446800	0.2955446645	0.0000000155
0.4	−0.0007852982	−0.0007852981	0.0000000000	0.3921056400	0.3921056085	0.0000000315
0.5	−0.0012213966	−0.0012213965	−0.0000000001	0.4877058000	0.4877057550	0.0000000450
0.6	−0.0017507640	−0.0017507638	−0.0000000002	0.5823547200	0.5823546642	0.0000000558
0.7	−0.0023721156	−0.0023721155	−0.0000000001	0.6760618700	0.6760618009	0.0000000691
0.8	−0.0030841869	−0.0030841867	−0.0000000002	0.7688366200	0.7688365361	0.0000000839
0.9	−0.0038857325	−0.0038857323	−0.0000000002	0.8606882100	0.8606881473	0.0000000627
1.0	−0.0047755265	−0.0047755261	−0.0000000004	0.9516258800	0.9516258196	0.0000000604
1.1	−0.0057523614	−0.0057523610	−0.0000000004	1.0416588000	1.0416586470	0.0000001530
1.2	−0.0068150489	−0.0068150481	−0.0000000008	1.1307957000	1.1307956328	0.0000000672
1.3	−0.0079624178	−0.0079624171	−0.0000000007	1.2190458000	1.2190456908	0.0000001092
1.4	−0.0091933161	−0.0091933154	−0.0000000007	1.3064177000	1.3064176460	0.0000000540
1.5	−0.0105066090	−0.0105066079	−0.0000000011	1.3929203000	1.3929202357	0.0000000643
1.6	−0.0119011770	−0.0119011771	0.0000000001	1.4785621000	1.4785621103	−0.0000000103
1.7	−0.0133759230	−0.0133759222	−0.0000000008	1.5633519000	1.5633518340	0.0000000660
1.8	−0.0149297600	−0.0149297595	−0.0000000005	1.6472980000	1.6472978859	0.0000001141
1.9	−0.0165616220	−0.0165616213	−0.0000000007	1.7304088000	1.7304086606	0.0000001394
2.0	−0.0182704570	−0.0182704565	−0.0000000005	1.8126926000	1.8126924692	0.0000001308

Table 3.11   Bond prices resulting from Runge–Kutta method of order 4

Bond (numerical)	Bond (analytical)	Difference
1.0000000000	1.0000000000	0.0000000
0.9900498986	0.9900498993	0.0000001
0.9801991870	0.9801991883	0.0000001
0.9704472401	0.9704472416	0.0000002
0.9607934147	0.9607934178	0.0000003
0.9512370566	0.9512370610	0.0000005
0.9417774961	0.9417775015	0.0000006
0.9324140506	0.9324140572	0.0000007
0.9231460261	0.9231460340	0.0000009
0.9139727212	0.9139727271	0.0000006
0.9048934161	0.9048934219	0.0000006
0.8959073804	0.8959073945	0.0000016
0.8870139063	0.8870139129	0.0000007
0.8782122273	0.8782122375	0.0000012
0.8695016165	0.8695016218	0.0000006
0.8608813069	0.8608813134	0.0000008
0.8523505552	0.8523505543	−0.0000001
0.8439085756	0.8439085818	0.0000007
0.8355546192	0.8355546292	0.0000012
0.8272879139	0.8272879260	0.0000015
0.8191076877	0.8191076988	0.0000014

with the help of a transformation $\mathbf{y} = \mathbf{V}\mathbf{z}$. For details of this transformation, see page 421 of Bickford (1990). Usually, however, numerical algorithms are applied directly to Equation (3.218). Most FE codes solve Equation (3.218) with the θ-method, as introduced in Section 3.2.6. The recurrence relation Equation (3.133) applied to systems $\dot{\mathbf{y}} = \mathbf{f}(\mathbf{x}, \mathbf{y})$ changes to:

$$\mathbf{y}_{n+1} = \mathbf{y}_n + h\left[(1 - \theta)\dot{\mathbf{y}}_n + \theta\dot{\mathbf{y}}_{n+1}\right] \qquad (3.220)$$

Multiplying Equation (3.220) by $\mathbf{M}$ leads to

$$\mathbf{M}\mathbf{y}_{n+1} = \mathbf{M}\mathbf{y}_n + \mathbf{M}h\left[(1 - \theta)\dot{\mathbf{y}}_n + \theta\dot{\mathbf{y}}_{n+1}\right] \qquad (3.221)$$

$$= \mathbf{M}\mathbf{y}_n + h(1 - \theta)\mathbf{M}\underbrace{\dot{\mathbf{y}}_n}_{=\mathbf{f}_n - \mathbf{k}\mathbf{y}_n} + h\theta\underbrace{\mathbf{M}\dot{\mathbf{y}}_{n+1}}_{=\mathbf{f}_{n+1} - \mathbf{k}\mathbf{y}_{n+1}} \qquad (3.222)$$

The recurrence formula is given by:

$$(\mathbf{M} + h\theta\mathbf{k})\mathbf{y}_{n+1} = \left[\mathbf{M} - (1 - \theta)h\mathbf{k}\right]\mathbf{y}_n + h\left[\theta\mathbf{f}_{n+1} + (1 - \theta)\mathbf{f}_n\right] \qquad (3.223)$$

Stability analysis is more complicated than for single initial value problems, since it also depends on the eigenvalues of $\mathbf{M}$ and $\mathbf{k}$. The interested reader is referred to Chapter 4 of Bickford (1990), where stability analysis is discussed for FE discretizations of parabolic PDEs self-adjoint in the spatial variable(s).

3.3   ORDINARY TWO-POINT BOUNDARY VALUE PROBLEMS

3.3.1   Introductory remarks

This section provides a quick introduction to two popular techniques of solving ordinary two-point boundary value problems: FDM and shooting methods. Both techniques are quite general and can be applied to linear and nonlinear problems. A third general solution technique for this kind of problem, the FEM, is discussed in Chapter 4. The problem to be solved is given by:

$$u''(x) + b_1(x)u'(x) + b_2(x)u(x) = b_3(x) \tag{3.224}$$

$$u(x^{\min}) = \gamma_1 \tag{3.225}$$

$$u(x^{\max}) = \gamma_2 \tag{3.226}$$

i.e. we do not discuss the treatment of Neumann and Robin BCs. Problems of this kind regularly result from economic applications of calculus of variations and optimal control. Usually, the independent variable is time. This tends to hide the fundamental difference from ordinary initial value problems. An initial value problem of order n is provided with n initial conditions, fixing the value of the ODE at hand at the starting point and the values of its first $n - 1$ derivatives. All numerical techniques use the information of these n initial conditions to roll out the solutions to later points in time. The user of any numerical technique has to decide up to what point in time the solution is to be computed. So the length of the domain of the ODE is not part of the problem, but under the control of the person solving this problem. For convenience, the user will continue the computations only to that point in time which is of interest to him. Two-point boundary value problems, however, have BCs specified at both ends of the domain, so that the domain is fixed. The types of BC allowed in order to provide a unique solution are discussed in Section A.3.3. This clear-cut distinction between initial value and boundary value problems is somewhat blurred by the following facts:

• Dynamic optimization problems in economics, i.e. applications of variational calculus and optimal control, usually deal with an infinite time horizon, which makes these problems at first sight look like initial value problems. However, they are two-point boundary value problems and a BC needs to be imposed on the problem at infinity. This condition at infinity is called the *transversality condition*.

• In finance, many initial value problems are backwards in time, i.e. the argument is not time but time to maturity. As an example, we might think of a bond. The value of the bond at maturity is known; the investor wants to know what the bond is worth today. Since time is backward, the numerical procedure starts at the final condition and is rolled backwards in time, usually up to today. This way, the length of the computational domain is fixed. However, no BC has to be applied at the point in time which is associated with today.

A further complication in the distinction of both problem classes can be found in the nature of shooting methods. Shooting methods replace the boundary value problems by a series of initial value problems. To understand the basic idea of a shooting method, the user must be aware of the difference between initial value and boundary value problems.

3.3.2   Finite difference methods

As with initial value problems, the basic idea of FDMs is to replace the derivatives in the ODE by FD approximations. First, the domain of the ODE, a closed interval, is subdivided into N intervals. For simplicity, we assume that they are of equal length h. Then, the following substitutions are necessary:

$$u(x) \to u(x_i) \tag{3.227}$$

$$u'(x) \to \frac{1}{2h}[u(x_{i+1}) - u(x_{i-1})] \tag{3.228}$$

$$u''(x) \to \frac{1}{h^2}[u(x_{i+1}) - 2u(x_i) + u(x_{i-1})] \tag{3.229}$$

Inserting these approximations in Equation (3.224) leads to:

$$\frac{1}{h^2}[u(x_{i+1}) - 2u(x_i) + u(x_{i-1})] + b_1(x_i)\frac{1}{2h}[u(x_{i+1}) - u(x_{i-1})] + b_2(x_i)u(x_i) = b_3(x_i) \tag{3.230}$$

for $1 \leq i \leq N - 1$. The BCs are integrated accordingly:

$$u(x_0) = \gamma_1 \tag{3.231}$$
$$u(x_N) = \gamma_2 \tag{3.232}$$

As a first example, we consider the problem of a pure Brownian motion leaving a prescribed region, see Section B.3.5:

$$\frac{1}{2}\frac{d^2u}{dx^2} = -1 \tag{3.233}$$
$$u(x^{\min}) = 0 \tag{3.234}$$
$$u(x^{\max}) = 0 \tag{3.235}$$

Instead of Equation (3.233) we consider the equivalent formulation:

$$u''(x) = -2 \tag{3.236}$$

The example is a special case of Equation (3.224) with:

$$b_1(x) = 0 \tag{3.237}$$
$$b_2(x) = 0 \tag{3.238}$$
$$b_3(x) = -2 \tag{3.239}$$

so that its approximate formulation reads:

$$\frac{1}{h^2}[u(x_{i-1}) - 2u(x_i) + u(x_{i+1})] = -2 \quad \forall \ 1 \leq i \leq N - 1 \tag{3.240}$$

$$u(x_0) = 0 \tag{3.241}$$
$$u(x_N) = 0 \tag{3.242}$$

Consequently, we have $N - 1$ unknowns and $N - 1$ equations. For $N = 4$, we get:

$$u(x_0) = 0 \tag{3.243}$$

$$\frac{1}{h^2}[u(x_0) - 2u(x_1) + u(x_2)] = -2 \tag{3.244}$$

$$\frac{1}{h^2}[u(x_1) - 2u(x_2) + u(x_3)] = -2 \tag{3.245}$$

$$\frac{1}{h^2}[u(x_2) - 2u(x_3) + u(x_4)] = -2 \tag{3.246}$$

$$u(x_4) = 0 \tag{3.247}$$

These equations may be written in matrix form:

$$\frac{1}{h^2}\begin{pmatrix} 1 & 0 & 0 & 0 & 0 \\ 1 & -2 & 1 & 0 & 0 \\ 0 & 1 & -2 & 1 & 0 \\ 0 & 0 & 1 & -2 & 1 \\ 0 & 0 & 0 & 0 & 1 \end{pmatrix}\begin{pmatrix} u(x_0) \\ u(x_1) \\ u(x_2) \\ u(x_3) \\ u(x_4) \end{pmatrix} = \begin{pmatrix} 0 \\ -2 \\ -2 \\ -2 \\ 0 \end{pmatrix} \qquad (3.248)$$

Since $x_0 = 0$ and $x_N = 2$ and, consequently, $h = 0.5$, the above system reduces to:

$$\begin{pmatrix} 4 & 0 & 0 & 0 & 0 \\ 4 & -8 & 4 & 0 & 0 \\ 0 & 4 & -8 & 4 & 0 \\ 0 & 0 & 4 & -8 & 4 \\ 0 & 0 & 0 & 0 & 4 \end{pmatrix}\begin{pmatrix} u(x_0) \\ u(x_1) \\ u(x_2) \\ u(x_3) \\ u(x_4) \end{pmatrix} = \begin{pmatrix} 0 \\ -2 \\ -2 \\ -2 \\ 0 \end{pmatrix} \qquad (3.249)$$

This system is solved by $(0, 0.75, 1, 0.75, 0)^T$. This solution and the analytical solution of the problem are both plotted in Figure 3.1. Note that the numerical solution consists of a list of points. In Figure 3.1, these points have been connected with linear segments. The matrix is symmetric because the problem under inspection is self-adjoint. If the ODE under inspection were nonlinear, its approximate solution would have been a system of nonlinear equations.

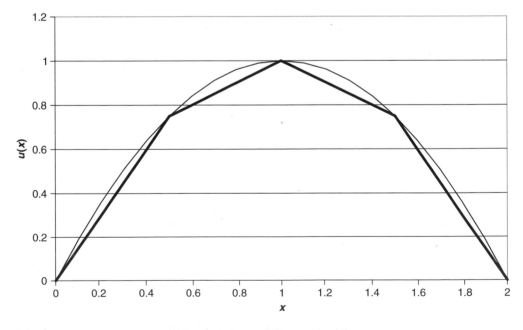

Figure 3.1   FD solution of the first exit time of Brownian motion

3.3.3 Shooting methods

The basic idea of all shooting methods is to replace the two-point boundary value problem:

$$u''(x) = f(x, u, u') \tag{3.250}$$

$$u(x^{\min}) = \gamma_1 \tag{3.251}$$

$$u(x^{\max}) = \gamma_2 \tag{3.252}$$

by an initial value problem:

$$u''(x) = f(x, u, u') \tag{3.253}$$

$$u(x^{\min}) = \gamma_1 \tag{3.254}$$

$$u'(x^{\min}) = M \tag{3.255}$$

Note that the above problem is nonlinear. The variable M is varied in such a manner to ensure that:

$$\lim_{n \to \infty} u(x^{\max}, M) = u(x^{\max}) = \gamma_2 \tag{3.256}$$

At first, some value for M has to be guessed. Then the problem in Equations (3.253) to (3.255) is solved with this guessed value for M. Next, the difference between the numerical value $u(x^{\max})$ and the value from the right-hand BC γ_2 is determined. Then, a second value for M has to be selected in such a way as to decrease the difference $u(x^{\max}) - \gamma_2$. This procedure is repeated until this difference falls under some prescribed threshold:

$$u(x^{\max}, M) \quad \gamma_2 \le \epsilon \tag{3.257}$$

The name of this method stems from the analogy to the procedure of firing at a stationary target. The first shot of the cannon is made with a guess of the appropriate angle. Then, for each next shot, the angle is varied in order to minimize the distance between the location of burst and the target. Instead of choosing M by trial and error, the process can be automated by combining the solver of the initial value problem with some root-finding technique. Popular choices are Newton-type algorithms. Alternatively, the solver of the initial value problem can be combined with an optimization technique that tries to minimize $[u(x^{\max}, M) - \gamma_2]^2$ by manipulating M.

Solely based on the large number of solvers for initial value problems, nonlinear systems and optimization problems, many variants of the shooting method can be set up easily. The shooting approach can be simplified somewhat for linear problems; see Section 11.2 of Burden and Faires (1998).

3.4 INITIAL BOUNDARY VALUE PROBLEMS

3.4.1 The explicit scheme

Since many textbooks on option pricing, such as Deutsch (2002), Hull (2000) and Wilmott (1998), provide thorough introductions to the approximate solution of PDEs arising in finance with FD, we will not repeat this material here. Instead, we will deliver a brief outline of the basic ideas of the most prominent FD methods. To avoid clutter, we employ the heat equation to which the Black–Scholes PDE (2.30) can be transformed. There are several (different) ways of transforming PDE (2.30) to $u_\tau = u_{xx}$; we present the one given by Seydel (2002):

$$S = Ke^x \tag{3.258}$$

$$t = T - \frac{2\tau}{\sigma^2} \tag{3.259}$$

$$q = \frac{2r}{\sigma^2} \tag{3.260}$$

$$V(S, t) = V(Ee^x, T - \frac{2r}{\sigma^2}) =: v(x, \tau) \tag{3.261}$$

$$v(x, \tau) =: E \exp\left(-0.5(q - 1)x - [0.25(q - 1)^2 + q]\tau\right) \tag{3.262}$$

The transformed IC for the call results in:

$$y(x, 0) = \max\left(e^{0.5x(q+1)} - e^{0.5x(q-1)}, 0\right) \tag{3.263}$$

An obvious disadvantage of this transformation is that the original domain $(S, t) \in [0, \infty) \times [0, T]$ becomes

$$-\infty < x < \infty, \quad 0 \leq \tau \leq 0.5\sigma^2 T \tag{3.264}$$

so that *two* BCs need to be approximated. For a call, this results in:

$$\lim_{x \to -\infty} y(x, \tau) = 0 \tag{3.265}$$

$$\lim_{x \to \infty} y(x, \tau) = \exp\left(0.5(q + 1)x + 0.25(q + 1)^2\tau\right) \tag{3.266}$$

As a first step, the (x, τ)-domain has to be discretized. We assume that it is possible to truncate the domain so that we can approximate the PDE on a compact spatial domain $[x^{\min}, x^{\max}]$. We also assume that it is possible to state Dirichlet BCs at both $x^{\min}$ and $x^{\max}$. The time-step size is given by k. The interval between two nodes in the spatial direction is given by:

$$h = \frac{x^{\max} - x^{\min}}{m} \tag{3.267}$$

with m being the number of spatial steps. The grid points, also called nodes, are (x_i, τ_j) with:

$$x_i = ih \quad \forall \quad i = 0, 1, \ldots, m \tag{3.268}$$

$$\tau_j = jk \quad \forall \quad j = 0, 1, \ldots \tag{3.269}$$

We approximate both u_τ and u_{xx} by their differences:

$$\frac{\partial u}{\partial \tau}(x_i, \tau_j) \approx \frac{u(x_i, \tau_j + k) - u(x_i, \tau_j)}{k} := \frac{w_{i,j+1} - w_{i,j}}{k} \tag{3.270}$$

$$\frac{\partial^2 u}{\partial x^2}(x_i, \tau_j) \approx \frac{u(x_i + h, \tau_j) - 2u(x_i, \tau_j) + u(x_i - h, \tau_j)}{h^2}$$

$$:= \frac{w_{i+1,j} - 2w_{i,j} + w_{i-1,j}}{h^2} \tag{3.271}$$

so that $w_{i,j}$ approximates $u(x_i, \tau_j)$. The local truncation error is of order $O(k + h^2)$, i.e. linear in the time discretization and quadratic in the spatial discretization. At interior grid points the

PDE $u_\tau = u_{xx}$ is valid, so that there the following approximation holds:

$$\frac{w_{i,j+1} - w_{i,j}}{k} - \frac{w_{i+1,j} - 2w_{i,j} + w_{i-1,j}}{h^2} = 0 \tag{3.272}$$

Solving the above difference equation for $w_{i,j+1}$ gives:

$$w_{i,j+1} = \left(1 - \frac{2k}{h^2}\right) w_{ij} + \frac{k}{h^2}(w_{i+1,j} + w_{i-1,j}) \tag{3.273}$$

The above difference equation is valid for each $i = 1, 2, \ldots, m$ and $j = 1, 2, \ldots$. The values for $w_{i,0}$ are taken from the IC associated with the PDE $u_\tau = u_{xx}$. Accordingly, the values for $w_{0,j}$ and $w_{m,j}$ are taken from the BCs. The explicit nature of this algorithm becomes clearer when stated in matrix form. Define the discretized IC:

$$\mathbf{w}^{(0)} = (w_{1,0}, \ldots, w_{m-1,0})^\mathsf{T} \tag{3.274}$$

Additionally, we need:

$$\mathbf{w}^{(j)} = (w_{1,j}, \ldots, w_{m-1,j})^\mathsf{T} \quad \bar{\mathbf{w}}^{(j)} = (w_{0,j}, \mathbf{w}^{(j)}, w_{m,j})^\mathsf{T} \tag{3.275}$$

$$\mathbf{A} = \begin{pmatrix} \lambda & (1-2\lambda) & \lambda & 0 & \cdots & 0 & 0 \\ 0 & \lambda & (1-2\lambda) & \lambda & \ddots & \vdots & \vdots \\ \vdots & 0 & \ddots & \ddots & \ddots & 0 & \vdots \\ \vdots & \vdots & \ddots & \ddots & \ddots & \lambda & 0 \\ 0 & 0 & \cdots & 0 & \lambda & (1-2\lambda) & \lambda \end{pmatrix} \tag{3.276}$$

with $\lambda = \frac{k}{h^2}$. The approximate solution to the initial boundary value problem is given by:

$$\mathbf{w}^{(j)} = \mathbf{A}\bar{\mathbf{w}}^{(j-1)} \quad \text{for each} \quad j = 1, 2, \ldots \tag{3.277}$$

This method is called the *forward difference method* because the time axis is discretized with forward differences. It is also called an *explicit method* because $\mathbf{w}^{(j)}$ can be explicitly stated in terms of $\mathbf{w}^{(j-1)}$. This advantage comes at the cost of severe limitations on the step size in τ: $0 < \lambda \le 0.5$. For $\lambda \ge 0.5$ the method becomes unstable.

3.4.2   The implicit scheme

To overcome the disadvantage of small steps in the explicit method, the *implicit method* or *backward differences method* discretizes the time axis with backward differences:

$$\frac{w_{i,j} - w_{i,j-1}}{k} - \frac{w_{i+1,j} - 2w_{i,j} + w_{i-1,j}}{h^2} = 0 \tag{3.278}$$

Equation (3.278) relates the time level j to the time level $j - 1$. It is not possible to find an approximation explicitly, since the information from the BC at time level j is needed already in $j - 1$. This becomes more obvious in the following representation. Let $\lambda = \frac{k}{h^2}$, as in Section 3.4.1. Then, Equation (3.278) can be rewritten as:

$$(1 + 2\lambda)w_{i,j} - \lambda(w_{i+1,j} - w_{i-1,j}) = w_{i,j-1} \tag{3.279}$$

for $j = 1, 2, \ldots$ and $i = 1, 2, \ldots, m$. This method is stable for all λ. The local truncation error of this method is also of order $O(k + h^2)$. This results in the fact that time intervals need to be chosen much smaller than spatial intervals. This disadvantage is overcome by the Crank–Nicolson method, to be described in Section 3.4.3.

3.4.3   The Crank–Nicolson method

The basic idea of the method is to combine explicit and implicit methods to arrive at an approach that possesses a local truncation error of $O(k^2 + h^2)$, so that the number of steps in time and space can be chosen to be of the same dimension. Adding Equations (3.272) and (3.278) we arrive at:

$$\frac{w_{i,j+1} - w_{i,j}}{k} - \frac{1}{2}\left[\frac{w_{i+1,j} - 2w_{i,j} + w_{i-1,j}}{h^2} + \frac{w_{i+1,j+1} - 2w_{i,j+1} + w_{i-1,j+1}}{h^2}\right] = 0$$

(3.280)

The Crank–Nicolson method is stable for any step size. However, it has problems with discontinuities in the initial condition, or its first derivative, resulting in oscillating values for u, u_x and higher spatial derivatives. This is a serious problem in financial applications, because most payoffs, or their first derivatives, possess discontinuities. How this problem can be overcome is discussed in Section 5.2.1.

3.4.4   Integrating early exercise

As shown in Section 2.3, early exercise turns the pricing problem into a moving boundary problem. The notion of this type of problem is probably best explained with the help of a drink. Think of a gin and tonic with ice cubes. We see in Section A.3.5 that the distribution of temperature can be modeled with the following heat equation:

$$u_t = \frac{k}{c\rho}u_{xx}$$

(3.281)

where k, c, and ρ are material parameters which are different for both materials. Until the ice cube has melted completely, different temperatures are observed within and outside the ice cube. The temperature distributions within and outside the ice cube both obey the heat Equation (3.281) – with different material parameters. Since the ice cube is melting, the interface between the liquid and the ice moves as time progresses. The interface is represented by the BC; its location is not known *a priori* and needs to be computed as part of the problem. This is the crucial difference from problems with boundaries changing their position in time, as arising in the pricing of barrier options with time-varying knock-out levels (Kunitomo and Ikeda, 1992). In the pricing of this special type of barrier option, the location of the barrier is known *a priori* and can be used to compute the solution. Moving boundary problems were first studied by J. Stefan in the context of melting ice in the late 19th century, so they are sometimes called Stefan problems. The early exercise feature divides the space–time domain of an American option into two regions, one in which exercising the option is optimal, and another in which keeping the option is optimal. The former is called the *early exercise region,* while the latter is called the *continuity region*; see Figure 3.2. It is possible to transform the problem of pricing an American option into a Stefan problem, as delineated in Chapter 6 of Wilmott *et al.* (1996). Furthermore,

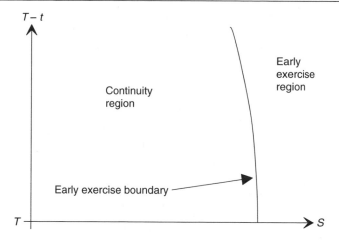

Figure 3.2 Early exercise boundary of a call with dividend yield

it is possible to show that the pricing of an American option constitutes a well-posed problem, even in the case of several underlyings (Villeneuve, 1999a, 1999b). The treatment of early exercise features in the context of the FDM is discussed in Wilmott (1998).

3.5 NOTES

Two-point boundary value problems
More extensive introductions to the numerical treatment of this kind of problem are given in many textbooks on numerical analysis (Burden and Faires, 1998; Press *et al.*, 1992). Economic problems of this kind with infinite time horizons often require more sophisticated methods such as reverse shooting (Judd, 1998, Section 10.7).

Solving pricing PDEs with FD
FD were introduced into finance by Brennan and Schwartz (1977, 1978). There are various FD methods to deal with several spatial variables. All suffer from the fact that the computational burden grows exponentially with the number of spatial variables. Due to this *curse of dimension*, problems involving more than three variables are usually solved with MC. Modern developments in the application of the FDM to problems from finance include nonuniform grids for the spatial variables (Izvorski, 1998) and automatic time stepping (Zvan *et al.*, 2000, Section 4.1). Parallel computing has been demonstrated by Dembo and Nelken (1991) for convertible bonds. For modern FD schemes and their parallel processing, see Evans (1997) .

Further PDE-based numerical techniques applicable to option-pricing problems
There is a wealth of numerical methods for parabolic problems which, at least theoretically, can be applied to problems arising in the pricing and hedging of contingent claims. All the methods can be classified as being fully discrete methods or semidiscrete methods. Semidiscrete methods usually discretize first in space but leave the time variable continuous. This is best explained with the help of an example (Heath, 2002, p. 453f). Consider the heat equation

$$u_t = k u_{xx}, \quad 0 \leq x \leq 1, \quad t \geq 0 \tag{3.282}$$

with the following initial and boundary conditions:

$$u(0, x) = f(x), \quad u(t, 0) = 0, \quad u(t, 1) = 0 \tag{3.283}$$

First, the spatial variable is discretized at $n + 2$ mesh points $x_i = ih$ $(i = 0, \ldots, n + 1)$ with $h = 1/(n + 1)$. At each mesh point, an FD approximation of u_{xx} is given by:

$$u_{xx}(t, x_i) \approx \frac{u(t, x_{i+1}) - 2u(t, x_i) + u(t, x_{i-1})}{h^2}, \quad i = 1, \ldots, n \tag{3.284}$$

Leaving the time variable continuous, the following system of ordinary initial value problems is obtained:

$$\dot{y}_i = \frac{c}{h^2} \left(y_{i+1}(t) - 2y_i(t) + y_{i-1}(t) \right), \quad i = 1, \ldots, n \tag{3.285}$$

where $\dot{y}_i(t) \approx u(t, x_i)$. The BCs result in $y_0(t) = y_{n+1}(t) = 0$, while the IC provides the initial values for the above system:

$$y_i(0) = f(x_i), \quad i = 1, \ldots, n \tag{3.286}$$

The partial differential equation (3.282) is approximated with a system of ODEs. A quote from Heath (2002) provides a geometric interpretation:

> If we think of the solution $u(t, x)$ as a surface over the (t, x) plane, this method computes cross sections of that surface along a series of lines, each of which passes through one of the discrete spatial mesh points and runs parallel to the time axis.

This approach is called the *vertical method of lines*. A less popular approach is to discretize time first and the spatial variables in a second step. This method is called the *horizontal method of lines* or *Rothe's method*. All FE methods for time-dependent problems to be discussed in this book are semidiscretizations. In a first step, the spatial variables are discretized. The resulting stiff system of ODEs is solved with FD. This is not the only possible way of discretizing evolution problems with FE. An alternative is to use space–time elements (Eriksson *et al.*, 1996, Section 16.4) and to discretize the space–time domain in one step in a similar manner to the FDM in Section 3.4. This, of course, is a fully discrete method. However, most FE codes employ a semidiscrete approach.

Closely related to FE are *finite volume methods* (Angermann and Knabner, 2000, Chapter 8; Morton, 1996); for an application of a finite volume method to pricing problems see Forsyth *et al.* (2001). Further methods include *multigrid methods* (Oosterlee *et al.*, 2001); for an application to option pricing see Clarke and Parrott (1999), for *projection methods* see Judd (1998) and for *(pseudo) spectral methods*, see Zhang (1998a). Spectral methods have been employed in bond and option pricing in Ashcroft (1996) and Bos *et al.* (2002). For a combination of FD and MC, see Heath and Platen (2002).

Part II
Finite Elements

4

Static 1D Problems

4.1 BASIC FEATURES OF FINITE ELEMENT METHODS

The basic idea of FE is to approximate the solution of a given differential equation with a set of algebraically simple functions. The spatial domain of the differential equation is divided into subdomains called elements. For each element, the parameters of this function are usually different. The functions are equal with respect to the function type but different with respect to the values of the parameters. Each of these functions only has local support, i.e. outside a small number of elements it takes on the value zero. The elements are nonoverlapping and cover the domain on which the differential equation is defined. In the 1D case, the elements are intervals. FE can be used not only to find approximate solutions for differential equations, but for any equation of calculus, hence integral, integrodifferential[1] and variational equations can be solved as well. Following engineering jargon, we will consider problems to be 1D when they possess one spatial variable. Problems with one spatial variable plus time are called *dynamic* 1D problems, although they actually depend on two variables. Problems with one spatial variable which do not depend on time are called *static* 1D problems. Consequently, 1D static problems are ordinary differential equations, while 1D dynamic problems are partial differential equations. This convention follows the fact that most FE approaches treat time differently from spatial variables. The most common approach is to discretize space with FE and time with FD. We will follow this approach in this book.

4.2 THE METHOD OF WEIGHTED RESIDUALS – ONE-ELEMENT SOLUTIONS

Rather than being a single algorithm, FE is a collection of concepts and ideas. One of these concepts is the method of weighted residuals (MWR). This method has many variants; the most important will be discussed here. This discussion is cast into a one-element framework so that, for the moment, further complications arising from element assembly can be put aside. The MWR will be explained with the help of an example to which it is applied. This test problem is a *linear second order ordinary boundary value problem*, often also called a *linear two-point boundary value problem*. Its economic interpretation is the expected exit time of an Ornstein–Uhlenbeck process of a given region, for instance the interest rate breaching a given corridor:

$$\frac{\sigma^2}{2}\frac{d^2u}{dx^2} + a(b-x)\frac{du}{dx} = -1 \tag{4.1}$$

$$u(x^{\min}) = u(x^{\max}) = 0 \tag{4.2}$$

---

[1] This makes FE a possible candidate for pricing models with jumps.

Table 4.1   Data for the 1D test problem

Parameter	Value
σ	0.02
a	0.1
b	0.1
x_{min}	0.05
x_{max}	0.15

See Table 4.1 for the input data. The general solution of this problem is given by:

$$u(x) = c_1 + \int \left(c_2 e^{-\frac{2abx}{\sigma^2} + \frac{ax^2}{\sigma^2}} - \frac{e^{\frac{ab^2}{\sigma^2} - \frac{2abx}{\sigma^2} + \frac{ax^2}{\sigma^2}} \sqrt{\pi} \operatorname{Erf}\left[\frac{\sqrt{a}(-b+x)}{\sigma} \right]}{\sqrt{a}\sigma} \right) dx \qquad (4.3)$$

with:

$$\operatorname{Erf} = \frac{2}{\sqrt{\pi}} \int_0^x e^{-t^2} dt \qquad (4.4)$$

The specific solution, i.e. c_1 and c_2, can only be found by numerical means. This example will be used for the discussion of multi-element solutions as well.

A quadratic polynomial

First, we try to approximate the differential equation with a quadratic polynomial:

$$\tilde{u} = a_1 + a_2 x + a_3 x^2 \qquad (4.5)$$

Two of the a_i can be eliminated by inserting the boundary conditions $u(0.05) = u(0.15) = 0$:

$$\tilde{u}(0.05) = a_1 + a_2\, 0.05 + a_3\, 0.05^2 \overset{!}{=} 0 \qquad (4.6)$$

$$\tilde{u}(0.15) = a_1 + a_2\, 0.15 + a_3\, 0.15^2 \overset{!}{=} 0 \qquad (4.7)$$

This leads to:

$$\tilde{u}(x) = a_1 - \frac{80a_1}{3}x + \frac{400a_1}{3}x^2 \qquad (4.8)$$

This expression satisfies the BC Equation (4.2). Now, the only thing that needs to be determined to get an approximation for u is a_1. Notice that, up to this point, Equation (4.1) has not entered our computations at all, only the BCs were used. As a next step, it will be shown how to extract the information from the ODE (4.1) to determine the unknown parameter a_1. The basic idea of an MWR is to insert the approximation (4.8) into the original differential equation (4.1). In general, the resulting expression does not vanish. If it does vanish, the differential equation at hand has been solved analytically:

$$R(x, a_1) = \frac{\sigma^2}{2} \frac{d^2\tilde{u}}{dx^2} + a(b - x)\frac{d\tilde{u}}{dx} + 1 \qquad (4.9)$$

$$= 0.0002 \frac{d^2\tilde{u}}{dx^2} + 0.1(0.1 - x)\frac{d\tilde{u}}{dx} + 1 \qquad (4.10)$$

Using the derivatives:

$$\frac{d\tilde{u}}{dx} = \frac{800a_1}{3}x - \frac{80a_1}{3} \tag{4.11}$$

$$\frac{d^2\tilde{u}}{dx^2} = \frac{800a_1}{3} \tag{4.12}$$

the residual turns out to be a quadratic polynomial as well:

$$R(x, a_1) = -\frac{2000a_1x^2 - 400a_1x + 16a_1 - 75}{75} \tag{4.13}$$

Four different approaches of the MWR family will be outlined:

1. *The collocation method:*
   For each unknown parameter, one point from the domain has to be determined, i.e. here just one. At this point, the residual R is forced to take on the value zero, i.e. $R(x, a) = 0$. These points are called collocation points. They need to be taken from the domain, including the points on the boundary. The approximate solution is required to satisfy the PDE exactly at the collocation points by forcing the residual to vanish. This does not imply that the error in the numerical solution vanishes at the collocation points. At the collocation points, it is the error in the residual that is zero, not the error in the solution.[2] As a first attempt we arbitrarily choose $x = 0.1$, so that:

$$R(0.1, a_1) \overset{!}{=} 0 \tag{4.14}$$

$$\Longrightarrow \quad a_1 = -\frac{75}{4} \tag{4.15}$$

$$\Longrightarrow \quad \tilde{u}\left(x, -\frac{75}{4}\right) = -2500x^2 + 500x - \frac{75}{4} \tag{4.16}$$

This function is plotted in Figure 4.1. We will try a different collocation point: $x = 0.075$.

$$R(0.075, a_1) \overset{!}{=} 0 \tag{4.17}$$

$$\Longrightarrow \quad a_1 = -\frac{300}{11} \tag{4.18}$$

$$\Longrightarrow \quad \tilde{u}\left(x, -\frac{300}{11}\right) = -\frac{40\,000}{11}x^2 + \frac{8000}{11}x - \frac{300}{11} \tag{4.19}$$

A third collocation point $x = 0.125$ leads to the same approximation as $x = 0.075$.

2. *The subdomain method:*
   In a first attempt, we take the entire domain as the subdomain. The parameter a_1 is determined by solving:

$$\frac{1}{\Delta x}\int_{\Delta x} R(x, a_1)\,dx \overset{!}{=} 0 \tag{4.20}$$

$$\frac{1}{0.15 - 0.05}\int_{0.05}^{0.15} R\,dx \overset{!}{=} 0 \tag{4.21}$$

---

[2] Suppose, for example, the differential equation is $u''(x) = f(x)$. At each collocation point, $u''(x_i) = f(x_i)$, that is, the differential equation is satisfied exactly at this point or, stated another way, the residual $u''(x_i) - f(x_i)$ is zero. This does not imply that the approximate solution is exactly equal to the true solution at that point, only that its second derivative is equal to f at that point.

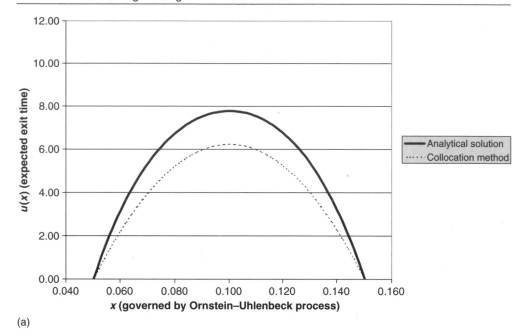

(a)

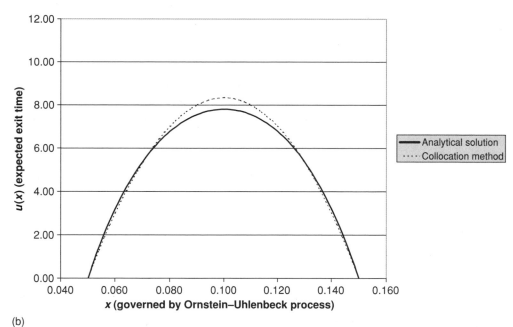

(b)

Figure 4.1 Collocation method: one element, quadratic trial function. The collocation point for (a) is $x = 0.1$, while for (b) $x = 0.075$ has been chosen

$$\Rightarrow a_1 \qquad\qquad = \frac{-225}{7} \approx -32.1429 \tag{4.22}$$

$$\Rightarrow \tilde{u}\left(x, \frac{-225}{7}\right) = -\frac{30\,000x^2}{7} + \frac{6000x}{7} - \frac{225}{7} \tag{4.23}$$

$$\approx -4285.71x^2 + 857.143x - 32.1428 \tag{4.24}$$

Another choice of the subdomain leads to the same result:

$$\frac{1}{0.10 - 0.05} \int_{0.05}^{0.10} R \, dx \stackrel{!}{=} 0 \tag{4.25}$$

$$\Rightarrow a_1 = \frac{-225}{7} \approx -32.1429 \tag{4.26}$$

A third choice of the subdomain gives:

$$\frac{1}{0.075 - 0.125} \int_{0.075}^{0.125} R \, dx \stackrel{!}{=} 0 \tag{4.27}$$

$$\Rightarrow a_1 \qquad\qquad = \frac{-900}{43} \approx -20.9302 \tag{4.28}$$

$$\Rightarrow \tilde{u}\left(x, \frac{-900}{43}\right) = -\frac{120\,000x^2}{43} + \frac{24\,000x}{43} - \frac{900}{43} \tag{4.29}$$

The subdomain method is illustrated in Figure 4.2.

3. *The least-squares method:*

$$\frac{\partial}{\partial a_1} \int_{0.05}^{0.15} [R(x, a_1)]^2 \, dx \stackrel{!}{=} 0 \tag{4.30}$$

$$\Rightarrow \qquad\qquad a_1 = \frac{-525}{23} \approx -22.8261 \tag{4.31}$$

$$\Rightarrow \tilde{u}\left(x, \frac{-523}{23}\right) = \frac{-70\,000x^2 + 14\,000x - 525}{23} \tag{4.32}$$

$$\approx -3043.48x^2 + 608.7x - 22.8261 \tag{4.33}$$

The least-squares method is illustrated in Figure 4.3(a).

4. *The Galerkin method:*

$$\tilde{u}(x, a_1) = a_1 - \frac{80a_1 x}{3} + \frac{400a_1 x^2}{3} \tag{4.34}$$

$$= a_1 \underbrace{(133.\bar{3}x^2 - 26.\bar{6}x + 1)}_{\phi_1} \tag{4.35}$$

$$\int_{0.05}^{0.15} R(x, a_1)\phi_1 \, dx \stackrel{!}{=} 0 \tag{4.36}$$

$$a_1 = -25 \tag{4.37}$$

$$\tilde{u}(x, a_1) = -\frac{10\,000x^2}{3} + \frac{2000x}{3} - 25 \tag{4.38}$$

$$= -3333.\bar{3} + 666.\bar{6}x - 25 \tag{4.39}$$

The Galerkin method is shown in Figure 4.3(b).

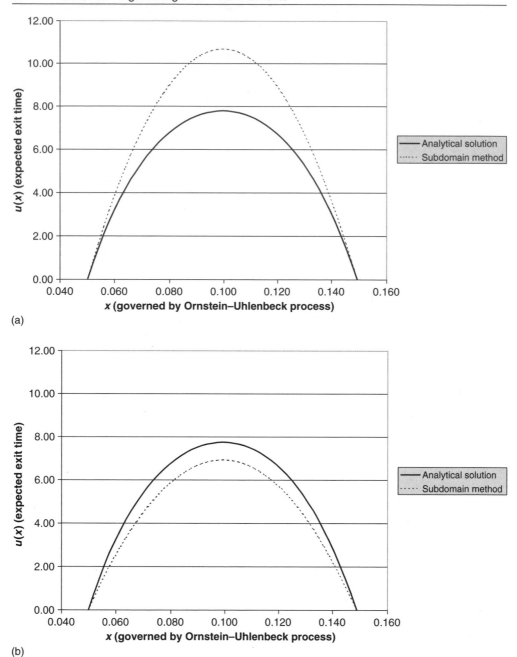

(a)

(b)

Figure 4.2   Subdomain method: one element, quadratic trial function. For (a), the entire domain has been used as the subdomain. For (b), the two center quarters of the domain have been used as subdomain

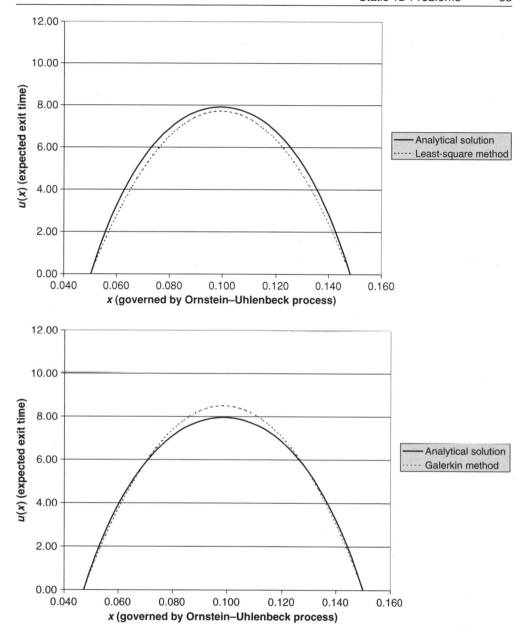

Figure 4.3 (a) Least-squares method; (b) Galerkin method: one element, quadratic trial function

A cubic polynomial

Next, we try to approximate the differential equation with a cubic polynomial:

$$\tilde{u} = a_1 + a_2 x + a_3 x^2 + a_4 x^3 \tag{4.40}$$

Again, two of the a_i can be eliminated by inserting the boundary conditions $u(0.05) = u(0.15) = 0$:

$$\tilde{u}(0.05) = a_1 + a_2\, 0.05 + a_3\, 0.05^2 + a_4\, 0.05^3 \overset{!}{=} 0 \tag{4.41}$$

$$\tilde{u}(0.15) = a_1 + a_2\, 0.15 + a_3\, 0.15^2 + a_4\, 0.15^3 \overset{!}{=} 0 \tag{4.42}$$

This leads to:

$$\tilde{u}(x) = a_4 x^3 + a_3 x^2 - \frac{x(80a_3 + 13a_4)}{400} + \frac{3(5a_3 + a_4)}{2000} \tag{4.43}$$

$$\tilde{u}'(x) = 3a_4 x^2 + 2a_3 x - \frac{80a_3 + 13a_4}{400} \tag{4.44}$$

$$\tilde{u}''(x) = 6a_4 x + 2a_3 \tag{4.45}$$

$$R(x, a_3, a_4) = 0.0002\tilde{u}''(x) + 0.1(0.1 - x)\tilde{u}'(x) + 1 \tag{4.46}$$

To determine the unknown a_i we employ the same four criteria as before:

1. *The collocation method:*
   For each undetermined parameter a_i, a point x_i in the domain has to be chosen. At each x_i, the residual is forced to be zero. These x_i are called *collocation points*. Since we do not have any additional information, they are distributed uniformly over the domain:

$$R(0.075, a_3, a_4) \overset{!}{=} 0 \tag{4.47}$$

$$R(0.125, a_3, a_4) \overset{!}{=} 0 \tag{4.48}$$

This gives the linear system:

$$a_3 = -\frac{73a_4 + 640\,000}{176} \tag{4.49}$$

$$a_3 = -\frac{163a_4 + 320\,000}{800} \tag{4.50}$$

resulting in:

$$a_4 = 0 \tag{4.51}$$

$$a_3 = -\frac{40\,000}{11} \tag{4.52}$$

$$\tilde{u}(x) = -\frac{40\,000x^2}{11} + \frac{8000x}{11} - \frac{300}{11} \tag{4.53}$$

$$= -3636.\overline{36}x^2 + 727.\overline{27}x - 27.\overline{27} \tag{4.54}$$

Notice that the coefficient for the cubic part has vanished.

2. *The subdomain method:*

For each parameter a_i, a different part of the domain is chosen. On this interval, the integral over the residual $R(\cdot)$ is forced to be zero:

$$\begin{cases} \frac{1}{\Delta x_1} \int_{\Delta x_1} R(x, a_3, a_4) \stackrel{!}{=} 0 \\ \frac{1}{\Delta x_2} \int_{\Delta x_2} R(x, a_3, a_4) \stackrel{!}{=} 0 \end{cases} \tag{4.55}$$

$$\Longleftrightarrow \begin{cases} \frac{1}{0.10-0.05} \int_{0.10}^{0.05} R(x, a_3, a_4) \stackrel{!}{=} 0 \\ \frac{1}{0.15-0.10} \int_{0.15}^{0.10} R(x, a_3, a_4) \stackrel{!}{=} 0 \end{cases} \tag{4.56}$$

$$\Longleftrightarrow \begin{cases} a_3 \stackrel{!}{=} -\frac{3(69a_4 + 1\,600\,000)}{1120} \\ a_3 \stackrel{!}{=} -\frac{3(31a_4 + 320\,000)}{224} \end{cases} \tag{4.57}$$

$$\Longleftrightarrow \begin{cases} a_3 = -\frac{30\,000}{7} \\ a_4 = 0 \end{cases} \tag{4.58}$$

$$\Longrightarrow \tilde{u} = \frac{-30\,000x^2 + 6000x - 225}{7} \approx -4285.71x^2 + 857.14x - 32.14 \tag{4.59}$$

Again, the cubic part has vanished.

3. *The least-squares method:*

$$\begin{cases} \frac{\partial}{\partial a_3} \int_{0.05}^{0.15} [R(x, a_3, a_4)]^2 \, \mathrm{d}x \stackrel{!}{=} 0 \\ \frac{\partial}{\partial a_4} \int_{0.05}^{0.15} [R(x, a_3, a_4)]^2 \, \mathrm{d}x \stackrel{!}{=} 0 \end{cases} \tag{4.60}$$

$$\Longleftrightarrow \begin{cases} 230a_3 + 69a_4 + 70\,000 = 0 \\ 19\,320a_3 + 6523a_4 + 58.8 \cdot 10^6 = 0 \end{cases} \tag{4.61}$$

$$\Longleftrightarrow \begin{cases} a_3 = -\frac{70\,000}{23} \approx -3043.5 \\ a_4 = 0 \end{cases} \tag{4.62}$$

$$\Longrightarrow \tilde{u}(x) = -\frac{6087}{2}x^2 + \frac{6087}{10}x - \frac{18\,261}{800} \tag{4.63}$$

$$\Longrightarrow \tilde{u} \approx -3043.5x^2 + 608.7x - 22.8262 \tag{4.64}$$

Again, the cubic part has vanished.

4. *The Galerkin method:*

First, $\tilde{u}$ has to be rearranged:

$$\tilde{u}(x, a_3, a_4) = a_4x^3 + a_3x^2 - \frac{x(80a_3 + 13a_4)}{400} + \frac{3(5a_3 + a_4)}{2000} \tag{4.65}$$

$$= a_3 \underbrace{\frac{400x^2 - 80x + 3}{400}}_{\phi_1} + a_4 \underbrace{\frac{(5x+1)(400x^2 - 80x + 3)}{2000}}_{\phi_2} \tag{4.66}$$

$$= a_3\phi_1 + a_4\phi_2 \tag{4.67}$$

Next, the Galerkin optimization criterion has to be applied:

$$\begin{cases} \int_{0.05}^{0.15} R\phi_1 \mathrm{d}x \stackrel{!}{=} 0 \\ \int_{0.05}^{0.15} R\phi_2 \mathrm{d}x \stackrel{!}{=} 0 \end{cases} \tag{4.68}$$

$$\Longleftrightarrow \begin{cases} 30a_3 + 9a_4 + 100\,000 = 0 \\ 12\,600a_3 + 3859a_4 + 42 \cdot 10^6 = 0 \end{cases} \tag{4.69}$$

$$\Longleftrightarrow \begin{cases} a_3 = -\frac{10\,000}{3} \approx -3333.\bar{3} \\ a_4 = 0 \end{cases} \tag{4.70}$$

$$\Longrightarrow \tilde{u}(x, a_3, a_4) = -\frac{10\,000x^2}{3} + \frac{2000x}{3} - 25 \tag{4.71}$$

$$= -3333.\bar{3} + 666.\bar{6}x - 25 \tag{4.72}$$

Again, the cubic part has vanished. The $\tilde{u}$ produced by using a quadratic and a cubic polynomial are the same.

Note that the cubic polynomial leads to solving an inhomogeneous linear system of two equations. As we will see, this can be generalized: a polynomial of order n leads to a system of $n - 1$ linear equations.

Polynomials of degree 4

Next, we try to approximate the differential equation with a polynomial of order 4:

$$\tilde{u} = a_1 + a_2x + a_3x^2 + a_4x^3 + a_5x^4 \tag{4.73}$$

Again, two of the a_i can be eliminated by inserting the boundary conditions $u(0.05) = u(0.15) = 0$:

$$\tilde{u}(0.05) = a_1 + a_2\,0.05 + a_3\,0.05^2 + a_4\,0.05^3 + a_4\,0.05^4 \overset{!}{=} 0 \tag{4.74}$$

$$\tilde{u}(0.15) = a_1 + a_2\,0.15 + a_3\,0.15^2 + a_4\,0.15^3 + a_4\,0.15^4 \overset{!}{=} 0 \tag{4.75}$$

This leads to:

$$\tilde{u}(x) = a_5x^4 + a_4x^3 + a_3x^2 - \underbrace{\frac{80a_3 + 13a_4 + 2a_5}{400}}_{a_2}x + \underbrace{\frac{3(400a_3 + 80a_4 + 13a_5)}{160\,000}}_{a_1} \tag{4.76}$$

$$\tilde{u}'(x) = 4a_5x^3 + 3a_4x^2 + 2a_3x - \frac{80a_3 + 13a_4 + 2a_5}{400} \tag{4.77}$$

$$\tilde{u}''(x) = 12a_5x^2 + 6a_4x + 2a_3 \tag{4.78}$$

$$R(x, a_3, a_4, a_5) = 0.0002\tilde{u}''(x) + 0.1(0.1 - x)\tilde{u}'(x) + 1 \tag{4.79}$$

$$= \frac{1}{40\,000}\left\{ a_3\left(-\frac{125}{64}x^2 + 25x - 1\right) \right.$$

$$+ a_4\,(12\,000x^3 - 1200x^2 - 178x + 13)$$

$$\left. + a_5\,(16\,000x^4 - 1600x^3 - 96x^2 - 20x + 2) \right\} - 1 \tag{4.80}$$

To determine the unknown a_i we employ the same four criteria as before:

1. *The collocation method:*
   The collocation points are distributed evenly over the domain:

$$R(0.075, a_3, a_4, a_5) \stackrel{!}{=} 0 \tag{4.81}$$

$$R(0.100, a_3, a_4, a_5) \stackrel{!}{=} 0 \tag{4.82}$$

$$R(0.125, a_3, a_4, a_5) \stackrel{!}{=} 0 \tag{4.83}$$

This is equivalent to:

$$\frac{8800a_3 + 1630a_4 + 167a_5}{32\,000\,000} = -1 \tag{4.84}$$

$$\frac{50a_3 + 15a_4 + 3a_5}{125\,000} = -1 \tag{4.85}$$

$$\frac{352a_3 + 146a_4 + 39a_5}{1\,280\,000} = -1 \tag{4.86}$$

The unique solution of this linear system is given by:

$$a_3 = \frac{707\,500}{-43} \approx -16\,453.49 \tag{4.87}$$

$$a_4 = \frac{4\,000\,000}{43} \approx 93\,023.26 \tag{4.88}$$

$$a_5 = \frac{-10\,000\,000}{-43} \approx -2.326 \cdot 10^5 \tag{4.89}$$

Also, we get:

$$a_1 = -\frac{6975}{172} \approx 40.55 \tag{4.90}$$

$$a_2 = \frac{61\,500}{43} \approx 1430.233 \tag{4.91}$$

Inserting these coefficients into Equation (4.73) gives us the approximate solution:

$$\tilde{u}(x) = -\frac{6975}{172} + \frac{61\,500}{43}x - \frac{707\,500}{43}x^2 + \frac{4\,000\,000}{43}x^3 - \frac{10\,000\,000}{43}x^4 \tag{4.92}$$

2. *The subdomain method:*
   We divide the domain into three nonoverlapping intervals of nonequal length. There is no need either to cover the entire domain or to choose nonoverlapping intervals.

$$\begin{cases} \frac{1}{0.075-0.05} \int_{0.05}^{0.075} R(x, a_3, a_4, a_5)\mathrm{d}x \stackrel{!}{=} 0 \\ \frac{1}{0.1-0.075} \int_{0.075}^{0.1} R(x, a_3, a_4, a_5)\mathrm{d}x \stackrel{!}{=} 0 \\ \frac{1}{0.15-0.1} \int_{0.1}^{0.15} R(x, a_3, a_4, a_5)\mathrm{d}x \stackrel{!}{=} 0 \end{cases} \tag{4.93}$$

$$\implies \begin{cases} a_1 = -\frac{6225}{137} \approx -45.438 \\ a_2 = \frac{226\,000}{137} \approx 1649.64 \\ a_3 = -\frac{2\,730\,000}{137} \approx 19\,927 \\ a_4 = \frac{16\,000\,000}{137} \approx 116\,788 \\ a_5 = -\frac{40\,000\,000}{137} \approx 291\,971 \end{cases} \tag{4.94}$$

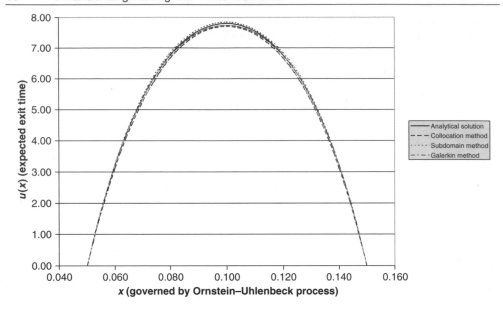

Figure 4.4   Several MWR methods: one element, trial function with polynomial of order $n = 4$

$$\Longrightarrow \tilde{u}(x) = -\frac{40\,000\,000}{137}x^4 + \frac{16\,000\,000}{137}x^3 \tag{4.95}$$

$$-\frac{2\,730\,000}{137}x^2 + \frac{226\,000}{137}x - \frac{6225}{137} \tag{4.96}$$

3. *The least-squares method:*
   In a similar fashion as above, the approximate solution is given by:

$$\tilde{u}(x) \approx -45.6789 + 1667.48x - 20\,320.9x^2 + 119\,835x^3 - 299\,587x^4 \tag{4.97}$$

4. *The Galerkin method:*
   In a similar fashion as above, the approximate solution is given by:

$$\tilde{u}(x) = -44.2729 + 1598.94x - 19\,150.1x^2 + 111\,554x^3 - 278\,884x^4 \tag{4.98}$$

Some plots for polynomials of order 4 are shown in Figure 4.4.

Polynomials of degree 5

All results are the same as for a polynomial of degree 4, i.e. all coefficients a_6 for the different methods vanish.

Polynomials of degree 6

The approximate solution $\tilde{u}$ is of the form:

$$\tilde{u}(x) = \sum_{i=1}^{N=7} a_i x^{i-1} \tag{4.99}$$

The coefficients for various methods can be found in Table 4.2.

Table 4.2 One-element solution: polynomial of degree 6

Coefficient	Collocation	Subdomain	Least-Squares	Galerkin
a_1	$-\dfrac{5\,103\,375}{92\,036}$	$-\dfrac{12\,392\,722\,175}{202\,462\,676}$	$\dfrac{205\,193\,526\,975}{3\,631\,135\,982}$	$\dfrac{6\,840\,675}{122\,366}$
a_2	$\dfrac{55\,022\,500}{23\,009}$	$\dfrac{417\,697\,943\,500}{151\,847\,007}$	$\dfrac{4\,473\,361\,218\,000}{1\,815\,567\,991}$	$\dfrac{148\,134\,000}{61\,183}$
a_3	$-\dfrac{960\,112\,500}{23\,009}$	$-\dfrac{7\,874\,300\,342\,500}{151\,847\,007}$	$-\dfrac{79\,408\,982\,490\,000}{1\,815\,567\,991}$	$-\dfrac{2\,601\,870\,000}{61\,183}$
a_4	$\dfrac{30\,550\,000\,000}{69\,027}$	$\dfrac{276\,351\,250\,000\,000}{455\,541\,021}$	$\dfrac{856\,028\,228\,000\,000}{1\,815\,567\,991}$	$\dfrac{27\,764\,000\,000}{61\,183}$
a_5	$-\dfrac{67\,125\,000\,000}{23\,009}$	$-\dfrac{674\,216\,875\,000\,000}{151\,847\,007}$	$-\dfrac{5\,710\,151\,370\,000\,000}{1\,815\,567\,991}$	$-\dfrac{183\,810\,000\,000}{61\,183}$
a_6	$\dfrac{250\,000\,000\,000}{23\,009}$	$\dfrac{2\,798\,590\,000\,000\,000}{151\,847\,007}$	$\dfrac{21\,420\,484\,800\,000\,000}{1\,815\,567\,991}$	$\dfrac{686\,400\,000\,000}{61\,183}$
a_7	$-\dfrac{1\,250\,000\,000\,000}{69\,027}$	$-\dfrac{15\,349\,250\,000\,000\,000}{455\,541\,021}$	$-\dfrac{35\,700\,808\,000\,000\,000}{1\,815\,567\,991}$	$-\dfrac{1\,144\,000\,000\,000}{61\,183}$

Table 4.3 One-element solution: polynomial of degree 8

Coefficient	Collocation	Subdomain	Least-Squares	Galerkin
a_1	$-\dfrac{22\,713\,739\,912\,175}{375\,944\,594\,176}$	$-\dfrac{6\,158\,796\,667\,512\,225}{99\,428\,778\,213\,632}$	$-\dfrac{179\,249\,870\,068\,225}{2\,909\,887\,748\,732}$	$-\dfrac{791\,939\,775}{12\,901\,028}$
a_2	$\dfrac{138\,104\,538\,619\,625}{48\,461\,607\,843}$	$\dfrac{1\,160\,624\,747\,579\,125}{388\,393\,664\,897}$	$\dfrac{6\,447\,800\,060\,152\,000}{2\,182\,415\,811\,549}$	$\dfrac{9\,464\,136\,000}{3\,225\,257}$
a_3	$-\dfrac{2\,900\,718\,158\,148\,125}{48\,461\,607\,843}$	$\dfrac{25\,395\,586\,871\,045\,625}{388\,393\,664\,897}$	$-\dfrac{139\,329\,207\,410\,360\,000}{2\,182\,415\,811\,549}$	$-\dfrac{87\,138\,920\,000}{1\,382\,253}$
a_4	$\dfrac{3\,717\,060\,810\,812\,500}{4\,405\,600\,713}$	$\dfrac{161\,427\,825\,461\,312\,500}{166\,454\,427\,813}$	$\dfrac{18\,248\,287\,036\,408\,000\,000}{19\,641\,742\,303\,941}$	$\dfrac{419\,848\,000\,000}{460\,751}$
a_5	$-\dfrac{134\,867\,011\,369\,531\,250}{16\,153\,869\,281}$	$-\dfrac{562\,410\,722\,230\,468\,750}{55\,484\,809\,271}$	$-\dfrac{62\,019\,991\,573\,540\,000\,000}{6\,547\,247\,434\,647}$	$-\dfrac{4\,252\,820\,000\,000}{460\,751}$
a_6	$\dfrac{920\,257\,011\,250\,000\,000}{16\,153\,869\,281}$	$\dfrac{4\,054\,920\,343\,750\,000\,000}{55\,484\,809\,271}$	$\dfrac{434\,001\,891\,065\,600\,000\,000}{6\,547\,247\,434\,647}$	$\dfrac{88\,774\,400\,000\,000}{1\,382\,253}$
a_7	$-\dfrac{1\,135\,886\,468\,750\,000\,000}{4\,405\,600\,713}$	$-\dfrac{57\,926\,565\,406\,250\,000\,000}{166\,454\,427\,813}$	$-\dfrac{5\,998\,093\,875\,488\,000\,000\,000}{19\,641\,742\,303\,941}$	$-\dfrac{407\,264\,000\,000\,000}{1\,382\,253}$
a_8	$\dfrac{33\,863\,143\,750\,000\,000\,000}{48\,461\,607\,843}$	$\dfrac{382\,620\,706\,250\,000\,000\,000}{388\,393\,664\,897}$	$\dfrac{5\,468\,692\,028\,800\,000\,000\,000}{6\,547\,247\,434\,647}$	$\dfrac{457\,600\,000\,000\,000}{569\,163}$
a_9	$-\dfrac{42\,411\,367\,187\,500\,000\,000}{48\,461\,607\,843}$	$-\dfrac{493\,363\,945\,312\,500\,000\,000}{388\,393\,664\,897}$	$-\dfrac{6\,835\,865\,036\,000\,000\,000\,000}{6\,547\,247\,434\,647}$	$-\dfrac{572\,000\,000\,000\,000}{569\,163}$

Polynomials of degree 7

All results are the same as for a polynomial of degree 6, i.e. all coefficients a_8 for the different methods vanish.

Polynomials of degree 8

The approximate solution $\tilde{u}$ is of the form:

$$\tilde{u}(x) = \sum_{i=1}^{N=9} a_i x^{i-1} \tag{4.100}$$

The coefficients for various methods can be found in Table 4.3.

Polynomials of degree 9

All results are the same as for a polynomial of degree 8, i.e. all coefficients a_{10} for the different methods vanish.

Table 4.4   One-element solution: polynomial of degree 10 (1)

Coefficient	Collocation	Subdomain
a_1	$-\dfrac{4\,284\,066\,132\,887\,006\,325}{66\,462\,501\,403\,959\,424}$	$-\dfrac{145\,229\,435\,725}{2\,290\,017\,792}$
a_2	$\dfrac{1\,700\,880\,051\,042\,270\,250}{519\,238\,292\,218\,433}$	$\dfrac{85\,109\,721\,125}{26\,836\,146}$
a_3	$-\dfrac{41\,495\,412\,886\,694\,688\,750}{519\,238\,292\,218\,433}$	$-\dfrac{2\,019\,611\,886\,875}{26\,836\,146}$
a_4	$\dfrac{2\,180\,379\,994\,569\,938\,875\,000}{1\,557\,714\,876\,655\,299}$	$\dfrac{17\,140\,878\,906\,250}{13\,418\,073}$
a_5	$-\dfrac{28\,652\,374\,541\,160\,045\,312\,500}{1\,557\,714\,876\,655\,299}$	$-\dfrac{656\,358\,935\,546\,875}{40\,254\,219}$
a_6	$\dfrac{93\,749\,078\,320\,140\,250\,000\,000}{519\,238\,292\,218\,433}$	$\dfrac{2\,097\,167\,187\,500\,000}{13\,418\,073}$
a_7	$-\dfrac{2\,034\,939\,182\,470\,468\,750\,000\,000}{1\,557\,714\,876\,655\,299}$	$-\dfrac{44\,774\,898\,437\,500\,000}{40\,254\,219}$
a_8	$\dfrac{3\,508\,072\,577\,587\,500\,000\,000\,000}{519\,238\,292\,218\,433}$	$\dfrac{76\,484\,375\,000\,000\,000}{13\,418\,073}$
a_9	$-\dfrac{12\,293\,800\,121\,640\,625\,000\,000\,000}{519\,238\,292\,218\,433}$	$-\dfrac{89\,160\,156\,250\,000\,000}{4\,472\,691}$
a_{10}	$\dfrac{78\,657\,599\,125\,000\,000\,000\,000\,000}{1\,557\,714\,876\,655\,299}$	$\dfrac{1\,718\,750\,000\,000\,000\,000}{40\,254\,219}$
a_{11}	$-\dfrac{77\,815\,601\,875\,000\,000\,000\,000\,000}{1\,557\,714\,876\,655\,299}$	$-\dfrac{1\,718\,750\,000\,000\,000\,000}{40\,254\,219}$

Table 4.5   One-element solution: polynomial of degree 10 (2)

Coefficient	Least-Squares	Galerkin
a_1	$-\dfrac{35\,084\,021\,533\,293\,655\,975}{551\,757\,216\,132\,728\,212}$	$-\dfrac{355\,929\,824\,575}{5\,603\,498\,964}$
a_2	$\dfrac{440\,158\,281\,462\,392\,194\,000}{137\,939\,304\,033\,182\,053}$	$\dfrac{13\,377\,692\,354\,000}{4\,202\,624\,223}$
a_3	$-\dfrac{31\,557\,464\,685\,059\,056\,510\,000}{413\,817\,912\,099\,546\,159}$	$-\dfrac{15\,181\,229\,570\,000}{200\,124\,963}$
a_4	$\dfrac{4\,869\,264\,200\,206\,655\,288\,000\,000}{3\,724\,361\,208\,895\,915\,431}$	$\dfrac{259\,244\,024\,000\,000}{200\,124\,963}$
a_5	$-\dfrac{20\,959\,057\,133\,688\,723\,140\,000\,000}{1\,241\,453\,736\,298\,638\,477}$	$-\dfrac{3\,332\,287\,660\,000\,000}{200\,124\,963}$
a_6	$\dfrac{203\,337\,050\,046\,121\,513\,600\,000\,000}{1\,241\,453\,736\,298\,638\,477}$	$\dfrac{32\,175\,478\,400\,000\,000}{200\,124\,963}$
a_7	$-\dfrac{4\,390\,053\,341\,896\,397\,968\,000\,000\,000}{3\,724\,361\,208\,895\,915\,431}$	$-\dfrac{76\,847\,888\,000\,000\,000}{66\,708\,321}$
a_8	$\dfrac{7\,569\,571\,892\,275\,702\,400\,000\,000\,000}{1\,241\,453\,736\,298\,638\,477}$	$\dfrac{8\,317\,379\,200\,000\,000\,000}{1\,400\,874\,741}$
a_9	$-\dfrac{26\,652\,431\,230\,766\,068\,000\,000\,000\,000}{1\,241\,453\,736\,298\,638\,477}$	$-\dfrac{29\,208\,244\,000\,000\,000\,000}{1\,400\,874\,741}$
a_{10}	$\dfrac{19\,100\,518\,183\,801\,600\,000\,000\,000\,000}{413\,817\,912\,099\,546\,159}$	$\dfrac{26\,873\,600\,000\,000\,000\,000}{600\,374\,889}$
a_{11}	$-\dfrac{19\,100\,518\,138\,801\,600\,000\,000\,000\,000}{413\,817\,912\,099\,546\,159}$	$-\dfrac{26\,873\,600\,000\,000\,000\,000}{600\,374\,889}$

Polynomials of degree 10

The approximate solution $\tilde{u}$ is of the form:

$$\tilde{u}(x) = \sum_{i=1}^{N=11} a_i x^{i-1} \tag{4.101}$$

The coefficients for various methods can be found in Tables 4.4 and 4.5.

Table 4.6 Condition numbers

Order of polynomial N	Collocation method	Galerkin method
2	2931.25	$2.3450 \cdot 10^7$
3	37254.3	$1.3628 \cdot 10^{10}$
4	$1.2386 \cdot 10^6$	$1.2022 \cdot 10^{13}$
5	$3.7569 \cdot 10^7$	$1.3163 \cdot 10^{16}$
10	$3.1711 \cdot 10^{15}$	$1.0677 \cdot 10^{32}$

Polynomials of degree n

So far, we have observed that the accuracy of the solution increases as the degree of the polynomial approximating the solution increases. Unfortunately, increasing the degree of the approximating polynomial not only increases the linear system to be solved, but makes the problem of solving the linear system increasingly ill-conditioned, as can be seen in Table 4.6. One way out of this dilemma is to partition the domain into parts and to look for functions with local support approximating the differential equation on this part only. These functions with local support are the finite elements to be introduced in this chapter.

A general form of the MWR

It was observed by Crandall (1956) that all of the above criteria and many more can be put into the general form:

$$\int_\Omega R(x, \mathbf{a}) W_1(x) \, \mathrm{d}x = 0 \tag{4.102}$$

$$\int_\Omega R(x, \mathbf{a}) W_2(x) \, \mathrm{d}x = 0 \tag{4.103}$$

$$\vdots$$

$$\int_\Omega R(x, \mathbf{a}) W_N(x) \, \mathrm{d}x = 0 \tag{4.104}$$

with $R(x, \mathbf{a})$ being the residual. The $W_i(x)$ are the weighting functions:

1. The *collocation method:* $W_i(x) = \delta(x - x_i)$ (with the Dirac delta function δ) so that

$$\int_\Omega R(x, \mathbf{a}) \, \delta(x - x_i) \, \mathrm{d}x = R(x_i; \mathbf{a}) \quad \forall i \tag{4.105}$$

2. The *subdomain method:* First we define the weighting function:

$$W_i(x) = \begin{cases} 0 & \text{for} \quad x^{\min} \le x_i \\ 1 & \text{for} \quad x_i \le x_{i+1} \\ 0 & \text{for} \quad x_{i+1} \le x^{\max} \end{cases} \tag{4.106}$$

so that $W_i(x)$ possesses only local support in the closed interval $\Delta x_i = [x^{\min}, x^{\max}]$. Consequently:

$$\int_\Omega R(x, \mathbf{a}) \, W_i(x) \, \mathrm{d}x = \int_{\Delta x_i} R(x, \mathbf{a}) \, \mathrm{d}x \quad \forall i \tag{4.107}$$

3. The *least-squares method:* Because of Leibnitz's rule, the following set of optimization criteria is to be met:

$$\frac{\partial}{\partial a_i} \int_\Omega [R(x, \mathbf{a})]^2 \; dx = 2 \int_\Omega R(x, \mathbf{a}) \underbrace{\frac{\partial}{\partial a_i} R(x, \mathbf{a})}_{=W_i(x)} \; dx \stackrel{!}{=} 0 \quad \forall i \tag{4.108}$$

4. The *Galerkin method:* The weighting function is given by $W_i(x) = \phi_i(x)$ so that:

$$\int_\Omega R(x, \mathbf{a})\phi_i(x)dx = 0 \quad \forall i \tag{4.109}$$

In the next sections, we will only discuss the collocation method and the method of Galerkin.

4.3 THE RITZ VARIATIONAL METHOD

The Ritz variational method is based on the variational formulation of the differential equation that is to be solved numerically. Assuming that the same discretization and degree of interpolation is used, the Ritz finite element method will turn out to be exactly the same as the Galerkin FEM for the class of problems considered in this section, which additionally have to be self-adjoint. Only for self-adjoint problems can an equivalent formulation in terms of calculus of variations be given. In science and engineering, most problems are self-adjoint, so that most FE textbooks discuss this case only; in economics and finance, problems are self-adjoint only by accident, so that the general case has to be investigated. The Ritz method is interesting for economists because the problems formulated as variational problems can be solved without constructing the Euler equation.

The Ritz method will be explained with the help of an example which is formulated as an ODE, so that the variational formulation has to be found first. We can not re-use the example from the previous section because it is not self-adjoint. The problem of a pure Brownian motion leaving a prescribed region can be formulated as a boundary value problem, see Section B.3.5:

$$\frac{1}{2}\frac{d^2u}{dx^2} = -1 \tag{4.110}$$

$$u(x^{\min}) = 0 \tag{4.111}$$

$$u(x^{\max}) = 0 \tag{4.112}$$

The equivalent variational formulation then is

$$V[u(x)] = \int_{x^{\min}}^{x^{\max}} \left[\frac{1}{4}(u')^2 - u\right] dx \tag{4.113}$$

$$u(x_{\min}) = 0 \tag{4.114}$$

$$u(x_{\max}) = 0 \tag{4.115}$$

The Ritz method has to be applied directly to the variational problem. The first step again is the construction of a trial function that meets the BCs. With a polynomial of order 3

$$\tilde{u}(x, a_1, \ldots, a_4) = a_1 + a_2x + a_3x^2 + a_4x^3 \tag{4.116}$$

and the two BCs $u(0) = 0$ and $u(2) = 0$, the trial function is given by:

$$\tilde{u}(x, a_3, a_4) = (-2a_3 - 4a_4)x + a_3x^2 + a_4x^3 \tag{4.117}$$

As in previous sections, the trial function meets the BCs for any choice of a_3 and a_4. The next step is to replace u by $\tilde{u}$ in the variational problem:

$$V[\tilde{u}(a_3, a_4)] = \int_0^2 \left[\frac{(\tilde{u}')^2}{4} - \tilde{u} \right] dx \tag{4.118}$$

Now, the extremum of $V[\tilde{u}(\cdot)]$ with respect to a_3 and a_4 is to be found. The necessary conditions are simply:

$$\frac{\partial V}{\partial a_i} \overset{!}{=} 0 \quad \forall\, i \tag{4.119}$$

This way, the dynamic variational problem has been reduced to an unconstrained static optimization problem. Inserting $\tilde{u}$ and $\tilde{u}'$ into Equation (4.118) and solving the integral leads to:

$$V[\tilde{u}(a_3, a_4)] = \frac{2\left[5a_3^2 + 10a_3(3a_4 + 1) + 6a_4(8a_4 + 5)\right]}{15} \tag{4.120}$$

Differentiating the above expression with respect to a_3 and a_4 and applying Equation (4.119) results in:

$$\frac{4(a_3 + 3a_4 + 1)}{3} \overset{!}{=} 0 \tag{4.121}$$

$$\frac{4(5a_3 + 16a_4 + 5)}{5} \overset{!}{=} 0 \tag{4.122}$$

so that:

$$a_4 = 0 \tag{4.123}$$
$$a_3 = -1 \tag{4.124}$$
$$a_2 = 2 \tag{4.125}$$
$$a_1 = 0 \tag{4.126}$$
$$\tilde{u} = 2x - x^2 \tag{4.127}$$

Here, it happens that the approximate solution equals the analytical solution $\tilde{u} = u$; compare this with Example 60 in Appendix B. It has to be emphasized that this is *not* the regular case. Usually, it holds that $\tilde{u} \neq u$. In this particular case at hand, analytical and approximate solutions coincide because u belongs to the trial function space.

As a next step, we demonstrate with the above example that the Ritz method and Galerkin's approach lead to the same approximate solution. As in Section 4.2, the trial function Equation (4.117) is inserted directly into the differential Equation (4.110), resulting in the following expression for the residual R:

$$R(x, a_3, a_4) = \frac{\tilde{u}_{xx}}{2} + 1 \tag{4.128}$$

$$= 0.5(2a_3 + 6a_4 x) + 1 \tag{4.129}$$

In order to be able to apply the Galerkin criterion, the trial functions ϕ_3 and ϕ_4, associated with a_3 and a_4, respectively, are needed:

$$\tilde{u}(x, a_3, a_4) = (-2a_3 - 4a_4)x + a_3 x^2 + a_4 x^3 \tag{4.130}$$

$$= a_3(x^2 - 2x) + a_4(x^3 - 4x) \tag{4.131}$$

$$= a_3 \phi_3 + a_4 \phi_4 \tag{4.132}$$

Applying the Galerkin criterion, $\int R\phi_i \, dx = 0$ for all i, results in:

$$\int_0^2 R\phi_3 \, dx = \int_0^2 [0.5(2a_3 + 6a_4x) + 1] (x^2 - 2x) \, dx \stackrel{!}{=} 0 \qquad (4.133)$$

$$\int_0^2 R\phi_4 \, dx = \int_0^2 [0.5(2a_3 + 6a_4x) + 1] (x^4 - 4x) \, dx \stackrel{!}{=} 0 \qquad (4.134)$$

Solving the above expression gives $a_3 = -1$ and $a_4 = 0$, so that the approximate solutions computed with Ritz's and Galerkin's methods are equal.

4.4   THE METHOD OF WEIGHTED RESIDUALS – A MORE GENERAL VIEW

This section provides a more general view of the MWR. The general idea of this method will be outlined without a specific problem at hand. Later sections will apply these basic concepts to many problems of financial or economic interest.

At a rather abstract level any differential equation can be written as:

$$L(u) = f \qquad (4.135)$$

with $L(\cdot)$ denoting a differential operator. For the purposes of this book it is supposed to depend on one, two or three spatial variables, but not time or time to maturity. It may be linear or nonlinear:

- $L(u) = u_{xx} + a_1(x)u_x + a_2u = f(x)$ will be discussed in this chapter.
- $L(u) = u_{xx} + u_{yy} = f(x, y)$ and more general problems in two spatial variables will be discussed in Chapter 6.
- $L(u) = u_{xx} + u_{yy} + u_{zz} = f(x, y, z)$ and more general problems in three spatial variables will be discussed in Chapter 8.
- Nonlinear operators will be treated in Chapter 10.

The first step of an FE approach is to subdivide the domain into nonoverlapping subdomains. Besides this, the finite elements need to cover the domain completely. In 1D problems, this reduces to dividing a straight line into intervals. In 2D and 3D problems, discretizing a domain can become much more complicated since the elements, usually triangles or rectangles, might not be able to fill a complicated domain exactly. However, in financial applications, usually only simple domains arise, which can be filled with elements without incurring a discretization error, (see Figure 6.10). In 2D, triangular and rectangular elements will be discussed; in 3D, only rectangular elements will be covered.

The approximate solution $\tilde{u}$ is to be of the following form:

$$\tilde{u} = \sum_{i=1}^{N+1} \tilde{u}_i \phi_i \qquad (4.136)$$

The ϕ_i are called *shape functions*, *basis functions* or *trial functions*. The choice of appropriate shape functions depends on various criteria, such as the order of the differential equation at hand and the order of the derivative of the approximate solution needed. The weights $\tilde{u}_i$ are to be determined by the numerical algorithm in such a way that $u \approx \tilde{u}$. This is made more precise by introducing the residual $R = L(\tilde{u}) - f$, which is to be minimized. This can be accomplished in various fashions, as outlined in Section 4.2. Since there are $N + 1$ unknowns, $N + 1$ equations need to be established to determine them:

$$\int_{\Omega} RW_j \, dx \stackrel{!}{=} 0 \quad \forall \; j = 1, \dots, N + 1 \tag{4.137}$$

It is instructive to take a closer look at Equation (4.137). Inserting the definition of the residual:

$$\int_{\Omega} (L(\tilde{u}) - f)W_j \, dx = 0 \quad \forall \; j = 1, \dots, N + 1 \tag{4.138}$$

$$\int_{\Omega} L(\tilde{u})W_j \, dx = \int_{\Omega} fW_j \, dx \quad \forall \; j = 1, \dots, N + 1 \tag{4.139}$$

For linear $L(\cdot)$ the LHS of Equation (4.139) simplifies to:

$$\int_{\Omega} L\left(\sum_i \tilde{u}_i \phi_i\right) W_j \, dx \tag{4.140}$$

$$= \int_{\Omega} \sum_i \tilde{u}_i L(\phi_i) \, W_j \, dx \tag{4.141}$$

$$= \sum_i \tilde{u}_i \underbrace{\int_{\Omega} L(\phi_i) \, W_j \, dx}_{=K_{ij}} \tag{4.142}$$

The RHS of Equation (4.139) can be rewritten as:

$$\mathbf{F} = \begin{pmatrix} \int_{\Omega} fW_1 \, dx \\ \vdots \\ \int_{\Omega} fW_N \, dx \\ \int_{\Omega} fW_{N+1} \, dx \end{pmatrix} \tag{4.143}$$

so that the FE solution of a linear differential equation reduces to solving a system of linear equations given by:

$$\mathbf{K}\bar{u} = \mathbf{F} \tag{4.144}$$

The discussion is not complete at this point. Nothing has been said about the BCs so far. In this book, several trial methods and two weighting functions will be discussed: the Galerkin method and the collocation method.

4.5 MULTI-ELEMENT SOLUTIONS

In this section, the approach of the previous section will be generalized to multi-element solutions. Having introduced the MWR as one building block of the FEM, a general seven-step procedure applicable to all kinds of differential equations arising in economics and finance will be presented. The seven-step procedure follows closely the approach of Bickford (1990). This seven-step procedure will be introduced with the help of a general two-point boundary value problem arising in economics which, furthermore, is not self-adjoint:

$$u''(x) + a_1(x)u'(x) + a_2(x)u(x) + f(x) = 0 \tag{4.145}$$

$$\alpha_1 u(x^{\min}) + \beta_1 u'(x^{\min}) = \gamma_1 \tag{4.146}$$

$$\alpha_2 u(x^{\max}) + \beta_2 u'(x^{\max}) = \gamma_2 \tag{4.147}$$

This states the most general case of two Robin BCs. Economic applications usually specialize to either two Dirichlet BCs or to one Dirichlet BC and one Neumann BC. The starting point of the seven-step procedure is the differential equation. It ends with the numerical solution of the discretized problem. It can be used with and without the element concept. The seven steps to be performed with each differential equation are:

1. Discretization.
2. Interpolation.
3. Elemental formulation.
4. Assembly.
5. Boundary conditions.
6. Solution.
7. Computation of derived variables.

First, we will derive a multi-element solution, following this recipe, for the problem given by Equations (4.145) to (4.147) using the Galerkin optimization criterion. Then, we will apply the collocation method.

4.5.1   The Galerkin method with linear elements

1. *Discretization:* Only the spatial variable is discretized with FE. Time or time to maturity in dynamic problems will be discretized with FD, as introduced in Section 3.2. The spatial domain of a 1D problem simply consists of a closed interval $[x_0, x_{N+1}]$. This interval is divided into nonoverlapping subintervals which do not need be of equal length (see Figure 4.5). These subintervals are called *elements*. The points on the border between two elements are called *nodes*, as in the context of FD. The collection of nodes and elements is called a *mesh*. The length of the interval $[x_i, x_{i+1}]$ is denoted by l_E^i. In the case of equidistant nodes, i.e. $l_E^i = l_E^j$ for all $i \neq j$, the element length is denoted simply by l_E.

2. *Interpolation:* Assuming that an approximate solution between two nodes $x_i \leq x_{i+1}$ can be represented by a straight line, linear interpolation is employed on this interval:

$$\tilde{u}_e(x) = ax + b \qquad (4.148)$$

Both a and b are to be chosen in a way that $\tilde{u}_e(x_i) = u_i$ and $\tilde{u}_e(x_{i+1}) = u_{i+1}$. This results in:

$$\tilde{u}_e(x) = \frac{x_{i+1} - x}{x_{i+1} - x_i} \tilde{u}_i + \frac{x - x_i}{x_{i+1} - x_i} \tilde{u}_{i+1} \qquad (4.149)$$

$$= N_i \tilde{u}_i + N_{i+1} \tilde{u}_{i+1} \qquad (4.150)$$

N_i and N_{i+1} are called *elemental linear interpolation functions*. On the entire domain $[x^{\min} = x_1, x^{\max} = x_{N+1}]$ the approximate function $\tilde{u}(x)$ is given by:

$$\tilde{u}(x) = \sum_{i=1}^{N+1} \tilde{u}_i n_i(x) \qquad (4.151)$$

Figure 4.5   Discretization of a 1D problem

with:

$$n_i(x_j) = 1 \quad \forall \quad i = j \tag{4.152}$$

$$n_i(x_j) = 0 \quad \forall \quad i \neq j \tag{4.153}$$

So that n_i equals one at node i and vanishes at every other node. Applying Equation (4.151) to x_j leads to:

$$\tilde{u}(x_j) = \sum_{i=1}^{N+1} \tilde{u}_i n_i(x_j) = \tilde{u}_j \tag{4.154}$$

This means that the coefficients $\tilde{u}_i$ are the approximate solution of the differential equation at x_i. Outside $[x_{i-1}, x_{i+1}]$ the function n_i is zero.

3. *Elemental formulation:* Substituting the approximate function Equation (4.151) into the boundary value problem given by Equation (4.145) leads to:

$$\tilde{u}'' + a_1(x)\tilde{u}' + a_2(x)\tilde{u} + f(x) = R(x, \tilde{u}_1, \ldots, \tilde{u}_{N+1}) \tag{4.155}$$

Again, as in Section 4.2, $R(\cdot)$ is the residual. The Galerkin criterion requires that the following expression vanishes:

$$\int_{x^{\min}}^{x^{\max}} R \, n_k(x) \, \mathrm{d}x \overset{!}{=} 0 \quad \forall \, k = 1, 2, \ldots, N + 1 \tag{4.156}$$

Equation (4.156) represents $N + 1$ linear algebraic equations, so that the $\tilde{u}_i$ can be uniquely determined. Substituting Equation (4.155) into Equation (4.156) leads to:

$$\int_{x^{\min}}^{x^{\max}} \left[\tilde{u}'' + a_1(x)\tilde{u}' + a_2(x)\tilde{u} + f(x) \right] n_k(x) \, \mathrm{d}x \overset{!}{=} 0 \quad \forall \, k = 1, 2, \ldots, N + 1 \tag{4.157}$$

To eliminate the second order derivative $\tilde{u}''$, the first term from the integral is integrated by parts:

$$\int_{x^{\min}}^{x^{\max}} \tilde{u}'' n_k = \left[\tilde{u}' n_k \right]_{x^{\min}}^{x^{\max}} - \int_{x^{\min}}^{x^{\max}} \tilde{u}' n_k' \, \mathrm{d}x \tag{4.158}$$

$$= n_k(x^{\max}) \left[\frac{\gamma_2 - \alpha_2 u(x^{\max})}{\beta_2} \right]$$

$$- n_k(x^{\min}) \left[\frac{\gamma_1 - \alpha_1 u(x^{\min})}{\beta_1} \right] - \int_{x^{\min}}^{x^{\max}} \tilde{u}' n_k' \, \mathrm{d}x \tag{4.159}$$

The above equation lacks mathematical rigor, since integration by parts is applied to a function with discontinuous derivatives. A rigorous resolution requires sophisticated mathematical tools. Alternatively, the derivation of the FE model can be set up differently so that Equation (4.158) can be avoided. For such a derivation, see Burnett (1987). Using the above result, Equation (4.157) reads:

$$n_k(x^{\max}) \left[\frac{\gamma_2 - \alpha_2 u(x^{\max})}{\beta_2} \right] - n_k(x^{\min}) \left[\frac{\gamma_1 - \alpha_1 u(x^{\min})}{\beta_1} \right]$$

$$- \int_{x^{\min}}^{x^{\max}} \left[n_k' \tilde{u}' - n_k a_1 \tilde{u}' - n_k a_2 \tilde{u} - n_k f \right] \mathrm{d}x \overset{!}{=} 0 \quad \forall k = 1, 2, \ldots, N + 1 \tag{4.160}$$

Next, we consider only the integral form of the above equation. Inserting Equation (4.151) into this integral gives:

$$\int_{x^{\min}}^{x^{\max}} \left[n_k' \left(\sum_{i=1}^{N+1} \tilde{u}_i n_i' \right) - n_k a_1 \left(\sum_{i=1}^{N+1} \tilde{u}_i n_i' \right) - n_k a_2 \left(\sum_{i=1}^{N+1} \tilde{u}_i n_i \right) - n_k f \right] dx$$

$$= \sum_{i=1}^{N+1} \left[\int_{x^{\min}}^{x^{\max}} (n_k' n_i' - a_1 n_k n_i' - a_2 n_k n_i) \, dx \right] \tilde{u}_i - \int_{x^{\min}}^{x^{\max}} n_k f \, dx \qquad (4.161)$$

Using this result leads to:

$$\sum_{i=1}^{N+1} \left[\int_{x^{\min}}^{x^{\max}} (n_k' n_i' - a_1 n_k n_i' - a_2 n_k n_i) \, dx \right] \tilde{u}_i = \int_{x^{\min}}^{x^{\max}} n_k f \, dx$$

$$+ n_k(x^{\max}) \left[\frac{\gamma_2 - \alpha_2 u(x^{\max})}{\beta_2} \right] - n_k(x^{\min}) \left[\frac{\gamma_1 - \alpha_1 u(x^{\min})}{\beta_1} \right] \quad \forall k = 1, 2, \ldots, N+1$$

$$(4.162)$$

Introducing the Kronecker delta, δ_{ij}, the above equation can be rewritten as:

$$\sum_{i=1}^{N+1} \left[\int_{x^{\min}}^{x^{\max}} (n_k' n_i' - a_1 n_k n_i' - a_2 n_k n_i) \, dx \right] \tilde{u}_i + \delta_{kN+1} \frac{\alpha_2}{\beta_2} \tilde{u}_{N+1} - \delta_{k1} \frac{\alpha_1}{\beta_1} \tilde{u}_1$$

$$= \int_{x^{\min}}^{x^{\max}} n_k f \, dx + \delta_{kN+1} \frac{\gamma_2}{\beta_2} - \delta_{k1} \frac{\gamma_1}{\beta_1} \quad \forall k = 1, 2, \ldots, N+1 \qquad (4.163)$$

with

$$\delta_{ij} = \begin{cases} 0 & \text{if } i \neq j \\ 1 & \text{if } i = j \end{cases} \qquad (4.164)$$

This accounts for the fact that:

$$n_k(x^{\min}) = \begin{cases} 0 & \forall k = 2, \ldots, N+1 \\ 1 & k = 1 \end{cases} \qquad (4.165)$$

$$n_k(x^{\max}) = \begin{cases} 0 & \forall k = 1, \ldots, N \\ 1 & k = N+1 \end{cases} \qquad (4.166)$$

All terms resulting from the BCs drop out for all k with $1 < k < N$ because of the Dirac delta functions. The system (4.163) can be written more compactly as:

$$\sum_{i=1}^{N+1} K_{ki} \tilde{u}_i = F_k \quad \forall \ k = 1, 2, \ldots, N+1 \qquad (4.167)$$

or, in an even more compact manner:

$$\mathbf{K}\tilde{\mathbf{u}} = \mathbf{F} \qquad (4.168)$$

where:

$$K_{ki} = \int_{x^{\min}}^{x^{\max}} (n_k' n_i' - a_1 n_k n_i' - a_2 n_k n_i) \, dx + \delta_{kN+1} \frac{\alpha_2}{\beta_2} - \delta_{k1} \frac{\alpha_1}{\beta_1}$$

$$F_k = \int_{x^{\min}}^{x^{\max}} n_k f \, dx + \delta_{kN+1} \frac{\gamma_2}{\beta_2} - \delta_{k1} \frac{\gamma_1}{\beta_1} \qquad (4.169)$$

Because of the local support of each finite element, many terms drop out. For all k with $1 < k < N$, each equation of the system (4.167) can be written as:

$$\int_{x_{k-1}}^{x_k} \left(n_k' n_{k-1}' - a_1 n_k n_{k-1}' - a_2 n_k n_{k-1} \right) \, dx \, \tilde{u}_{k-1} + \tag{4.170}$$

$$\int_{x_{k-1}}^{x_{k+1}} \left(n_k' n_k' - a_1 n_k n_k' - a_2 n_k n_k \right) \, dx \, \tilde{u}_k + \tag{4.171}$$

$$\int_{x_k}^{x_{k+1}} \left(n_k' n_{k+1}' - a_1 n_k n_{k+1}' - a_2 n_k n_{k+1} \right) \, dx \, \tilde{u}_{k+1} = \int_{x_{k-1}}^{x_{k+1}} n_k f \, dx \tag{4.172}$$

Ignoring the BCs for a short moment, the LHS of the above equation can be written as:

$$\mathbf{K} = \begin{pmatrix} K_{11} & K_{12} & 0 & 0 & \cdots & 0 & 0 & 0 \\ K_{21} & K_{22} & K_{23} & 0 & \cdots & 0 & 0 & 0 \\ 0 & K_{32} & K_{33} & K_{34} & \cdots & 0 & 0 & 0 \\ \vdots & \vdots & \vdots & \vdots & \ddots & \vdots & \vdots & \vdots \\ 0 & 0 & 0 & 0 & \cdots & K_{N,N-1} & K_{N,N} & K_{N,N+1} \\ 0 & 0 & 0 & 0 & \cdots & 0 & K_{N+1,N} & K_{N+1,N+1} \end{pmatrix} \tag{4.173}$$

The matrix $\mathbf{K}$ is a tridiagonal matrix and consequently increasingly sparse for growing N. Only for self-adjoint differential equations is it symmetric.

4. *Assembly:* It is instructive to write out the system (4.167) in full for $N = 2$:

$$\int_{x_1}^{x_2} F_{11} \, dx \, \tilde{u}_1 + \int_{x_1}^{x_2} F_{12} \, dx \, \tilde{u}_2 \qquad\qquad\qquad - \frac{\alpha_1}{\beta_1} \tilde{u}_1 = \int_{x_1}^{x_2} n_1 f \, dx - \frac{\gamma_1}{\beta_1}$$

$$\int_{x_1}^{x_2} F_{21} \, dx \, \tilde{u}_1 + \int_{x_1}^{x_2} F_{22} \, dx \, \tilde{u}_2 + \int_{x_2}^{x_3} F_{22} \, dx \, \tilde{u}_2 + \int_{x_2}^{x_3} F_{23} \, dx \, \tilde{u}_3 = \int_{x_1}^{x_2} n_2 f \, dx + \int_{x_2}^{x_3} n_2 f \, dx$$

$$\int_{x_2}^{x_3} F_{32} \, dx \, \tilde{u}_2 + \int_{x_2}^{x_3} F_{33} \, dx \, \tilde{u}_3 + \frac{\alpha_2}{\beta_2} \tilde{u}_3 = \frac{\gamma_2}{\beta_2} + \int_{x_2}^{x_3} n_3 f \, dx$$

with $F_{ik} = n_i' n_k' - a_1 n_i n_k' - a_2 n_i n_k$.

The first terms of the first two lines represent the first element. Accordingly, the third and fourth terms of the second line and the first two terms of the third line arise from the second element. This makes it possible to view the system (4.167) from an elemental perspective instead of a global nodal view. The elemental stiffness matrices $\mathbf{k}_{ei}$ associated with the two elements are given by:

$$\mathbf{k}_{e1} = \begin{pmatrix} \int_{x_1}^{x_2} F_{11} \, dx & \int_{x_1}^{x_2} F_{12} \, dx \\ \int_{x_1}^{x_2} F_{21} \, dx & \int_{x_1}^{x_2} F_{22} \, dx \end{pmatrix} \tag{4.174}$$

$$\mathbf{k}_{e2} = \begin{pmatrix} \int_{x_2}^{x_3} F_{11} \, dx & \int_{x_2}^{x_3} F_{12} \, dx \\ \int_{x_2}^{x_3} F_{21} \, dx & \int_{x_2}^{x_3} F_{22} \, dx \end{pmatrix} \tag{4.175}$$

Notice that the only differences between these elemental stiffness matrices are the limits of integration. Defining an elemental interpolation vector $\mathbf{N}$:

$$\mathbf{N} = \begin{pmatrix} N_1 \\ N_2 \end{pmatrix} \quad \text{with} \quad \mathbf{N}' = \begin{pmatrix} N_1' \\ N_2' \end{pmatrix} \tag{4.176}$$

the elemental stiffness matrices can be rewritten as:

$$\mathbf{k}_{e1} = \int_{x_1}^{x_2} \left(\mathbf{N}' \mathbf{N}'^T - a_1 \mathbf{N} \mathbf{N}'^T - a_2 \mathbf{N} \mathbf{N}^T \right) \, dx \tag{4.177}$$

$$\mathbf{k}_{e2} = \int_{x_2}^{x_3} \left(\mathbf{N}' \mathbf{N}'^T - a_1 \mathbf{N} \mathbf{N}'^T - a_2 \mathbf{N} \mathbf{N}^T \right) \, dx \tag{4.178}$$

Only for elements of equal length and constant a_1 and a_2 are the elemental stiffness matrices the same.

In the case of two elements, each elemental stiffness matrix is expanded to the global size as follows:

$$\mathbf{k}_{g1} = \begin{pmatrix} \int_{x_1}^{x_2} F_{11}\,dx & \int_{x_1}^{x_2} F_{12}\,dx & 0 \\ \int_{x_1}^{x_2} F_{21}\,dx & \int_{x_1}^{x_2} F_{22}\,dx & 0 \\ 0 & 0 & 0 \end{pmatrix} \tag{4.179}$$

$$\mathbf{k}_{g2} = \begin{pmatrix} 0 & 0 & 0 \\ 0 & \int_{x_2}^{x_3} F_{11}\,dx & \int_{x_2}^{x_3} F_{12}\,dx \\ 0 & \int_{x_2}^{x_3} F_{21}\,dx & \int_{x_2}^{x_3} F_{22}\,dx \end{pmatrix} \tag{4.180}$$

$$\mathbf{k}_G = \mathbf{k}_{g1} + \mathbf{k}_{g2} \tag{4.181}$$

In a similar way, the load matrix is decomposed into elemental load matrices:

$$\mathbf{f}_{e1} = \begin{pmatrix} \int_{x_1}^{x_2} N_1\,f\,dx \\ \int_{x_1}^{x_2} N_2\,f\,dx \end{pmatrix} = \int_{x_1}^{x_2} \mathbf{N}\,f\,dx \tag{4.182}$$

$$\mathbf{f}_{e2} = \begin{pmatrix} \int_{x_2}^{x_3} N_1\,f\,dx \\ \int_{x_2}^{x_3} N_2\,f\,dx \end{pmatrix} = \int_{x_2}^{x_3} \mathbf{N}\,f\,dx \tag{4.183}$$

The integration into the global system is accomplished by extending the elemental load matrices:

$$\mathbf{f}_{g1} = \begin{pmatrix} \int_{x_1}^{x_2} N_1\,f\,dx \\ \int_{x_1}^{x_2} N_2\,f\,dx \\ 0 \end{pmatrix} \qquad \mathbf{f}_{g2} = \begin{pmatrix} 0 \\ \int_{x_2}^{x_3} N_1\,f\,dx \\ \int_{x_2}^{x_3} N_2\,f\,dx \end{pmatrix} \tag{4.184}$$

$$\mathbf{f}_G = \mathbf{f}_{g1} + \mathbf{f}_{g2} \tag{4.185}$$

At global level, the BCs can be represented as:

$$\mathbf{BT}_G = \begin{pmatrix} -\frac{\alpha_1}{\beta_1} & 0 & 0 \\ 0 & 0 & 0 \\ 0 & 0 & \frac{\alpha_2}{\beta_2} \end{pmatrix} \tag{4.186}$$

$$\mathbf{bt}_G = \begin{pmatrix} -\frac{\gamma_1}{\beta_1} \\ 0 \\ \frac{\gamma_2}{\beta_2} \end{pmatrix} \tag{4.187}$$

The final set of equations is given by $\mathbf{K}_G \bar{\mathbf{u}}_G = \mathbf{F}_G$, with the global stiffness matrix and global load matrix given by:

$$\mathbf{K}_G = \mathbf{k}_G + \mathbf{BT}_G \tag{4.188}$$

$$\mathbf{F}_G = \mathbf{f}_G + \mathbf{bt}_G \tag{4.189}$$

In the general case of N elements, the elemental stiffness and load matrices are given by:

$$\mathbf{k}_{ei} = \int_{x_i}^{x_{i+1}} \left(\mathbf{N}'\mathbf{N}'^\mathrm{T} - a_1\mathbf{N}\mathbf{N}'^\mathrm{T} - a_2\mathbf{N}\mathbf{N}^\mathrm{T} \right) \mathrm{d}x \tag{4.190}$$

$$\mathbf{f}_{ei} = \int_{x_i}^{x_{i+1}} \mathbf{N} f \, \mathrm{d}x \tag{4.191}$$

For constant a_1 and a_2, the elemental stiffness matrix can be simplified. Using:

$$\mathbf{N} = \begin{pmatrix} \frac{x_{i+1}-x}{x_{i+1}-x_i} \\ \frac{x-x_i}{x_{i+1}-x_i} \end{pmatrix} = \frac{1}{l_E^i} \begin{pmatrix} x_{i+1}-x \\ x-x_i \end{pmatrix} \tag{4.192}$$

$$\mathbf{N}' = \frac{1}{l_E^i} \begin{pmatrix} -1 \\ 1 \end{pmatrix} \tag{4.193}$$

inserting these expressions into Equation (4.190) and solving the integral results in:

$$\mathbf{k}_{ei} = \frac{1}{l_E^i} \begin{pmatrix} 1 & -1 \\ -1 & 1 \end{pmatrix} - \frac{a_1}{2} \begin{pmatrix} -1 & 1 \\ -1 & 1 \end{pmatrix} - \frac{a_2 \, l_E^i}{6} \begin{pmatrix} 2 & 1 \\ 1 & 2 \end{pmatrix} \tag{4.194}$$

For nonconstant parameters $a_k = a_k(x)$, $k = 1, 2$, either the integral in Equation (4.190) needs to be solved – possibly numerically (see Section A.2) – or it is possible to insert an average value into Equation (4.194):

$$\bar{a}_k = \frac{a_k(x_i) + a_k(x_{i+1})}{2} \qquad k = 1, 2 \tag{4.195}$$

Nonconstant $f = f(x)$ have to be treated in a similar fashion.
For a multi-element solution, the final set of equations from an elemental perspective is given by:

$$\mathbf{K}_G = \sum_{i=1}^{N} \mathbf{k}_{gi} + \mathbf{BT}_G \tag{4.196}$$

$$\mathbf{F}_G = \sum_{i=1}^{N} \mathbf{f}_{gi} + \mathbf{bt}_G \tag{4.197}$$

5. *Boundary conditions:* The case of one or two Dirichlet conditions cannot be put into the above framework because this would result in a division by zero in Equation (4.186) and/or Equation (4.187). We will discuss one Dirichlet condition at $x_1 = x^{\min}$ and a Neumann or Robin condition at $x_{N+1} = x^{\max}$. The extension to a Dirichlet condition at $x_{N+1} = x^{\max}$ or Dirichlet conditions at both ends of the domain is straightforward. The set of linear equations given by $\mathbf{K}\bar{\mathbf{u}} = \mathbf{F}$ has to be augmented to reflect $u(x_1) = \frac{\gamma_1}{\alpha_1}$. Equation (4.168) can be rewritten as:

$$\left[\begin{array}{ccccc} K_{11} & K_{12} & 0 & \dots & 0 & 0 \\ K_{21} & K_{22} & K_{23} & \dots & 0 & 0 \\ \vdots & \vdots & \vdots & \ddots & \vdots & \vdots \\ 0 & 0 & 0 & \dots & K_{N+1,N} & K_{N+1,N+1} \end{array} \middle| \begin{array}{c} F_1 \\ F_2 \\ \vdots \\ F_{N+1} \end{array} \right] \tag{4.198}$$

The condition $u(x_1) = \frac{\gamma_1}{\alpha_1}$ is enforced by changing the first line of Equation (4.198) to:

$$
\begin{bmatrix}
1 & 0 & 0 & \cdots & 0 & 0 & \Big| & \frac{\gamma_1}{\alpha_1} \\
K_{21} & K_{22} & K_{23} & \cdots & 0 & 0 & \Big| & F_2 \\
\vdots & \vdots & \vdots & \ddots & \vdots & \vdots & \Big| & \vdots \\
0 & 0 & 0 & \cdots & K_{N+1,N} & K_{N+1,N+1} & \Big| & F_{N+1}
\end{bmatrix}
\tag{4.199}
$$

6. *Solution:* The solution of linear systems arising from FE discretizations is a vast field. Many textbooks on FE analysis devote entire chapters to this topic. By not doing so in this book, it is not intended to convey the impression that this topic is of minor importance. On the contrary: it is of utmost importance in real-life computing, since all computing time is spent on solving these systems. Since there are several excellent treatments of this topic, only some quite general comments on the proper choice of a linear solver will be made here. The solver should be capable of taking advantage of symmetry and sparseness. Gaussian elimination with at least partial pivoting has proven to be very reliable in FE analysis. Besides this, the computation of the condition number is useful.

7. *Computation of derived variables:* Since the approximation of the unknown function u is obtained with the help of linear elements, differentiating the numerical solution leads to constant values for u' within each element and discontinuities at the border between two elements. This is one of the reasons for introducing higher order elements in Sections 4.5.2 and 4.5.3.

Example 4
As a first simple example, a two-element model will be constructed using the seven-step procedure outlined above.

$$
u''(x) = -2, \quad u(0) = 0, \quad u'(2) = -2
\tag{4.200}
$$

This is a self-adjoint problem with a Dirichlet BC at $x = 0$ and a Neumann BC at $x = 2$. The analytical solution is given by $u(x) = -(x-1)^2 + 1$. In terms of Equation (4.145) we have $a_1 = 0$, $a_2 = 0$, and $f = 2$.

1. *Discretization: Two elements of equal length imply $x_1 = 0$, $x_2 = 1$ and $x_3 = 2$. The element length is $l_E = 1$.*

2. *Interpolation: Linear elements are employed, so that the interpolation function is given by Equation (4.151).*

3. *Elemental formulation: The elemental stiffness matrix Equation (4.190) reduces to:*

$$
\mathbf{k}_{ei} = \int_{x_i}^{x_{i+1}} \mathbf{N}' \mathbf{N}'^T dx
\tag{4.201}
$$

The elemental load matrix Equation (4.191) reads:

$$
\mathbf{f}_{ei} = \int_{x_i}^{x_{i+1}} 2\mathbf{N} \, dx
\tag{4.202}
$$

Since $l_E = 1$:

$$
\mathbf{N} = \begin{pmatrix} x_{i+1} - x \\ x - x_i \end{pmatrix}, \quad \mathbf{N}' = \begin{pmatrix} -1 \\ 1 \end{pmatrix}
\tag{4.203}
$$

Inserting these expressions into the elemental load matrix and elemental mass matrix, we get:

$$\mathbf{k}_{ei} = \int_{x_i}^{x_{i+1}} \begin{pmatrix} -1 \\ 1 \end{pmatrix} (-1, 1) \, dx \tag{4.204}$$

$$= \begin{pmatrix} 1 & -1 \\ -1 & 1 \end{pmatrix} \tag{4.205}$$

$$\mathbf{f}_{ei} = 2 \int_{x_i}^{x_{i+1}} \begin{pmatrix} x_{i+1} - x \\ x - x_i \end{pmatrix} dx = \begin{pmatrix} 1 \\ 1 \end{pmatrix} \tag{4.206}$$

4. *Assembly: Next, the results of the previous step are assembled:*

$$\mathbf{F}_G = \mathbf{f}_{g1} + \mathbf{f}_{g2} + \mathbf{b}\mathbf{t}_G \tag{4.207}$$

$$= \begin{pmatrix} 1 \\ 1 \\ 0 \end{pmatrix} + \begin{pmatrix} 0 \\ 1 \\ 1 \end{pmatrix} + \begin{pmatrix} 0 \\ 0 \\ -2 \end{pmatrix} = \begin{pmatrix} 1 \\ 2 \\ -1 \end{pmatrix} \tag{4.208}$$

$$\mathbf{K}_G = \mathbf{k}_{g1} + \mathbf{k}_{g2} + \mathbf{B}\mathbf{T}_G \tag{4.209}$$

$$= \begin{pmatrix} 1 & -1 & 0 \\ -1 & 1 & 0 \\ 0 & 0 & 0 \end{pmatrix} + \begin{pmatrix} 0 & 0 & 0 \\ 0 & 1 & -1 \\ 0 & -1 & 1 \end{pmatrix} + \begin{pmatrix} 0 & 0 & 0 \\ 0 & 0 & 0 \\ 0 & 0 & 0 \end{pmatrix} \tag{4.210}$$

$$= \begin{pmatrix} 1 & -1 & 0 \\ -1 & 2 & -1 \\ 0 & -1 & 1 \end{pmatrix} \tag{4.211}$$

5. *Boundary conditions: The unconstrained system is given by:*

$$\left[\begin{array}{ccc|c} 1 & -1 & 0 & 1 \\ -1 & 2 & -1 & 2 \\ 0 & -1 & 1 & -1 \end{array} \right] \tag{4.212}$$

Since $u''(x) = -2$ constitutes a self-adjoint problem, $\mathbf{K}_G$ is symmetric. The above system is constrained by the BC $u(0) = 0$, so that it has to be modified to:

$$\left[\begin{array}{ccc|c} 1 & 0 & 0 & 0 \\ -1 & 2 & -1 & 2 \\ 0 & -1 & 1 & -1 \end{array} \right] \tag{4.213}$$

6. *Solution: The above system is solved by:*

$$\tilde{\mathbf{u}} = \begin{pmatrix} \tilde{u}_1 \\ \tilde{u}_2 \\ \tilde{u}_3 \end{pmatrix} = \begin{pmatrix} 0 \\ 1 \\ 0 \end{pmatrix} \tag{4.214}$$

7. *Computation of derived variables: The first derivative of the numerical solution is constant for each element, as shown in Figure 4.6. The approximation of the derived variable is quite poor. It becomes better as the number of elements is increased, as shown later in Figure 4.7. A further problem arises when the second derivative is of interest. Obviously, the second derivative of a linear approximation function vanishes everywhere. The only way out of this problem is to employ shape functions of a higher order. This approach is investigated in Section 4.5.2.*

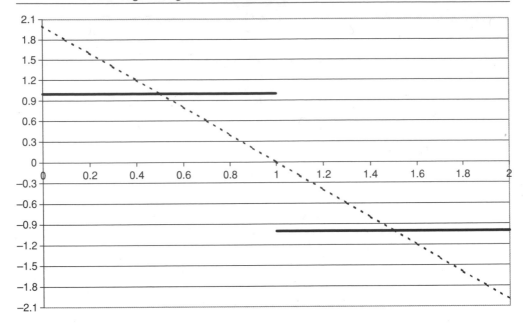

Figure 4.6 First derivative of a two-element solution

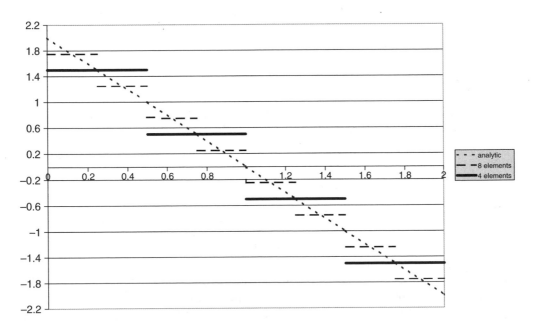

Figure 4.7 First derivatives of four-element and eight-element solutions

As a next step, the number of elements is increased in order to get an impression of how the numerical solution converges to the analytical.

1. Discretization: The domain is divided into 4, 8, 16 and 32 elements of equal length.

2. Interpolation: The same interpolation function as in the two-element solution is to be used.

3. Elemental formulation: The elemental formulation depends on the element length l_E, which varies since the number of elements is increased. The elemental stiffness matrix Equation (4.190) reduces to:

$$\mathbf{k}_{ei} = \int_{x_i}^{x_{i+1}} \mathbf{N}' \mathbf{N}'^{\mathrm{T}} \mathrm{d}x \qquad (4.215)$$

The elemental load matrix Equation (4.191) reads:

$$\mathbf{f}_{ei} = \int_{x_i}^{x_{i+1}} 2\mathbf{N} \, \mathrm{d}x \qquad (4.216)$$

with:

$$\mathbf{N} = \begin{pmatrix} \frac{x_{i+1}-x}{x_{i+1}-x_i} \\ \frac{x-x_i}{x_{i+1}-x_i} \end{pmatrix}, \quad \mathbf{N}' = \frac{1}{l_E}\begin{pmatrix} -1 \\ 1 \end{pmatrix} \qquad (4.217)$$

so that:

$$\mathbf{k}_{ei} = \frac{1}{l_F}\begin{pmatrix} 1 & -1 \\ -1 & 1 \end{pmatrix} \qquad (4.218)$$

$$\mathbf{f}_{ei} = \begin{pmatrix} l_E \\ l_E \end{pmatrix} \qquad (4.219)$$

4. Assembly: For $N = 4$, assembly in a completely analogous fashion to that above results in:

$$\begin{pmatrix} 2 & -2 & 0 & 0 & 0 \\ -2 & 4 & -2 & 0 & 0 \\ 0 & -2 & 4 & -2 & 0 \\ 0 & 0 & -2 & 4 & -2 \\ 0 & 0 & 0 & -2 & 2 \end{pmatrix} \tilde{\mathbf{u}} = \begin{pmatrix} \frac{1}{2} \\ 1 \\ 1 \\ 1 \\ -\frac{3}{2} \end{pmatrix} \qquad (4.220)$$

and for $N = 8$:

$$\begin{pmatrix} 4 & -4 & 0 & 0 & 0 & 0 & 0 & 0 & 0 \\ -4 & 8 & -4 & 0 & 0 & 0 & 0 & 0 & 0 \\ 0 & -4 & 8 & -4 & 0 & 0 & 0 & 0 & 0 \\ 0 & 0 & -4 & 8 & -4 & 0 & 0 & 0 & 0 \\ 0 & 0 & 0 & -4 & 8 & -4 & 0 & 0 & 0 \\ 0 & 0 & 0 & 0 & -4 & 8 & -4 & 0 & 0 \\ 0 & 0 & 0 & 0 & 0 & -4 & 8 & -4 & 0 \\ 0 & 0 & 0 & 0 & 0 & 0 & -4 & 8 & -4 \\ 0 & 0 & 0 & 0 & 0 & 0 & 0 & 4 & -4 \end{pmatrix} \tilde{\mathbf{u}} = \begin{pmatrix} 0.25 \\ 0.5 \\ 0.5 \\ 0.5 \\ 0.5 \\ 0.5 \\ 0.5 \\ 0.5 \\ -1.75 \end{pmatrix} \qquad (4.221)$$

5. *Boundary conditions: The BCs are integrated in a similar fashion as in the two-element example. For N = 4:*

$$
\left[
\begin{array}{ccccc|c}
1 & 0 & 0 & 0 & 0 & 0 \\
-2 & 4 & -2 & 0 & 0 & 1 \\
0 & -2 & 4 & -2 & 0 & 1 \\
0 & 0 & -2 & 4 & -2 & 1 \\
0 & 0 & 0 & -2 & 2 & -1.5
\end{array}
\right]
\tag{4.222}
$$

and for N = 8:

$$
\left[
\begin{array}{ccccccccc|c}
1 & 0 & 0 & 0 & 0 & 0 & 0 & 0 & 0 & 0 \\
-4 & 8 & -4 & 0 & 0 & 0 & 0 & 0 & 0 & 0.5 \\
0 & -4 & 8 & -4 & 0 & 0 & 0 & 0 & 0 & 0.5 \\
0 & 0 & -4 & 8 & -4 & 0 & 0 & 0 & 0 & 0.5 \\
0 & 0 & 0 & -4 & 8 & -4 & 0 & 0 & 0 & 0.5 \\
0 & 0 & 0 & 0 & -4 & 8 & -4 & 0 & 0 & 0.5 \\
0 & 0 & 0 & 0 & 0 & -4 & 8 & -4 & 0 & 0.5 \\
0 & 0 & 0 & 0 & 0 & 0 & -4 & 8 & -4 & 0.5 \\
0 & 0 & 0 & 0 & 0 & 0 & 0 & 4 & -4 & -1.75
\end{array}
\right]
\tag{4.223}
$$

It is obvious at first sight that the coefficient matrix $\mathbf{K}_G$ becomes more and more sparse for growing N.

6. *Solution: The four-element solution is given by:*

$$
\bar{\mathbf{u}} = (0, \frac{3}{4}, 1, \frac{3}{4}, 0)^T
\tag{4.224}
$$

The eight-element solution is given by:

$$
\bar{\mathbf{u}} = (0, \frac{7}{16}, \frac{3}{4}, \frac{15}{16}, 1, \frac{15}{16}, \frac{3}{4}, \frac{7}{16}, 0)^T
\tag{4.225}
$$

7. *Computation of derived variables: Here, the same comments as for the two-element solution apply. The first derivatives of the solutions with four and eight elements are plotted in Figure 4.7.* ☐

Example 5
The next example is not self-adjoint and also possesses a Robin BC. An analytical solution is available so that the numerical results can be compared.

$$
u'' - u' - 2u = 4x^2
\tag{4.226}
$$

$$
u(0) = 0
\tag{4.227}
$$

$$
u'(2) + u(2) = 478.3829
\tag{4.228}
$$

This problem is solved by:

$$
u(x) = c_1 e^{-x} + c_2 e^{2x} - 2x^2 + 2x - 3
\tag{4.229}
$$

$$
c_1 = 0
\tag{4.230}
$$

$$
c_2 = 3
\tag{4.231}
$$

1. *Discretization: As a first step, the domain [0, 2] is subdivided into two elements. Then, the mesh will be refined by using four and eight elements. All elements are to be of equal length l_E.*

2. *Interpolation: The same approach as in the previous example is chosen.*

3. *Elemental formulation: In view of Equation (4.145), the example has the coefficients $a_1 = -1, a_2 = -2,$ and $f = -4x^2$ so that the elemental stiffness matrix and the elemental load matrix are given by:*

$$\mathbf{k}_{ei} = \frac{1}{l_E} \begin{pmatrix} 1 & -1 \\ -1 & 1 \end{pmatrix} + \frac{1}{2} \begin{pmatrix} -1 & 1 \\ -1 & 1 \end{pmatrix} + \frac{2l_E}{6} \begin{pmatrix} 2 & 1 \\ 1 & 2 \end{pmatrix} \tag{4.232}$$

$$\mathbf{f}_{ei} = -\frac{4}{l_E} \int_{x_i}^{x_{i+1}} x^2 \begin{pmatrix} \frac{x_{i+1} - x}{x - x_i} \end{pmatrix} dx \tag{4.233}$$

$$= -\frac{l_E}{3} \begin{pmatrix} x_{i+1}^2 + 2x_i x_{i+1} + 3x_i^2 \\ 3x_{i+1}^2 + 2x_i x_{i+1} + x_i^2 \end{pmatrix} \tag{4.234}$$

For the two-element solution, this results in:

$$\mathbf{k}_{ei} = \frac{1}{6} \begin{pmatrix} 7 & -1 \\ -7 & 13 \end{pmatrix} \tag{4.235}$$

$$\mathbf{f}_{e1} = \frac{1}{6} \begin{pmatrix} -2 \\ -6 \end{pmatrix} \tag{4.236}$$

$$\mathbf{f}_{e2} = \frac{1}{6} \begin{pmatrix} -22 \\ -34 \end{pmatrix} \tag{4.237}$$

For the four-element solution, this results in:

$$\mathbf{k}_{ei} = \frac{1}{6} \begin{pmatrix} 11 & -8 \\ -14 & 17 \end{pmatrix} \tag{4.238}$$

$$\mathbf{f}_{e1} = \begin{pmatrix} -0.042 \\ -0.125 \end{pmatrix} \tag{4.239}$$

$$\mathbf{f}_{e2} = \begin{pmatrix} -0.458 \\ -0.708 \end{pmatrix} \tag{4.240}$$

$$\mathbf{f}_{e3} = \begin{pmatrix} -1.375 \\ -1.792 \end{pmatrix} \tag{4.241}$$

$$\mathbf{f}_{e4} = \begin{pmatrix} -2.792 \\ -3.375 \end{pmatrix} \tag{4.242}$$

Note that the elemental stiffness matrix Equation (4.238) is not symmetric.

4. *Assembly: Next, the results of the previous step are assembled. For the two-element solution, this results in:*

$$\mathbf{F}_G = \mathbf{f}_{g1} + \mathbf{f}_{g2} + \mathbf{bt}_G \tag{4.243}$$

$$= \frac{1}{6} \begin{pmatrix} -2 \\ -6 \\ 0 \end{pmatrix} + \frac{1}{6} \begin{pmatrix} 0 \\ -22 \\ -34 \end{pmatrix} + \begin{pmatrix} 0 \\ 0 \\ -478.3829 \end{pmatrix} = \frac{1}{6} \begin{pmatrix} -2 \\ -28 \\ 2836.2974 \end{pmatrix} \tag{4.244}$$

$$\mathbf{K}_G = \mathbf{k}_{g1} + \mathbf{k}_{g2} + \mathbf{BT}_G \tag{4.245}$$

$$= \frac{1}{6} \begin{pmatrix} 7 & -1 & 0 \\ -7 & 13 & 0 \\ 0 & 0 & 0 \end{pmatrix} + \frac{1}{6} \begin{pmatrix} 0 & 0 & 0 \\ 0 & 7 & -1 \\ 0 & -7 & 13 \end{pmatrix} + \begin{pmatrix} 0 & 0 & 0 \\ 0 & 0 & 0 \\ 0 & 0 & 1 \end{pmatrix} \tag{4.246}$$

$$= \frac{1}{6} \begin{pmatrix} 7 & -1 & 0 \\ -7 & 20 & -1 \\ 0 & -7 & 19 \end{pmatrix} \tag{4.247}$$

For the four-element solution, only the results are reported:

$$\mathbf{F}_G = \mathbf{f}_{g1} + \mathbf{f}_{g2} + \mathbf{f}_{g3} + \mathbf{f}_{g4} + \mathbf{bt}_G \tag{4.248}$$

$$= \frac{1}{6} \begin{pmatrix} -0.042 \\ -0.583 \\ -2.083 \\ -4.583 \\ 475.008 \end{pmatrix} \tag{4.249}$$

$$\mathbf{K}_G = \mathbf{k}_{g1} + \mathbf{k}_{g2} + \mathbf{k}_{g3} + \mathbf{k}_{g4} + \mathbf{BT}_G \tag{4.250}$$

$$= \frac{1}{6} \begin{pmatrix} 11 & -8 & 0 & 0 & 0 \\ -14 & 28 & -8 & 0 & 0 \\ 0 & -14 & 28 & -8 & 0 \\ 0 & 0 & -14 & 28 & -8 \\ 0 & 0 & 0 & -14 & 23 \end{pmatrix} \tag{4.251}$$

The eight-element solution is constructed accordingly.

5. *Boundary conditions: Since the BC is on the LHS, the constraint is enforced by replacing the first line:*

$$\left[\begin{array}{ccc|c} 1 & 0 & 0 & 0 \\ -1.167 & 3.333 & -0.166 & -4.667 \\ 0 & -1.167 & 3.167 & 472.716 \end{array} \right] \tag{4.252}$$

and for the four-element model:

$$\left[\begin{array}{ccccc|c} 1 & 0 & 0 & 0 & 0 & 0 \\ -2.333 & 4.667 & -1.333 & 0 & 0 & -0.583 \\ 0 & -2.333 & 4.667 & -1.333 & 0 & -2.083 \\ 0 & 0 & -2.333 & 4.667 & -1.333 & -4.583 \\ 0 & 0 & 0 & -2.333 & 3.833 & 475.008 \end{array} \right] \tag{4.253}$$

6. *Solution: The solutions are given by*

$$\bar{\mathbf{u}} = \begin{pmatrix} 0 \\ 6.178 \\ 151.555 \end{pmatrix} \quad \bar{\mathbf{u}} = \begin{pmatrix} 0 \\ 4.634 \\ 16.657 \\ 51.751 \\ 155.416 \end{pmatrix} \tag{4.254}$$

for the two-element and four-element model, respectively. These solutions are plotted in Figure 4.8. Note that the solution is not exact for the last node on the RHS. Boundary

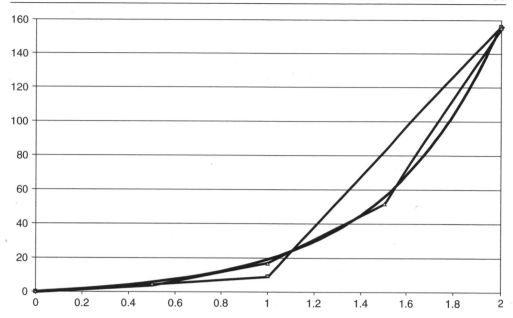

Figure 4.8 Two- and four-element solutions

conditions involving the term u_x are not satisfied identically. As for interior points, Robin and Neumann conditions are only satisfied approximately. As the mesh becomes finer, this error also becomes smaller.

7. *Computation of derived variables: Due to the linear shape functions, the derivative of the numerical solution shows jumps at the nodes of the elements.*   □

4.5.2 The Galerkin method with quadratic trial functions

1. *Discretization:* The discretization is the same as in the case with linear trial functions. The only restriction is that the number of nodes must be odd. This reflects the fact that there are three nodes associated with each element in order to uniquely pin down a quadratic function. Assume n elements: when using linear shape functions this leads to $n + 1$ nodes. Since each element requires one additional node somewhere in the middle of the element, the total number of nodes is $2n + 1$.

2. *Interpolation:* The approximate solution is of the form:

$$\tilde{u}_e(x) = \alpha + \beta x + \gamma x^2 \qquad (4.255)$$

The shape function $\tilde{u}_e(x)$ passes through three points $(x_i, \tilde{u}_i)$, $(x_{i+1}, \tilde{u}_{i+1})$ and $(x_{i+2}, \tilde{u}_{i+2})$. This leads to:

$$\tilde{u}_i = \alpha + \beta x_i + \gamma x_i^2 \qquad (4.256)$$

$$\tilde{u}_{i+1} = \alpha + \beta x_{i+1} + \gamma x_{i+1}^2 \qquad (4.257)$$

$$\tilde{u}_{i+2} = \alpha + \beta x_{i+2} + \gamma x_{i+2}^2 \qquad (4.258)$$

With the above three equations it is possible to determine α, β and γ. Inserting these results into Equation (4.255) yields:

$$\tilde{u}_e(x) = N_i(x)\tilde{u}_i + N_{i+1}(x)\tilde{u}_{i+1} + N_{i+2}(x)\tilde{u}_{i+2} \tag{4.259}$$

$$= \tilde{\mathbf{u}}_e^T \mathbf{N} \tag{4.260}$$

$$= \mathbf{N}^T \tilde{\mathbf{u}}_e \tag{4.261}$$

with:

$$N_i(x) = \frac{(x - x_{i+1})(x - x_{i+2})}{(x_i - x_{i+1})(x_i - x_{i+2})} \tag{4.262}$$

$$N_{i+1}(x) = \frac{(x - x_i)(x - x_{i+2})}{(x_{i+1} - x_i)(x_{i+1} - x_{i+2})} \tag{4.263}$$

$$N_{i+2}(x) = \frac{(x - x_i)(x - x_{i+1})}{(x_{i+2} - x_i)(x_{i+2} - x_{i+1})} \tag{4.264}$$

The above expressions are quadratic Lagrange interpolating polynomials. The derivatives are given by:

$$N_i'(x) = \frac{2x - x_{i+1} - x_{i+2}}{(x_i - x_{i+1})(x_i - x_{i+2})} \tag{4.265}$$

$$N_{i+1}'(x) = \frac{2x - x_i - x_{i+2}}{(x_{i+1} - x_i)(x_{i+1} - x_{i+2})} \tag{4.266}$$

$$N_{i+2}'(x) = \frac{2x - x_i - x_{i+1}}{(x_{i+2} - x_i)(x_{i+2} - x_{i+1})} \tag{4.267}$$

3. *Elemental formulation:* The derivation of the elemental formulation can be accomplished along the lines of the derivation for linear elements. The form of the elemental matrices is independent of the type of interpolation. The interpolation matrices and vectors, however, are different in size. As derived in Section 4.5.1, the elemental matrices and vectors are given by:

$$\mathbf{k}_{ei} = \int_{x_i}^{x_{i+2}} \left(\mathbf{N}'\mathbf{N}'^T - a_1\mathbf{N}\mathbf{N}'^T - a_2\mathbf{N}\mathbf{N}^T\right) dx \tag{4.268}$$

$$\mathbf{f}_{ei} = \int_{x_i}^{x_{i+2}} \mathbf{N}f \, dx \tag{4.269}$$

Since the ODE to be solved is not self-adjoint, the matrices need not be symmetric.

4. *Assembly:* The process of assembling the elemental matrices is completely analogous to the case with linear FE. The elemental matrices are 3×3. In a two-element problem, assembling two matrices of this size yields a 5×5 matrix. Only the lower right element of the first matrix and the upper left element of the second matrix overlap and need to be added.

5. *Boundary conditions:* Both Neumann and Dirichlet BCs can be treated in just the same way as with linear elements.

6. *Solution:* The result of the FE discretization with quadratic shape functions is a system of linear equations, for which the same comments hold as for linear shape functions. Note that the band structure of the matrix is somewhat different because the elemental matrices are 3×3.

7. *Computation of derived variables:* Because of the quadratic shape function, the first derivative of the approximate solution is a linear function. The problem of large jumps across the borders of the elements, as with linear shape functions, is mitigated; however it does not vanish. How to obtain a solution with a continuous first derivative is shown in Section 4.5.3.

Example 6
As a first simple example, a one-element model will be constructed using the seven-step procedure outlined above.

$$u''(x) = -2, \quad u(0) = 0, \quad u(2) = 0 \tag{4.270}$$

This is a self-adjoint problem with two Dirichlet BCs. The analytical solution is given by $u(x) = -(x-1)^2 + 1$. In terms of Equation (4.145), we have $a_1 = 0$, $a_2 = 0$ and $f = -2$.

1. *Discretization: Since there is only one element, this issue reduces to placing one node between the two which are located on the boundaries of the domain. For simplicity, we choose the middle, so that the three nodes are given by $x_1 = 0$, $x_2 = 1$ and $x_3 = 2$.*

2. *Interpolation: Quadratic interpolation is employed.*

3. *Elemental formulation: Because $a_1 = 0$ and $a_2 = 0$, Equation (4.268) reduces to:*

$$\mathbf{k}_{ei} = \int_{x_i}^{x_{i+2}} \mathbf{N}' \mathbf{N}'^T dx \tag{4.271}$$

$$= \begin{pmatrix} \int_{x_i}^{x_{i+2}} N_i' N_i' \, dx & \int_{x_i}^{x_{i+2}} N_i' N_{i+1}' \, dx & \int_{x_i}^{x_{i+2}} N_i' N_{i+2}' \, dx \\ \int_{x_i}^{x_{i+2}} N_{i+1}' N_i' \, dx & \int_{x_i}^{x_{i+2}} N_{i+1}' N_{i+1}' \, dx & \int_{x_i}^{x_{i+2}} N_{i+1}' N_{i+2}' \, dx \\ \int_{x_i}^{x_{i+2}} N_{i+2}' N_i' \, dx & \int_{x_i}^{x_{i+2}} N_{i+2}' N_{i+1}' \, dx & \int_{x_i}^{x_{i+2}} N_{i+2}' N_{i+2}' \, dx \end{pmatrix} \tag{4.272}$$

This matrix is symmetric since the ODE under inspection is self-adjoint. Since we are interested in a one-element solution, $i = 0$, $x_i = x_0$, $x_{i+1} = 1$ and $x_{i+2} = 2$

$$\mathbf{k}_{e0} = \begin{pmatrix} \frac{7}{6} & -\frac{4}{3} & \frac{1}{6} \\ -\frac{4}{3} & \frac{8}{3} & -\frac{4}{3} \\ \frac{1}{6} & -\frac{4}{3} & \frac{7}{6} \end{pmatrix} \tag{4.273}$$

The elemental load matrix reads:

$$\mathbf{f}_{ei} = \int_{x_i}^{x_{i+2}} f \mathbf{N} \, dx \tag{4.274}$$

$$= 2 \int_{x_i}^{x_{i+2}} \mathbf{N} \, dx \tag{4.275}$$

$$\Rightarrow \quad \mathbf{f}_{e0} = 2 \begin{pmatrix} N_0 \\ N_1 \\ N_2 \end{pmatrix} = \begin{pmatrix} \frac{2}{3} \\ \frac{8}{3} \\ \frac{2}{3} \end{pmatrix} \tag{4.276}$$

4. *Assembly: Since there are no Neumann BCs, $\mathbf{K}_G = \mathbf{k}_{e0}$ and $\mathbf{F}_G = \mathbf{f}_{e0}$.*

5. *Boundary conditions: Dirichlet BCs are integrated via constraints, as explained in Section 4.5.1. The linear system given by*

$$
\left[
\begin{array}{ccc|c}
\frac{7}{6} & -\frac{4}{3} & \frac{1}{6} & \frac{2}{3} \\
\\
-\frac{4}{3} & \frac{8}{3} & -\frac{4}{3} & \frac{8}{3} \\
\\
\frac{1}{6} & -\frac{4}{3} & \frac{7}{6} & \frac{2}{3}
\end{array}
\right]
\tag{4.277}
$$

needs to be modified to accommodate the BC $u(0) = u(2) = 0$:

$$
\left[
\begin{array}{ccc|c}
1 & 0 & 0 & 0 \\
\\
-\frac{4}{3} & \frac{8}{3} & -\frac{4}{3} & \frac{8}{3} \\
\\
0 & 0 & 1 & 0
\end{array}
\right]
\tag{4.278}
$$

6. *Solution: The unique solution to the above system is obviously $\bar{\mathbf{u}} = (0, 1, 0)^{\mathrm{T}}$. Although this solution vector is exactly the same as for linear elements, Equation (4.214), it bears a different interpretation. The solution with linear elements is a polygon passing through the points $(0, 0)$, $(1, 1)$ and $(2, 1)$. The solution with one quadratic element is a parabola passing through the same points. This solution coincides with the analytical solution. This issue is nicely described by Bickford (1990):*

> The emergence of the exact solution from the finite element model is an inherent property of the class of problems considered in this chapter, namely that if for *some* choice of the parameters $\tilde{u}_i$, the exact solution is contained within the approximate solution assumed for the finite element model, the method will select the values of the $\tilde{u}_i$ that correspond to that exact solution. When the exact solution is not contained within the approximate finite element solution, the solution will choose the *best* values of the unknowns $\tilde{u}_i$.

The meaning of best depends on the FE method used. Several optimization criteria resulting from the method of weighted residuals have been presented in Section 4.2.

7. *Computation of derived variables: Since the numerical and analytical solutions are the same, their derivatives coincide as well.*

A model with two or more elements will produce the same solution. We will show this with a two-element example which also shows the assembly process for quadratic shape functions.

1. *Discretization: The domain is discretized in two elements of equal length with nodes at both ends of the elements and in the middle.*

2. *Interpolation: Quadratic interpolation is employed.*

3. *Elemental formulation: The elemental stiffness and load matrices are given by:*

$$
\mathbf{k}_{ei} = \frac{1}{3}
\begin{pmatrix}
7 & -8 & 1 \\
-8 & 16 & -8 \\
1 & -8 & 7
\end{pmatrix}
\tag{4.279}
$$

$$
\mathbf{f}_{ei} = \frac{1}{3}
\begin{pmatrix}
1 \\
4 \\
1
\end{pmatrix}
\tag{4.280}
$$

4. *Assembly: Assembling* $\mathbf{k}_{e1}$, $\mathbf{k}_{e2}$, $\mathbf{f}_{e1}$ *and* $\mathbf{f}_{e2}$ *results in the linear system:*

$$\frac{1}{3}\left[\begin{array}{ccccc|c} 7 & -8 & 1 & 0 & 0 & 1 \\ -8 & 16 & -8 & 0 & 0 & 4 \\ 1 & -8 & 7+7 & -8 & 1 & 1+1 \\ 0 & 0 & -8 & 16 & -8 & 4 \\ 0 & 0 & 1 & -8 & 7 & 1 \end{array}\right] \tag{4.281}$$

5. *Boundary conditions: To integrate the BCs, system (4.281) has to be modified:*

$$\frac{1}{3}\left[\begin{array}{ccccc|c} 3 & 0 & 0 & 0 & 0 & 0 \\ -8 & 16 & -8 & 0 & 0 & 4 \\ 1 & -8 & 14 & -8 & 1 & 2 \\ 0 & 0 & -8 & 16 & -8 & 4 \\ 0 & 0 & 0 & 0 & 3 & 0 \end{array}\right] \tag{4.282}$$

6. *Solution: This system is solved by* $\bar{\mathbf{u}} = (0, \frac{3}{4}, 1, \frac{3}{4}, 0)^\mathrm{T}$. *The same comments as for the one-element solution apply.*

7. *Computation of derived variables: Here, the same comments as for the one-element solution apply as well.* □

4.5.3 The collocation method with cubic Hermite trial functions

Although the Galerkin approach is the most popular FEM at present, the collocation method represents a valid alternative for applications in finance and economics. First of all, its relatively simple nature makes it easy to understand. As a direct consequence, it is easier to implement than a Galerkin method. For instance, no integration is required. One of the reasons for the popularity of the Galerkin approach in physics and engineering is the equivalence of this method and the approach by Ritz when applied to self-adjoint operators. In economics and finance, however, self-adjoint operators occur only by chance.

Here, only the case of an ODE, as given by Equation (4.145), and two essential (Dirichlet) boundary conditions will be discussed. The integration of one natural (Neumann) condition is straightforward; see Section 5.4 of Sewell (2000). Since the collocation method is simpler, one step in our seven-step procedure drops out:

1. *Discretization:* For the distribution of the nodes on the domain, the same comments as in the previous sections apply. In addition to the nodes, collocation points have to be chosen as in Section 4.2, i.e. they may be located anywhere in the element, including the nodes, not necessarily following a particular pattern, see Figure 4.9. Note that the node numbering is different from Figure 4.5. The accuracy of the collocation method depends heavily on the location of the collocation points. The optimal choices for the collocation points are the

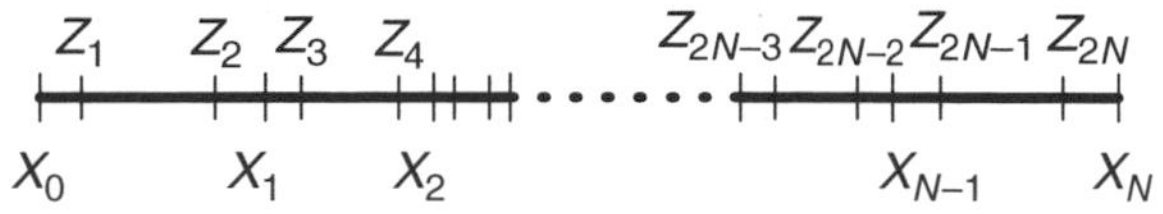

Figure 4.9 Nodes and collocation points

Gauss points:

$$z_{2i,2i+1} = x_i + \beta_{1,2}(x_{i+1} - x_i) \qquad (4.283)$$

$$\beta_{1,2} = 0.5 \pm \frac{0.5}{\sqrt{3}} \qquad (4.284)$$

This choice leads to an order of convergence of four $O(h^4)$, i.e. halving the element length h results in a reduction of the approximation error by factor 16. This approach is often called *orthogonal collocation*.

2. *Interpolation:* The collocation method is generally used either with shape functions representing a higher degree polynomial and possessing global support, or with shape functions with higher order continuity and local support. The shape function used by a collocation method to solve a second order problem must have continuous first derivatives, because the collocation equations supply information about the second derivative of the differential equation. If the first derivatives were not continuous, this information would be useless. A popular choice satisfying this requirement is Hermite polynomials (see Figures 4.10 and 4.11). Hermite polynomials are cubic splines that are C^∞ within the elements and C^1 across the element borders:

$$
\begin{aligned}
H_k(x) &= 3 \left(\frac{x - x_{k-1}}{x_k - x_{k-1}} \right)^2 - 2 \left(\frac{x - x_{k-1}}{x_k - x_{k-1}} \right)^3 \quad \text{for } x_{k-1} \leq x \leq x_k \\
&= 3 \left(\frac{x_{k+1} - x}{x_{k+1} - x_k} \right)^2 - 2 \left(\frac{x_{k+1} - x}{x_{k+1} - x_k} \right)^3 \quad \text{for } x_k \leq x \leq x_{k+1} \\
&= 0 \quad \text{elsewhere}
\end{aligned}
\qquad (4.285)
$$

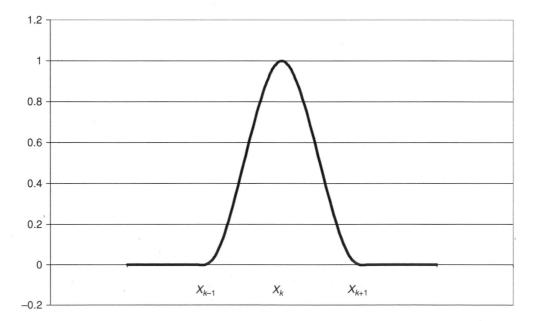

Figure 4.10 Cubic Hermite shape function $H_k(x)$

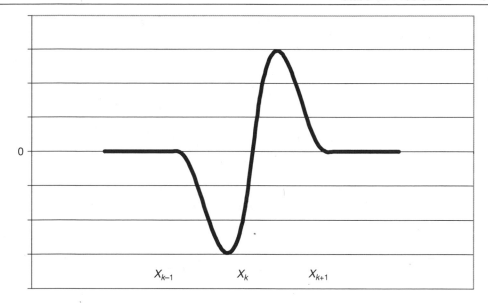

Figure 4.11   Cubic Hermite shape function $S_{k(x)}$

$$S_k(x) = -\frac{(x - x_{k-1})^2}{(x_k - x_{k-1})} + \frac{(x - x_{k-1})^3}{(x_k - x_{k-1})^2} \quad \text{for } x_{k-1} \le x \le x_k$$

$$= \frac{(x_{k+1} - x)^2}{(x_{k+1} - x_k)} - \frac{(x_{k+1} - x)^3}{(x_{k+1} - x_k)^2} \quad \text{for } x_k \le x \le x_{k+1}$$

$$= 0 \quad \text{elsewhere} \tag{4.286}$$

These piecewise cubics are designed such that the following hold:

$$H_k(x_i) = \delta_{ki} \tag{4.287}$$

$$H_k'(x_i) = 0 \tag{4.288}$$

$$S_k(x_i) = 0 \tag{4.289}$$

$$S_k'(x_i) = \delta_{ki} \tag{4.290}$$

3. *Elemental formulation:* An elemental formulation is advantageous in the Galerkin method when the problem requires irregular spacing of the nodes and a transformation to local coordinates becomes useful. The collocation method can deal directly with irregular spacing so this point is irrelevant. Also, the next two steps – *Boundary conditions* and *Assembly* – need to be interchanged.

4. *Assembly:* The interpolation function $\tilde{u}(x)$ is given by:

$$\tilde{u}(x) = \sum_{i=1}^{N+1} b_{2i-1} S_{i-1} + b_{2i} H_{i-1} \tag{4.291}$$

$$= b_1 S_0 + b_2 H_0 + b_3 S_1 + b_4 H_1 + b_5 S_2 + b_6 H_2 + \ldots$$

$$+ b_{2N-1} S_{N-1} + b_{2N} H_{N-1} + b_{2N+1} S_N + b_{2N+2} H_N \tag{4.292}$$

There are $2N + 2$ coefficients to be determined, so $2N + 2$ equations are needed. $2N$ of these come from the collocation points.

$$L(\tilde{u}(z_k)) = 0 \quad \forall \ k = 1, \ldots, 2N \tag{4.293}$$

$$\Leftrightarrow \ \tilde{u}''(z_k) + a_1(z_k)\tilde{u}'(z_k) + a_2(z_k)\tilde{u}(z_k) = f(z_k) \quad \forall \ k = 1, \ldots, 2N \tag{4.294}$$

with:

$$\tilde{u}(z_k) = \sum_{i=1}^{N+1} b_{2i-1}S_{i-1}(z_k) + b_{2i}H_{i-1}(z_k) \tag{4.295}$$

The last two equations come from the BCs.

5. *Boundary conditions:* In the case of two essential BCs, the basis functions H_0 and H_N are reserved to integrate the BCs. The left-hand BC is associated with b_2:

$$b_2 H_0(x^{min}) = \frac{\gamma_1}{\alpha_1} \tag{4.296}$$

$$\Leftrightarrow b_2 = \frac{\gamma_1}{\alpha_1} \quad \left(\text{since } H_0(x^{min}) = 1\right) \tag{4.297}$$

The right-hand BC is associated with b_{2N+2}:

$$b_{2N+2} H_N(x^{max}) = b_{2N+2} = \frac{\gamma_2}{\alpha_2} \tag{4.298}$$

6. *Solution:* The resulting linear system is not symmetric, not even for self-adjoint problems. If the collocation points are taken in ascending order and the shape functions are ordered as delineated above, the linear system is banded with half-bandwidth 2.

7. *Computation of derived variables:* Since the numerical solution is a cubic polynomial, its derivatives can be computed analytically. The first three derivatives, $\tilde{u}'(x)$, $\tilde{u}''(x)$ and $\tilde{u}'''(x)$, are of use since $\tilde{u}^{(4)}(x) \equiv 0$.

Example 7 (Example 6 revisited)
Consider the following test problem once more:

$$u_{xx} = -2 \tag{4.299}$$

$$u(0) = 0 \tag{4.300}$$

$$u(2) = 0 \tag{4.301}$$

1. *Discretization: We will derive a two-element solution with both elements being of equal length, so that $x_0 = 0$, $x_1 = 1$ and $x_2 = 2$. Besides this, we need four collocation points z_i ($i = 1, 2, 3, 4$) for which the following conditions hold:*

$$z_{2i,2i+1} = x_i + \beta_{1,2}(x_{i+1} - x_i) \tag{4.302}$$

$$\beta_{1,2} = 0.5 \pm \frac{0.5}{\sqrt{3}} \tag{4.303}$$

$$z_0 = \frac{1}{2} - \frac{\sqrt{3}}{6} \approx 0.211324 \tag{4.304}$$

$$z_1 = \frac{1}{2} + \frac{\sqrt{3}}{6} \approx 0.788675 \tag{4.305}$$

$$z_2 = \frac{3}{2} - \frac{\sqrt{3}}{6} \approx 1.211324 \tag{4.306}$$

$$z_3 = \frac{3}{2} + \frac{\sqrt{3}}{6} \approx 1.788675 \tag{4.307}$$

2. *Interpolation: The numerical solution is to be of the following form:*

$$\tilde{u}(x) = a_1 S_0(x) + a_2 H_0(x) + a_3 S_1(x) + a_4 H_1(x) + a_5 S_2(x) + a_6 H_2(x) \tag{4.308}$$

with S_i and H_i as defined in Equations (4.285)–(4.286). Since $i = 0, 1, 2$ we get:

$$S_0(x) = \begin{cases} (1-x)^2 - (1-x)^3 & 0 \le x \le 1 \\ 0 & 1 < x \le 2 \end{cases} \tag{4.309}$$

$$S_1(x) = \begin{cases} -x^2 + x^3 & 0 \le x \le 1 \\ (2-x)^2 - (2-x)^3 & 1 < x \le 2 \end{cases} \tag{4.310}$$

$$S_2(x) = \begin{cases} 0 & 0 \le x \le 1 \\ -(x-1)^2 + (x-1)^3 & 1 < x \le 2 \end{cases} \tag{4.311}$$

$$H_0(x) = \begin{cases} 3(1-x)^2 - 2(1-x)^3 & 0 \le x \le 1 \\ 0 & 1 < x \le 2 \end{cases} \tag{4.312}$$

$$H_1(x) = \begin{cases} 3x^2 - 2x^3 & 0 \le x \le 1 \\ 3(2-x)^2 - 2(2-x)^3 & 1 < x < 2 \end{cases} \tag{4.313}$$

$$H_2(x) = \begin{cases} 0 & 0 \le x \le 1 \\ 3(x-1)^2 - 2(x-1)^3 & 1 < x \le 2 \end{cases} \tag{4.314}$$

Accordingly:

$$S_0''(x) = \begin{cases} 6x - 4 & 0 \le x \le 1 \\ 0 & 1 < x \le 2 \end{cases} \tag{4.315}$$

$$S_1''(x) = \begin{cases} 6x - 2 & 0 \le x \le 1 \\ 6x - 10 & 1 < x \le 2 \end{cases} \tag{4.316}$$

$$S_2''(x) = \begin{cases} 0 & 0 \le x \le 1 \\ 6x - 8 & 1 < x \le 2 \end{cases} \tag{4.317}$$

$$H_0''(x) = \begin{cases} 12x - 6 & 0 \le x \le 1 \\ 0 & 1 < x \le 2 \end{cases} \tag{4.318}$$

$$H_1''(x) = \begin{cases} 6 - 12x & 0 \le x \le 1 \\ 12x - 18 & 1 < x \le 2 \end{cases} \tag{4.319}$$

$$H_2''(x) = \begin{cases} 0 & 0 \le x \le 1 \\ -12x + 18 & 1 < x \le 2 \end{cases} \tag{4.320}$$

3. *Assembly: Inserting $\tilde{u}(x)$ into the differential equation (4.299) we get:*

$$\tilde{u}_{xx}(x) = a_1 S_0''(x) + a_2 H_0''(x) + a_3 S_1''(x) + a_4 H_1''(x) + a_5 S_2''(x) + a_6 H_2''(x) = -2 \tag{4.321}$$

The four collocation points provide four equations:

$$\tilde{u}_{xx}(z_0) = a_1 S_0''(z_0) + a_2 H_0''(z_0) + a_3 S_1''(z_0) + a_4 H_1''(z_0)$$
$$= (-1 - \sqrt{3})a_1 - 2\sqrt{3}a_2 + (1 - \sqrt{3})a_3 + 2\sqrt{3}a_4 = -2 \tag{4.322}$$

$$\tilde{u}_{xx}(z_1) = a_1 S_0''(z_1) + a_2 H_0''(z_1) + a_3 S_1''(z_1) + a_4 H_1''(z_1)$$
$$= (-1 + \sqrt{3})a_1 + 2\sqrt{3}a_2 + (1 + \sqrt{3})a_3 - 2\sqrt{3}a_4 = -2 \tag{4.323}$$

$$\tilde{u}_{xx}(z_2) = a_3 S_1''(z_2) + a_4 H_1''(z_2) + a_5 S_2''(z_2) + a_6 H_2''(z_2)$$
$$= (1 + \sqrt{3})a_3 - 2\sqrt{3}a_4 + (1 - \sqrt{3})a_5 + 2\sqrt{3}a_6 = -2 \tag{4.324}$$

$$\tilde{u}_{xx}(z_3) = a_3 S_1''(z_3) + a_4 H_1''(z_3) + a_5 S_2''(z_3) + a_6 H_2''(z_3)$$
$$= (-1 + \sqrt{3})a_3 + 2\sqrt{3}a_4 + (1 + \sqrt{3})a_5 - 2\sqrt{3}a_6 = -2 \tag{4.325}$$

Now we have four equations with six unknowns. To make the system complete, we take the BCs.

4. *Boundary conditions: The left-hand BC is associated with a_2, while the right-hand BC is associated with a_6. In a similar fashion as in the Galerkin approach, the BCs are integrated into the linear system:*

$$a_2 = \frac{\gamma_1}{\alpha_1} = 0 \tag{4.326}$$

$$a_6 = \frac{\gamma_2}{\alpha_2} = 0 \tag{4.327}$$

5. *Solution: In matrix form this looks like:*

$$\begin{pmatrix} 0 & 1 & 0 & 0 & 0 & 0 \\ -1-\sqrt{3} & -2\sqrt{3} & 1-\sqrt{3} & 2\sqrt{3} & 0 & 0 \\ -1+\sqrt{3} & 2\sqrt{3} & 1+\sqrt{3} & -2\sqrt{3} & 0 & 0 \\ 0 & 0 & 1+\sqrt{3} & -2\sqrt{3} & 1-\sqrt{3} & 2\sqrt{3} \\ 0 & 0 & -1+\sqrt{3} & 2\sqrt{3} & 1+\sqrt{3} & -2\sqrt{3} \\ 0 & 0 & 0 & 0 & 0 & 1 \end{pmatrix} \begin{pmatrix} a_1 \\ a_2 \\ a_3 \\ a_4 \\ a_5 \\ a_6 \end{pmatrix} = \begin{pmatrix} 0 \\ -2 \\ -2 \\ -2 \\ -2 \\ 0 \end{pmatrix} \tag{4.328}$$

The solution of this system is:

$$\mathbf{a} = \begin{pmatrix} 2 \\ 0 \\ 0 \\ 1 \\ -2 \\ 0 \end{pmatrix} \tag{4.329}$$

These values for the a_i are to be inserted into Equation (4.308). For the first element, this yields:

$$\tilde{u}(x) = 2S_0(x) + 0H_0(x) + 0S_1(x) + 1H_1(x) \tag{4.330}$$
$$= 2\left[(1+x)^2 - (1-x)^3\right] + \left[3x^2 - 2x^3\right] \tag{4.331}$$
$$= -(x-1)^2 \tag{4.332}$$

For the second element, we get:

$$\tilde{u}(x) = 0S_1(x) + 1H_1(x) - 2S_2(x) + 0H_1(x) \tag{4.333}$$

$$= \left[3(2-x)^2 - 2(2-x)^3\right] - 2\left[-(x-1)^2 + (x-1)^3\right] \tag{4.334}$$

$$= -(x-1)^2 \tag{4.335}$$

The numerical solution coincides with the analytical. This result was to be expected on the same basis as for quadratic shape functions in Example 3.

6. *Computation of derived variables: Since the numerical solution and the analytical solution are identical, so are all their derivatives.* $\quad\square$

4.6 CASE STUDIES

4.6.1 The Evans model of a monopolist

This section serves as a numerical illustration to Section 2.1. Using the data from Chiang (1992), given in Table 4.7, a very accurate numerical solution can be achieved with three elements of equal length, as shown in Table 4.8.

Table 4.7 Data for the Evans model

Parameter	Value
a	160
b	8
h	100
α	0.1
β	0
γ	1000

4.6.2 First exit time of a geometric Brownian motion

We next compute numerically the expected exit time of a geometric Brownian motion

$$dY = aY\,dt + \sigma Y\,dX \tag{4.336}$$

with $a = 0.1$ and $\sigma = 0.2$ leaving the region [20, 60]. In practice, such a problem arises when one wants to know how long – on average – it takes for a given asset to leave a certain corridor for the first time (see Table 4.9). This information can be useful, for instance, in assessing the risk of a double barrier option.

Relabeling the independent variable as x, this problem can be formulated as an ODE (see Section B.3.5):

$$axu_x + \frac{\sigma^2 x^2}{2}u_{xx} = -1 \tag{4.337}$$

with:
$$u(x^{\min}) = 0 \tag{4.338}$$

$$u(x^{\max}) = 0 \tag{4.339}$$

Table 4.8 Results for the Evans model

Time t	Analytical solution	Numerical solution
0.0	11.4444	11.4444
0.1	11.6390	11.6391
0.2	11.8340	11.8341
0.3	12.0295	12.0295
0.4	12.2254	12.2254
0.5	12.4218	12.4218
0.6	12.6186	12.6187
0.7	12.8161	12.8161
0.8	13.0140	13.0141
0.9	13.2126	13.2127
1.0	13.4118	13.4119
1.1	13.6117	13.6117
1.2	13.8122	13.8122
1.3	14.0134	14.0135
1.4	14.2154	14.2155
1.5	14.4182	14.4182
1.6	14.6217	14.6217
1.7	14.8261	14.8261
1.8	15.0313	15.0313
1.9	15.2374	15.2374
2.0	15.4444	15.4444

Table 4.9 First exit time of a geometric Brownian motion

Asset price	Analytical solution	Numerical solution	Absolute difference
20.0	0.00000000	0.00000000	0.00000000
25.0	5.41981140	5.41981128	0.00000012
30.0	6.08946690	6.08946721	−0.00000031
35.0	5.42660530	5.42660555	−0.00000025
40.0	4.37095240	4.37095254	−0.00000014
45.0	3.22515940	3.22515949	−0.00000009
50.0	2.09472720	2.09472725	−0.00000005
55.0	1.01618150	1.01618148	0.00000002
60.0	0.00000000	0.00000000	0.00000000

Inserting the data gives:

$$0.1xu_x + \frac{0.04x^2}{2}u_{xx} = -1 \tag{4.340}$$

with:

$$u(20) = 0 \tag{4.341}$$
$$u(60) = 0 \tag{4.342}$$

The discretization consists of 99 equal-size collocation elements with a cubic Hermite basis function. The analytical solution to the problem (4.337) to (4.339) is given by:

$$u(x) = \frac{1}{\frac{\sigma^2}{2} - a}\left(\ln\left(\frac{x}{x^{\min}}\right) - \frac{1 - \left(\frac{x}{x^{\min}}\right)^{1-2a/\sigma^2}}{1 - \left(\frac{x^{\max}}{x^{\min}}\right)^{1-2a/\sigma^2}}\ln\left(\frac{x^{\min}}{x^{\max}}\right)\right) \tag{4.343}$$

4.6.3 The steady-state distribution of the Ornstein–Uhlenbeck process

The following case study shows that it is highly recommendable to investigate whether *a priori* statements about the behavior of the solution can be made. The Ornstein–Uhlenbeck process is given by:

$$dY = a(b - Y)\,dt + \sigma\,dX \tag{4.344}$$

Its steady-state distribution $p_\infty(y)$ is given by

$$-\sigma^2 p_\infty'' - ap_\infty + a(b - y)p_\infty' = 0 \tag{4.345}$$

$$\lim_{y \to -\infty} p_\infty(y) = 0 \tag{4.346}$$

$$\lim_{y \to \infty} p_\infty(y) = 0 \tag{4.347}$$

For details, see Appendix B. We will use the values of a, b and σ given in Table 4.10. The conventional approach to finding a numerical solution for this problem is to reduce the domain to a compact one by cutting off its edges at infinity:

$$p_\infty(y^{\min}) = 0 \tag{4.348}$$

$$p_\infty(y^{\max}) = 0 \tag{4.349}$$

Presupposing that no errors have occurred, any numerical routine should compute that $p_\infty(y) = 0$, which is somewhat surprising since $p_\infty(y)$ represents a cumulative distribution function so that the integral under p_∞ should take on the value one. What has happened?

By cutting off the domain as outlined above, the problem has been turned into a homogeneous two-point boundary value problem with homogeneous boundary conditions. We know from Theorem 16 (Appendix A), however, that problems of this kind only possess a trivial solution, so the numerical output is correct. In order to investigate the reasons for the failure we use the analytical solution Equation (B.101) to derive the exact boundary conditions:

$$p(y_\infty = -0.3) = p(y_\infty = 0.8) = 0.00146569 \tag{4.350}$$

The result of the approach is given in Figure 4.12. This result was achieved with 50 equidistant elements using a collocation finite element method with Hermite polynomials of order 3 as basis functions. The area under the curve is approximately 0.9995842, which indicates that this might be a solution, based on the fact that it is a density function. The difference to 1 results from three components:

- the domain being cut off at both ends;
- Equation (4.350) only having a limited number of digits;
- numerical inaccuracies.

Table 4.10 Data for the Ornstein–Uhlenbeck process

Parameter	Value
a	4.0
b	0.25
σ	0.4

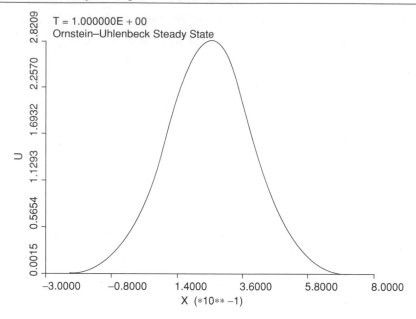

Figure 4.12   Steady-state distribution of an Ornstein–Uhlenbeck process

The influence of the first component can be estimated easily. We now know that it is not the numerical solution that has failed but the *ad hoc* technique of cutting off parts of the previously infinite domain. Similar problems arise frequently in option pricing. There, it is usually possible to come up with some artificial boundary conditions based on economic reasoning.

This problem exhibits another phenomenon: it is extremely sensitive to the boundary conditions for small σ (say $\sigma = 0.02$). Notice that with $p = 0$ on the boundary, the exact solution is $p(x) = 0$, while changing p to the correct values at the boundary changes the exact solution to something far from 0. This problem is extremely tough, as it stands, to solve numerically. Similar problems will be discussed in Section 4.6.4. Last but not least, one needs to emphasize that all difficulties arising in this case study do not result from the numerical method but from the problem at hand.

4.6.4   Convection-dominated problems

Recently, various publications dealing with convection-dominated problems in finance have appeared (for instance Seydel, 2002). First, we will present the numerical problems arising in convection-dominated problems with the help of an example. Consider the following boundary value problem (Knabner and Angermann, 2000, Chapter 9):

$$-ku'' + u' = 0 \tag{4.351}$$

For small k (and consequently large Péclet numbers) the above differential equation loses its elliptic character and starts to resemble a hyperbolic problem. Then, Equation (4.351) is called *convection-dominated* because the convection term u' dominates the character of the problem. Although Equation (4.351) is still elliptic by definition, techniques for elliptic problems tend

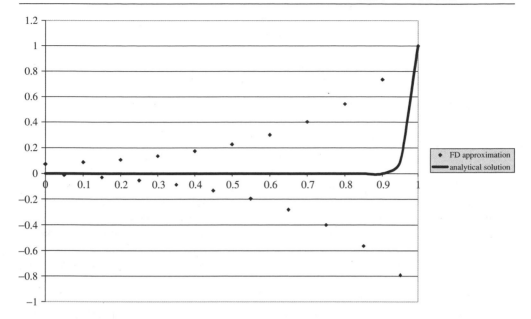

Figure 4.13 Oscillating FD approximation of Equation (4.351) with $k = 0.002$ and $h = 0.5$

to fail for decreasing k. Together with the BCs $u(0) = 0$ and $u(1) = 1$ this problem is solved by:

$$u(x) = \frac{1 - \exp(x/k)}{1 - \exp(1/k)} \qquad (4.352)$$

The analytical solution is plotted in Figure 4.13. Consider an FD approximation as explained in Section 3.3.2. The domain is discretized into $M + 1$ intervals of equal length h.

$$u_0 = 0 \qquad (4.353)$$

$$-k\frac{u_{i-1} - 2u_i + u_{i+1}}{h^2} + \frac{u_{i+1} - u_{i-1}}{2h} = 0 \qquad i = 1, \ldots, M \qquad (4.354)$$

$$u_{M+1} = 1 \qquad (4.355)$$

Equation (4.354) can be rewritten as:

$$\left(-\frac{2k}{h} - 1\right) u_{i-1} + \frac{4k}{h} u_i + \left(-\frac{2k}{h} + 1\right) u_{i+1} = 0 \qquad i = 1, \ldots, M \qquad (4.356)$$

The above expression can be solved explicitly with the *ansatz* $u_i = \lambda^i$, resulting in:

$$u_i = \frac{1 - \left(\frac{2k+h}{2k-h}\right)^i}{1 - \left(\frac{2k+h}{2k-h}\right)^{M+1}} \qquad (4.357)$$

For $2k < h$ the above expression oscillates, as shown in Figure 4.13. Only for $2k > h$ do the oscillations vanish. For a problem with a Péclet number of 10^7 this results in a maximal interval length h of 2×10^{-7}, thus requiring at least 2×10^7 intervals to solve problem (4.351) on the domain [0, 1]. This step size is prohibitively small. However, this also shows that problems

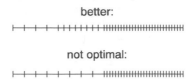

Figure 4.14 Grid refinement for 1D problems

which are only mildly convection-dominated can still be solved by simply increasing the number of intervals. It is advisable to increase the element length gradually, as in Figure 4.14. Problems from finance of practical interest usually are only mildly convection-dominated, so the effect of a finer FE discretization is explored by solving the above problem with various meshes. Figure 4.15 shows a collocation FE approximation with cubic elements on a regular grid with 9 and 99 elements. The coarse discretization exhibits an oscillation similar to the FD approach in Figure 4.13.

In equity and FX option pricing convection-dominated problems arise because of low asset prices and/or low volatilities, see for instance Example 37 in Appendix A. The volatilities have to be much smaller than usually observed in the market for this phenomenon to appear; see Figure 4.16. As in this example, the oscillations are often not obvious in the approximate function, but in its first derivative. Here, at first sight the price of the option itself has no obvious error but its Delta is obviously corrupted. Consequently, a serious approximation of Gamma cannot be obtained on this basis. Several models for pricing continuously sampled Asian options in the Black–Scholes framework can lead to convection-dominated problems (Zvan *et al.*, 1998c). However, since there are no options of this type in the market, option premiums resulting from these models can only be used as approximations for Asian options with many observations. Besides, there are simple PDE approaches for this type of problem which are not convection-dominated; see Section 5.2.3.

While the above issues are of little practical relevance, one-factor interest rate models of the Vasicek type (Vasicek, 1977) are usually convection-dominated for typical market data. Assuming that the short rate r is governed by:

$$\mathrm{d}r = u(r, t)\,\mathrm{d}t + w(r, t)\,\mathrm{d}X \qquad (4.358)$$

the pricing PDE is given by:

$$\frac{\partial V}{\partial t}(r, t) + \frac{1}{2}w(r, t)^2\frac{\partial^2 V}{\partial r^2}(r, t) + [u(r, t) - \lambda w(r, t)]\frac{\partial V}{\partial r}(r, t) - rV(r, t) = 0 \qquad (4.359)$$

with λ being the market price of risk. This includes the popular Hull–White model (Hull and White, 1990) with $w = \sqrt{\beta(t)}$ (volatility of r) and $u = [\eta(t) - \gamma(t)r]$ (η mean reversion level, γ mean reversion speed). For typical interest rates between 0% and 10%, as observed in Western Europe, the US and especially Japan, the diffusion term becomes small. Models of this type are popular in risk control where the output of a model is not as closely observed as on a trading desk, so that an erroneous output due to a large convection term might not be detected immediately. Besides, the situation is often aggravated by using Crank–Nicolson on nondifferentiable payoffs. The problem can usually be cured by refining the grid modestly and taking a different numerical scheme.

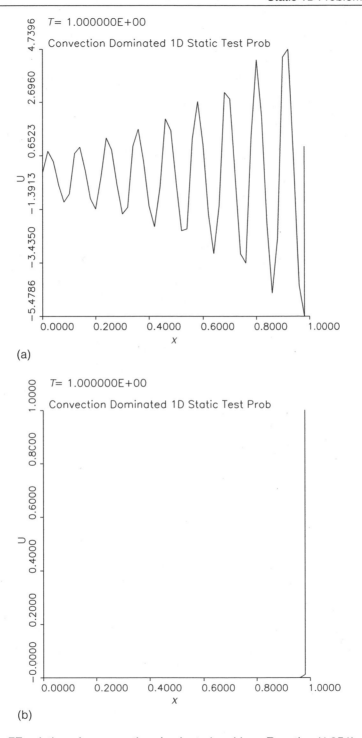

(a)

(b)

Figure 4.15 FE solution of a convection-dominated problem, Equation (4.351): (a) nine elements; (b) 99 elements

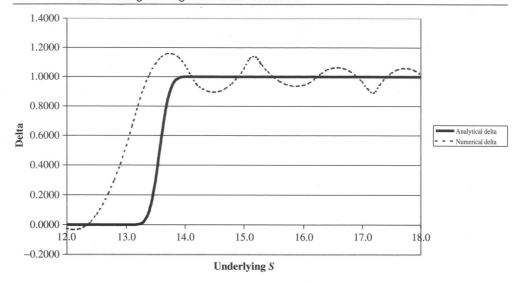

Figure 4.16   Spurious oscillations resulting from low volatility ($\sigma = 0.01$)

4.7   CONVERGENCE

In this section a brief and heuristic explanation shall be given on how the number of elements and the type of shape functions influence the accuracy of the solution and its derivatives. For the rest, only the errors resulting from the approximate character of the FE approach are of interest; it is assumed that all integrals and all linear systems are solved exactly. Also, no rounding errors occur. In the case of a linear shape function, the FE approximation for the element $[x_i, x_{i+1}]$ is given by:

$$\tilde{u}_e(x) = a + b(x - x_i) \tag{4.360}$$

This can be interpreted as the first two terms of a Taylor expansion around x. The Taylor series of the exact solution is given by:

$$u(x) = \sum_{n=0}^{\infty} \frac{u^{(n)}(x_i)}{n!}(x - x_i)^n \tag{4.361}$$

The symbol $e_{m,n}$ is to denote the error in the nth derivative of an FE solution with polynomials of order m as shape functions. The difference between an FE solution with linear elements, as in Equation (4.360), and the analytical solution is given by:

$$e_{1,0} = \sum_{n=0}^{\infty} \frac{u^{(n)}(x_i)}{n!}(x - x_i)^n - [a + b(x - x_i)] \tag{4.362}$$

$$= c_0 + c_1(x - x_i) + \frac{u''(\xi)(x_{i+1} - x_i)^2}{2} - a - b(x - x_i) \quad x_i \le \xi \le x_{i+1} \tag{4.363}$$

because of the Lagrange remainder formula. For elements with a sufficiently small length $h = x_{i+1} - x_i$ it should hold that $c_0 \approx a$ and $c_1 \approx b$ so that:

$$e_{1,0}(x) = \frac{u''(\xi)h^2}{2} = O(h^2) \tag{4.364}$$

i.e. the error is proportional to h^2. This implies that halving the element length reduces the error by approximately a factor of four for small enough h. A similar statement can be arrived at for the error of the first derivative:

$$e_{1,1} = u'(x) - \tilde{u}'_e(x) \tag{4.365}$$

$$= \sum_{n=1}^{\infty} \frac{f^{(n)}}{n!}(x - x_i)^{n-1} - b \tag{4.366}$$

$$= c_1 + u'(\xi)(x - x_i) - b \tag{4.367}$$

Since $b \approx c_1$ it follows that $e_{1,1} = u'(\xi)h = O(h)$, implying that halving the mesh length also halves the error in the first derivative for a linearly interpolated FE solution. Also, with the help of the Taylor series, it can be stated that:

$$e_{p,0} = O(h^{p+1}) \tag{4.368}$$

This means that halving the element length while using an interpolation polynomial of order p reduces the error by a factor $p + 1$. Generally, the mth derivative possesses only accuracy of order $p + 1 - m$:

$$e_{p,m} = O(h^{p+1-m}) \tag{4.369}$$

Although the above heuristic analysis gives some guidelines on efficient ways of refining a grid, the user of a numerical method is usually interested in error bounds. However, (tight) bounds for (pointwise) error for nonself-adjoint problems are scarce. One way out of this problem is to transform the problem to a self-adjoint problem and use error bounds available for this type of problem. Unfortunately, not all problems can be transformed in such a way. Two other methods often used in practice to indicate convergence (although not guaranteeing it) are:

1. Also based on the Taylor argument above, it is obvious that increasing the number of elements decreases the (pointwise) error. Doubling the number of elements several times should usually decrease the error in such a way that each additional FE run produces an approximate solution hardly different from its predecessor.

2. In option pricing, for constant input parameters, such as volatility, interest rates etc., analytical solutions are available. These can be used to compare the accuracy of a numerical solution.

4.8 NOTES

The reader interested in the history of WRM is referred to Kaliakin (2002) and the references given there. A detailed treatment of the least-squares FEM can be found in Jiang (1998). This method is often considered to be superior in dealing with convection-dominated problems like those in Section 4.6.4. The collocation method is a lot easier to set up than the Galerkin method. This is even more true for problems with more than one spatial variable. Good textbook references for the collocation method include Sewell (2000) and Lapidus and Pinder (1981).The Galerkin approach is the most popular in engineering due to its relationship with the Ritz method.

5
Dynamic 1D Problems

5.1 DERIVATION OF ELEMENT EQUATIONS

5.1.1 The Galerkin method

The differential equation to be solved in this section is a linear parabolic partial differential equation forward in time:

$$\frac{\partial u}{\partial t} = a_0(x)\frac{\partial^2 u}{\partial x^2} + a_1(x)\frac{\partial u}{\partial x} + a_2(x)u + f(x) \tag{5.1}$$

Denoting the derivative with respect to time with a dot ($\dot{u}$) and the derivative with respect to x with a prime (u'), Equation (5.1) reads:

$$\dot{u} = a_0 u'' + a_1 u' + a_2 u + f \tag{5.2}$$

For reasons that become clear later, Equation (5.2) is rewritten as:

$$\dot{u} = (a_0 u')' + (a_1 - a_0')u' + a_2 u + f \tag{5.3}$$

Equations (5.3) and (5.1) are equivalent because of the chain rule applied to the second order term:

$$(a_0 u')' - a_0' u' = a_0 u'' \tag{5.4}$$

The BCs are given by:

$$\alpha_1 u(x^{\min}) + \beta_1 u'(x^{\min}) = \gamma_1 \tag{5.5}$$
$$\alpha_2 u(x^{\max}) + \beta_2 u'(x^{\max}) = \gamma_2 \tag{5.6}$$

To render the problem well-posed, an IC is needed as well:

$$u(t_0, x) = u_0(x) \tag{5.7}$$

The α_i, β_i, and γ_i are allowed to be functions of time. The BCs are not allowed to change their location in space as time progresses, see Figure 5.1. In financial applications, this is usually not a problem since the locations of the BCs do not change as time progresses. This property is often not inherent to engineering applications. Think of a metal rod which changes its length as heat is induced. A financial problem with BCs moving their position is the pricing of options with knock-out barriers with varying knock-out levels; see Ikeda and Kunitomo (1992). The time–space domain of the parabolic problem is not compact by itself, since time might step forward forever. For numerical purposes, however, the computations have to come to an end somewhere, so the user of any numerical techniques for parabolic problems has to specify a point in time up to which the solution is to be found. In option pricing this is usually a trivial task since time goes backward. The contract matures at some later point in time. This FC is

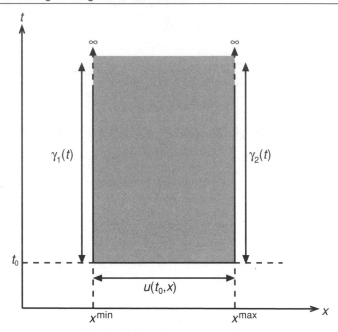

Figure 5.1   Domain of Equation (5.1) and associated IC and BCs

the starting point for the computations which are calculated up to that point of time for which a theoretical price is wanted. Next, we employ the seven-step procedure to the above problem.

1. *Discretization:* Only the spatial domain is discretized in the same way as for static 1D problems. At this point, time is not discretized. The specific form of the IC has some implication for the choice of a good mesh, especially when the IC is not continuous or has discontinuous derivatives, as is often the case in option pricing. This important issue is discussed in further detail at the end of Section 5.2.1.

2. *Interpolation:* The interpolation function is given by:

$$\tilde{u}(x, t) = \sum_{i=1}^{N+1} \tilde{u}_i(t) n_i(x) \tag{5.8}$$

The nodal values are now functions of time instead of constants as in the static problem. This means that the spatial variable is discretized, while the time variable is not. Note the difference to the FD approach to parabolic problems which discretizes both time and space. The approach of separating time and space is known as *semidiscretization* or the *method of Kantorovich.*[1]

3. *Elemental formulation:* First we define the residual as in previous sections:

$$R(x, t, \tilde{u}_1, \ldots, \tilde{u}_{N+1}) = (a_0 \tilde{u}')' + (a_1 - a_0')\tilde{u}' + a_2 \tilde{u} + f - \tilde{u}_t \tag{5.9}$$

---

[1] This method was propagated by the late Nobel prize winner L. V. Kantorovich in a publication in 1937 which was later translated and published (Kantorovich and Krylov, 1964).

The Galerkin criterion requires:

$$\int_{x^{\min}}^{x^{\max}} R\, n_k \, \mathrm{d}x \stackrel{!}{=} 0 \quad \forall\, k = 1, \ldots, N+1 \tag{5.10}$$

$$\Leftrightarrow \int_{x^{\min}}^{x^{\max}} \left[(a_0 \tilde{u}')' + (a_1 - a_0')\tilde{u}' + a_2 \tilde{u} + f - \tilde{u}_t \right] n_k \, \mathrm{d}x \stackrel{!}{=} 0 \quad \forall\, k = 1, \ldots, N+1 \tag{5.11}$$

Integrating the first term to eliminate the second order derivative:

$$\int_{x^{\min}}^{x^{\max}} (a_0 \tilde{u}')' \, n_k \, \mathrm{d}x = \left[a_0 \tilde{u}' n_k \right]_{x^{\min}}^{x^{\max}} - \int_{x^{\min}}^{x^{\max}} a_0 \tilde{u}' \, n_k' \, \mathrm{d}x \tag{5.12}$$

$$= \left[a_0 \tilde{u}'(x^{\max}) n_k(x^{\max}) \right] - \left[a_0 \tilde{u}'(x^{\min}) n_k(x^{\min}) \right] - \int_{x^{\min}}^{x^{\max}} a_0 \tilde{u}' \, n_k' \, \mathrm{d}x$$

$$= a_0 n_k(x^{\max}) \left[\frac{\gamma_2 - \alpha_2 u(x^{\max})}{\beta_2} \right] - a_0 n_k(x^{\min}) \left[\frac{\gamma_1 - \alpha_1 u(x^{\min})}{\beta_1} \right]$$

$$- \int_{x^{\min}}^{x^{\max}} a_0 \tilde{u}' \, n_k' \, \mathrm{d}x$$

Inserting the above result in Equation (5.11) leads to:

$$\int_{x^{\min}}^{x^{\max}} \left[a_0 \tilde{u}' n_k' - (a_1 - a_0')\tilde{u}' n_k - a_2 \tilde{u} n_k - f n_k + n_k \dot{\tilde{u}} \right] \mathrm{d}x$$

$$\stackrel{!}{=} a_0 n_k(x^{\max}) \left[\frac{\gamma_2 - \alpha_2 u(x^{\max})}{\beta_2} \right] - a_0 n_k(x^{\min}) \left[\frac{\gamma_1 - \alpha_1 u(x^{\min})}{\beta_1} \right]$$
$$\forall\, k = 1, \ldots, N+1 \tag{5.13}$$

Next we consider only the LHS. Inserting Equation (5.8) into Equation (5.13) results in:

$$\int_{x^{\min}}^{x^{\max}} \left[a_0 n_k' \left(\sum_{i=1}^{N+1} \tilde{u}_i n_i' \right) - (a_1 - a_0')n_k \left(\sum_{i=1}^{N+1} \tilde{u}_i n_i' \right) - a_2 n_k \left(\sum_{i=1}^{N+1} \tilde{u}_i n_i \right) - f n_k + n_k \left(\sum_{i=1}^{N+1} \dot{\tilde{u}}_i n_i \right) \right] \mathrm{d}x$$
$$= \sum_{i=1}^{N+1} \left[\int_{x^{\min}}^{x^{\max}} \left(a_0 n_k' n_i' - (a_1 - a_0')n_k n_i' - a_2 n_k n_i \right) \mathrm{d}x \right] \tilde{u}_i + \sum_{i=1}^{N+1} \left[\int_{x^{\min}}^{x^{\max}} (n_k n_i) \, \mathrm{d}x \right] \dot{\tilde{u}}_i - \int_{x^{\min}}^{x^{\max}} n_k f \, \mathrm{d}x$$

Inserting the above result into Equation (5.13) leads to:

$$\sum_{i=1}^{N+1} \left[\int_{x^{\min}}^{x^{\max}} \left(a_0 n_k' n_i' - (a_1 - a_0')n_k n_i' - a_2 n_k n_i \right) \mathrm{d}x \right] \tilde{u}_i + a_0 \delta_{kN+1} \frac{\alpha_2}{\beta_2} \tilde{u}_{N+1} - a_0 \delta_{k1} \frac{\alpha_1}{\beta_1} \tilde{u}_1$$

$$+ \sum_{i=1}^{N+1} \left[\int_{x^{\min}}^{x^{\max}} (n_k n_i) \, \mathrm{d}x \right] \dot{\tilde{u}}_i = \int_{x^{\min}}^{x^{\max}} n_k f \, \mathrm{d}x + a_0 \delta_{kN+1} \frac{\gamma_2}{\beta_2} - a_0 \delta_{k1} \frac{\gamma_1}{\beta_1}$$
$$\forall\, k = 1, \ldots, N+1 \tag{5.14}$$

$$\Leftrightarrow \mathbf{A}\bar{\mathbf{u}} + \mathbf{B}\dot{\bar{\mathbf{u}}} = \mathbf{q} \tag{5.15}$$

with:

$$A_{ki} = \int_{x^{min}}^{x^{max}} \left(a_0 n'_k n'_i - (a_1 - a'_0) n_k n'_i - a_2 n_k n_i \right) dx + a_0 \delta_{kN+1} \frac{\alpha_2}{\beta_2} \tilde{u}_{N+1} - a_0 \delta_{k1} \frac{\alpha_1}{\beta_1} \tilde{u}_1 \quad (5.16)$$

$$B_{ki} = \int_{x^{min}}^{x^{max}} (n_k n_i) \, dx \quad (5.17)$$

$$q_k = \int_{x^{min}}^{x^{max}} n_k f \, dx + a_0 \delta_{kN+1} \frac{\gamma_2}{\beta_2} - a_0 \delta_{k1} \frac{\gamma_1}{\beta_1} \quad (5.18)$$

4. *Assembly:* Similar to the static problem given by Equation (4.168), the dynamic problem (5.15) consists of:

$$\mathbf{A} = \sum_e \mathbf{K}_G + \mathbf{BT} \quad (5.19)$$

$$\mathbf{B} = \sum_e \mathbf{m}_G \quad (5.20)$$

$$\mathbf{q} = \sum_e \mathbf{q}_G + \mathbf{bt} \quad (5.21)$$

With:

$$\mathbf{k}_e = \int_{x^{min}}^{x^{max}} \left(a_0 \mathbf{N}' \mathbf{N}'^T - (a_1 - a'_0) \mathbf{N} \mathbf{N}'^T - a_2 \mathbf{N} \mathbf{N}^T \right) dx \quad (5.22)$$

$$\mathbf{m}_e = \int_{x^{min}}^{x^{max}} \mathbf{N} \mathbf{N}^T \, dx \quad (5.23)$$

$$\mathbf{q}_e = \int_{x^{min}}^{x^{max}} \mathbf{N} f \, dx \quad (5.24)$$

$$\mathbf{BT}_G = \begin{pmatrix} -a_0 \frac{\alpha_1(t)}{\beta_1(t)} & 0 & \cdots & 0 & 0 \\ 0 & 0 & \cdots & 0 & 0 \\ \vdots & \vdots & \ddots & \vdots & \vdots \\ 0 & 0 & \cdots & 0 & 0 \\ 0 & 0 & \cdots & 0 & a_0 \frac{\alpha_2(t)}{\beta_2(t)} \end{pmatrix} \quad (5.25)$$

$$\mathbf{bt}_G = \begin{pmatrix} -a_0 \frac{\gamma_1(t)}{\beta_1(t)} \\ 0 \\ \vdots \\ 0 \\ a_0 \frac{\gamma_2(t)}{\beta_2(t)} \end{pmatrix} \quad (5.26)$$

In order to solve the system of initial value problems (5.15), a starting value $\bar{\mathbf{u}}(0) = \bar{\mathbf{u}}_0$ for each ODE is needed. These starting values are obtained by discretizing the initial condition Equation (5.7) associated with the PDE (5.2). Note that the dynamic problem contains the static problem as a special case: by setting $\mathbf{B}$ in Equation (5.15) to zero we arrive at Equation (4.168).

5. *Boundary conditions:* As in the static case, this approach cannot deal with Dirichlet conditions. The remedy is similar to the static case, i.e. the Dirichlet BC is enforced with a

constraint. Let us consider the case of one Dirichlet condition which is located at $x^{\min}$. Equation (5.15) written out in full reads:

$$a_{11}\tilde{u}_1 + a_{12}\tilde{u}_2 + a_{13}\tilde{u}_3 + \cdots + b_{11}\dot{\tilde{u}}_1 + b_{12}\dot{\tilde{u}}_2 + b_{13}\dot{\tilde{u}}_3 + \ldots = q_1(t) \quad (5.27)$$
$$a_{21}\tilde{u}_1 + a_{22}\tilde{u}_2 + a_{23}\tilde{u}_3 + \cdots + b_{21}\dot{\tilde{u}}_1 + b_{22}\dot{\tilde{u}}_2 + b_{23}\dot{\tilde{u}}_3 + \ldots = q_2(t) \quad (5.28)$$
$$a_{31}\tilde{u}_1 + a_{32}\tilde{u}_2 + a_{33}\tilde{u}_3 + \cdots + b_{31}\dot{\tilde{u}}_1 + b_{32}\dot{\tilde{u}}_2 + b_{33}\dot{\tilde{u}}_3 + \ldots = q_3(t) \quad (5.29)$$
$$\vdots$$

The first equation of the above system is replaced by the Dirichlet BC:

$$\alpha_1\tilde{u}_1 \qquad\qquad\qquad = \gamma_1(t) \quad (5.30)$$
$$a_{21}\tilde{u}_1 + a_{22}\tilde{u}_2 + a_{23}\tilde{u}_3 + \cdots + b_{21}\dot{\tilde{u}}_1 + b_{22}\dot{\tilde{u}}_2 + b_{23}\dot{\tilde{u}}_3 + \ldots = q_2(t) \quad (5.31)$$
$$a_{31}\tilde{u}_1 + a_{32}\tilde{u}_2 + a_{33}\tilde{u}_3 + \cdots + b_{31}\dot{\tilde{u}}_1 + b_{32}\dot{\tilde{u}}_2 + b_{33}\dot{\tilde{u}}_3 + \ldots = q_3(t) \quad (5.32)$$
$$\vdots$$

In short, the constrained set of equations can be written as:

$$\mathbf{M}\dot{\bar{u}} + \mathbf{K}\bar{u} = \mathbf{f} \quad (5.33)$$

6. *Solution:* The numerical solution of systems of ordinary initial value problems such as Equation (5.33) has already been discussed in Section 3.2.9. Note that systems resulting from FE discretizations tend to be stiff.

7. *Computation of derived variables:* The derived variables in the spatial dimension can be computed in a similar fashion to the static problem. The derivative with respect to time can only be computed with, FD, since the semidiscrete discretization of the parabolic PDE provided a continuous approximation only of the spatial variable.

Example 8 (A Black-Scholes transformation)
The PDE derived by Black and Scholes and introduced in Section 2.2 can be transformed to the heat equation, see Section 3.4:

$$v_\tau = v_{xx} \quad (5.34)$$

1. *Discretization: A four-element model with elements of equal length is to be investigated. The domain, an infinite strip, is truncated at $x^{\min}$ and $x^{\max}$.*

2. *Interpolation: Linear interpolation is to be used for all four elements. Consequently, there are five nodes labeled $1, \ldots, 5$.*

3. *Elemental formulation: The elemental matrices are given by:*

$$\mathbf{k_e} = \int_{x_i}^{x_{i+1}} \mathbf{N'N'}^T \, dx = \frac{4}{x^{\max} - x^{\min}}\begin{bmatrix} 1 & -1 \\ -1 & 1 \end{bmatrix} \quad (5.35)$$

$$\mathbf{m_e} = \int_{x_i}^{x_{i+1}} \mathbf{NN}^T \, dx = \frac{x^{\max} - x^{\min}}{24}\begin{bmatrix} 2 & 1 \\ 1 & 2 \end{bmatrix} \quad (5.36)$$

$$\mathbf{q_e} = \mathbf{0} \quad (5.37)$$

4. *Assembly: Assuming that Dirichlet BCs are available, $\mathbf{bt} = \mathbf{BT} = 0$ so that the assembly*

equations are given by:

$$\mathbf{Au} + \mathbf{B\dot{u}} = \mathbf{0} \tag{5.38}$$

$$\mathbf{A} = \sum_e \mathbf{k_G} = \frac{4}{x^{\max} - x^{\min}} \begin{pmatrix} 1 & -1 & 0 & 0 & 0 \\ -1 & 2 & -1 & 0 & 0 \\ 0 & -1 & 2 & -1 & 0 \\ 0 & 0 & -1 & 2 & -1 \\ 0 & 0 & 0 & -1 & 1 \end{pmatrix} \tag{5.39}$$

$$\mathbf{B} = \sum_e \mathbf{m_G} = \frac{x^{\max} - x^{\min}}{24} \begin{pmatrix} 2 & 1 & 0 & 0 & 0 \\ 1 & 4 & 1 & 0 & 0 \\ 0 & 1 & 4 & 1 & 0 \\ 0 & 0 & 1 & 4 & 4 \\ 0 & 0 & 0 & 1 & 2 \end{pmatrix} \tag{5.40}$$

The initial condition for system (5.38) is given by the discretization of Equation (3.266).

5. *Boundary conditions: The BCs at* $x^{\min}$ *and* $x^{\max}$ *result in* $u_1 = v(x^{\min}, \tau)$ *and* $u_5 = v(x^{\max}, \tau)$. *The constrained system reads:*

$$\frac{96}{(x^{\max} - x^{\min})^2} \begin{pmatrix} 2 & -1 & 0 \\ -1 & 2 & -1 \\ 0 & -1 & 2 \end{pmatrix} \begin{pmatrix} u_2 \\ u_3 \\ u_4 \end{pmatrix} + \begin{pmatrix} 4 & 1 & 0 \\ 1 & 4 & 1 \\ 0 & 1 & 4 \end{pmatrix} \begin{pmatrix} \dot{u}_2 \\ \dot{u}_3 \\ \dot{u}_4 \end{pmatrix} = \mathbf{0} \tag{5.41}$$

6. *Solution: With the BCs specified, the above stiff system can be solved analytically. More realistic examples with more elements can be solved with the techniques outlined in Section 3.2.9.*

7. *Computation of derived variables: Here, the same comments as above hold.* □

5.1.2 The collocation method

In this section we derive the collocation method for the parabolic PDE (5.2):

$$\dot{u} = a_0 u'' + a_1 u' + a_2 u + f \tag{5.42}$$

Here, only Dirichlet conditions will be considered, so that the BCs are given by:

$$\alpha_1 u(x^{\min}) = \gamma_1 \tag{5.43}$$

$$\alpha_2 u(x^{\max}) = \gamma_2 \tag{5.44}$$

To render the problem well-posed, an IC is needed as well:

$$u(t_0, x) = u_0(x) \tag{5.45}$$

In contrast to Section 4.5.3, this exposition is not restricted to local basis functions.

1. *Discretization:* At this point, the same comments as in Section 5.1.1 hold. When cubic Hermite basis functions are used, some care has to be employed in the discretization of the initial condition. Since some derivatives must be interpolated, if the initial values are not smooth (i.e. infinite derivatives), the resulting cubic interpolants may have undesired noise or large spikes.

2. *Interpolation:* As mentioned above, this section is not restricted to local basis functions (giving an FE method). Also, basis functions with global support can be employed. Examples

for basis functions with global support are Legendre or Chebyshev polynomials; employed in this setting this approach is called the *spectral method*. Trigonometric basis functions lead to a *pseudospectral method*. In a similar way as in the context of the Galerkin method, a separation of variables technique is employed:

$$\tilde{u}(x, t) = \sum_{i=1}^{N+1} \tilde{u}_i(t)\phi_i(x) \tag{5.46}$$

i.e. time and space are separated.

3. *Elemental formulation:* For basis functions with global support, this issue drops out since the entire domain consists only of one element. For basis functions with local support, i.e. FE methods, the extension from a 1D static problem to a 1D dynamic problem can be achieved in a similar fashion as shown for the Galerkin method.

4. *Assembly:* Inserting the derivatives of the basis function (5.46) into the parabolic PDE (5.2) leads to the following system of ODEs:

$$\sum_{i=1}^{N+1} \dot{\tilde{u}}_i(t)\phi_i(x_j) = a_0 \sum_{i=1}^{N+1} \tilde{u}_i(t)\phi_i''(x_j) + a_1 \sum_{i=1}^{N+1} \tilde{u}_i(t)\phi_i'(x_j) + a_2 \sum_{i=1}^{N+1} \tilde{u}_i(t)\phi_i(x_j) - f \tag{5.47}$$

$$= \sum_{i=1}^{N+1} [a_0\phi_i''(x_j) + a_1\phi_i'(x_j) + a_2\phi_i(x_j)]\tilde{u}_i(t) + f \tag{5.48}$$

for $j = 1, \ldots, N + 1$. In order to change this expression to matrix notation, we define the following $(N + 1) \times (N + 1)$ matrices:

$$\mathbf{M} = [\phi_i(x_j)]_{(ji)} \tag{5.49}$$

$$\mathbf{N} = \lfloor a_0\phi_i''(x_j) + a_1\phi_i'(x_j) + a_2\phi_i(x_j)]_{(ji)} \tag{5.50}$$

Equation (5.47) can then be rewritten as:

$$\mathbf{M}\dot{\bar{u}} = \mathbf{N}\bar{u} + \mathbf{f} \tag{5.51}$$

Note that the above system consists of initial value ODEs just as the semidiscretization of a parabolic PDE with the Galerkin method; compare with system (5.33).

5. *Boundary conditions:* The integration of the Dirichlet BC is the same as in Section 5.1.1. The first line and last line of the system (5.51) are replaced by the BCs, as in Equation (5.27) to (5.32).

6. *Solution:* The numerical solution of systems of stiff ordinary initial value problems, such as Equation (5.51), has been discussed in Section 3.2.9.

7. *Computation of derived variables:* Here, the same comments as for the Galerkin semidiscretization from Section 5.1.1 apply.

5.2   CASE STUDIES

5.2.1   Plain vanilla options

As a first case study we will start with a plain vanilla European call. The seemingly strange term *plain vanilla* distinguishes this type of option from options with additional features which are usually labeled *exotic*. They are called plain vanilla because they are about as common as vanilla ice cream. There is no definite distinction in what is exotic and what is vanilla.

So you sometimes run across descriptions like *plain vanilla Asian put*, which is to indicate that this instrument is an Asian put without any additional features. The notion of whether an instrument is vanilla or not also depends on the market place and time. While in the mid 1970s a simple American put was considered by some people as being highly exotic, nowadays capped swaptions are regarded as vanilla.

This first case study will be very detailed in order to highlight many features of the FEM used to price derivatives and the program used. In order to make the case studies independent of each other, there will be some repetition. However, many basics will only be discussed in this subsection. So, if you are only interested in one special product you might want to read this case study first. The problem can be stated as:

$$\frac{\partial V}{\partial t} + \frac{1}{2}\sigma^2 S^2 \frac{\partial^2 V}{\partial S^2} + rS\frac{\partial V}{\partial S} - rV = 0 \qquad (5.52)$$

$$V(S, T) = \max(0, S - E) \qquad (5.53)$$

$$V(0, t) = 0 \qquad (5.54)$$

In order to render the problem well-posed, we also need a condition which prohibits unlimited growth as S goes to infinity. This can be delivered by financial insight. The more the option goes into the money, the closer its value becomes to that of the underlying minus the discounted strike. This way, the value of the option cannot grow at a faster rate than the underlying. In order to be able to treat this problem numerically we have to turn the semi-infinite domain into a finite one. The easiest and most common approach to do this is to cut off the domain at some point. In our case, we will put a border on the domain where the option is deeply in the money. There, we know that the value of the option is that of the underlying minus the discounted strike:

$$V(S^{\max}, t) = S^{\max} - Ee^{-r(T-t)} \qquad (5.55)$$

As a second possibility, we can formalize that the option grows by one unit as the underlying grows by one unit:

$$\frac{\partial V(S^{\max}, t)}{\partial S} = 1 \qquad (5.56)$$

A third possibility is to specify the second derivative on the boundary:

$$\frac{\partial^2 V(S^{\max}, t)}{\partial S^2} = 0 \qquad (5.57)$$

As a first example we will redo an exercise from Hull's famous textbook, see Table 5.1 (Hull, 2000, p. 244). In our first PDE2D input file we will simply state the above problem to see what happens. Then we will modify the code for higher accuracy and measures of sensitivity, the *Greeks*.

Table 5.1   Data for a plain vanilla call

Parameter	Value
Strike price	40
Interest rate	0.1
Volatility	0.2
Maturity	0.5 year

Source: Hull, 2000

Table 5.2 Premium results for plain vanilla call (1)

Underlying	FE (100 × 100) premium	Analytical premium	Difference (%)
30	0.09458228	0.091410090	3.47028
33	0.41563677	0.411541770	0.99504
36	1.2225290	1.220152982	0.19473
39	2.6801168	2.680053687	0.00235
42	4.7586967	4.759422387	−0.00153
45	7.2874827	7.287820620	−0.00464
48	10.077581	10.07749335	0.00087
51	12.994461	12.994325000	0.00105
54	15.964607	15.964646460	−0.00025

The reported errors and differences here and in following sections are based on more significant digits than are shown in the tables. *Difference* is defined as relative deviation:

$$\text{difference} = \frac{\text{result}_{FE} - \text{result}}{\text{result}} \quad [\%] \qquad (5.58)$$

For the rest of the book *difference* is not necessarily the same as *error*, since finite elements will also be compared to other numerical methods. For the computation of both *difference* and *error* all available digits have been used, although normally not all are reported here.

The results in Table 5.2 were achieved by using Equation (5.55) with $S^{max} = 100$. First, we want to investigate how the choice of type and location of the BC influences the price. For this reason, Equations (5.56) and (5.57) are also applied with various values of S^{max}. The results for $S = 42$ are reported in Table 5.3. To make the results for different values of S^{max} comparable, the number of elements has to be increased as S^{max} grows. On the RHS of the domain, elements are added until the influence on the price vanishes.

The results in Table 5.2 are based on a rather arbitrary choice of the number of time-steps and nodes. There are several ways to increase accuracy:

1. Increase the number of nodes or (equivalently) elements;
2. Increase the number of time-steps;
3. Enforce a tighter FE net in areas of high curvature, around points where the payoff function or its first derivative is not differentiable, or around areas of interest;
4. Use adaptivity in time and/or space;
5. Employ different shape functions.

All of these approaches will be explored on the next couple of pages. They all carry advantages and disadvantages.

1. Simply increasing both the number of elements and time-steps leads to the results reported in Table 5.4. This brute force approach ignores the fact that the FE approach is of $O(dt)$ in time (Backward Euler) and $O(dx^4)$ in space.
2. Since this FE approach is of order $O(dt)$ (Backward Euler) or $O(dt^2)$ (Crank–Nicolson) in time and $O(dx^4)$ in the spatial variables, it seems worthwhile to increase only the number of time-steps to 500 and 1000 with both methods. The results are reported in Tables 5.5 and 5.6. The performance of the Crank–Nicolson method is much better. However, this comes at the price of possible oscillations caused by the payoff not being differentiable at the strike.

Table 5.3 Premium results for a plain vanilla call (2)

S^{max}	Elements	Analytical	BC Equation (5.55)	(%)	BC Equation (5.56)	(%)	BC Equation (5.57)	(%)
45.0	45	4.759422387	4.601901	−3.310%	4.953426	4.076%	3.7665852	−20.860%
46.0	46	4.759422387	4.677904	−1.713%	4.854858	2.005%	4.3031742	−9.586%
47.0	47	4.759422387	4.718858	−0.852%	4.803043	0.917%	4.5574401	−4.244%
48.0	48	4.759422387	4.739493	−0.419%	4.777449	0.379%	4.6726616	−1.823%
49.0	49	4.759422387	4.749262	−0.213%	4.756640	−0.058%	4.7226227	−0.773%
50.0	50	4.759422387	4.753626	−0.122%	4.760263	0.018%	4.7433783	−0.337%
60.0	60	4.759422387	4.756642	−0.058%	4.756640	−0.058%	4.7566396	−0.058%
70.0	70	4.759422387	4.756640	−0.058%	4.756640	−0.058%	4.7566398	−0.058%
80.0	80	4.759422387	4.756640	−0.058%	4.756640	−0.058%	4.7566398	−0.058%
90.0	90	4.759422387	4.756640	−0.058%	4.756640	−0.058%	4.7566398	−0.058%
100.0	100	4.759422387	4.756640	−0.058%	4.756640	−0.058%	4.7566398	−0.058%

Table 5.4   Premium results for a plain vanilla call (3)

Underlying	FE (3200 × 3200)	Difference to analytical solution (%)	FE (25600 × 25600)	Difference to analytical solution (%)
30	0.0914	−0.0909	0.0914	0.01131
33	0.4116	−0.0207	0.4116	0.00256
36	1.2202	−0.0004	1.2202	0.00003
39	2.6800	0.0028	2.6800	−0.00035
42	4.7593	0.0018	4.7594	−0.00023
45	7.2878	0.0007	7.2878	−0.00009
48	10.0775	0.0002	10.0775	−0.00002
51	12.9943	0.0001	12.9943	−0.00001
54	15.9646	0.0000	15.9646	0.00000

Table 5.5   Constant time stepping (Crank–Nicolson)

Underlying	FE	Difference to analytical solution (%)	FE	Difference to analytical solution (%)
30	0.09141	0.001749	0.09141	0.001812
33	0.41154	−0.000066	0.41154	−0.000010
36	1.22015	−0.000162	1.22015	−0.000154
39	2.68005	−0.000100	2.68005	−0.000104
42	4.75942	−0.000042	4.75942	−0.000046
45	7.28782	−0.000007	7.28782	−0.000009
48	10.07749	0.000006	10.07749	0.000006
51	12.99433	0.000008	12.99433	0.000008
54	15.96465	0.000003	15.96465	0.000003
Time-steps	500		1000	

Table 5.6   Constant time stepping (Backward Euler)

Underlying	FE	Difference to analytical solution (%)	FE	Difference to analytical solution (%)
30	0.09194	0.579128	0. 09168	0.290671
33	0.41208	0.130410	0. 41181	0.065228
36	1.22017	0.001378	1.22016	0.000608
39	2.67956	−0.018387	2.67981	−0.009246
42	4.75886	−0.011743	4.75914	−0.005895
45	7.28749	−0.004596	7.28765	−0.002303
48	10.07737	−0.001234	10.07743	−0.000619
51	12.99429	−0.000285	12.99431	−0.000139
54	15.96461	−0.000210	15.96463	−0.000103
Time-steps	500		1000	

Table 5.7 Results from a refined grid

Underlying	FE (120 × 100) premium	Analytical premium	Difference (%)
30	0.09404	0.09141009	2.872094
33	0.41422	0.41154177	0.650350
36	1.22025	1.220152982	0.007919
39	2.67760	2.680053687	−0.091397
42	4.75664	4.759422387	−0.058444
45	7.28615	7.28782062	−0.022860
48	10.07687	10.07749335	−0.006186
51	12.99413	12.994325	−0.001478
54	15.96447	15.96464646	−0.001093

3. In this example, at the strike the payoff function is not differentiable. Besides this, this is the area with the most curvature. The equidistant mesh from above is replaced by a grid in which all intervals between 40 and 60 are halved (see Table 5.7). This implies that the number of elements is increased by 20. Notice that the strike is still located on the border between two elements. The accuracy of the approximation has increased, but much less than with increasing the number of time-steps.

4. How is adaptivity in time realized in PDE2D? This is explained in Sewell (2000):

> The user can either specify a constant, user-chosen, stepsize dt, or request adaptive stepsize control. If adaptive stepsize control is chosen, then each time step, two steps of size $dt/2$ are taken, and that solution is compared with the result when one step of size dt is taken. If the difference between the two answers is less than a user-supplied tolerance (for each variable), the time step dt is accepted (and the next step dt is doubled, if the agreement is too 'good'); otherwise dt is halved and the process is repeated. When the step is accepted, an extrapolation is done using the answer obtained using one step of size $dt(U_1(t_{n+1}))$ and two steps of size $dt/2(U_2(t_{n+1}))$. When the backward Euler method is used to compute both answers, the extrapolated value is $U(t_{n+1}) = 2U_2(t_{n+1}) - U_1(t_{n+1})$, which increases the order of the truncation error of the backward Euler method from $O(dt)$ to $O(dt^2)$ while preserving its stability of stiff systems. When the Crank–Nicolson method is used, the extrapolated value is $U(t_{n+1}) = 0.5U_2(t_{n+1}) + 0.5U_1(t_{n+1})$, which preserves the $O(dt^2)$ truncation error of the Crank–Nicolson method, while greatly improving its stability on stiff systems.

The main difference between the approach used in PDE2D and the RKF method introduced in Section 3.2.8 is that PDE2D uses the same integration method with different step sizes, while RKF uses two different ODE solvers. Applying the adaptive versions of the backward Euler method and the Crank–Nicolson method with an error tolerance of 0.01 leads to numerical solutions with approximately 100 time-steps, as in Tables 5.8 and 5.9, while greatly improving accuracy compared to fixed time stepping. By increasing the error tolerance, the number of time-steps increases and the difference to the numerical solutions vanishes.

5. The influence of different shape functions is explored with the help of rainbow options in Section 7.2.1.

Bermudan exercise

All options discussed so far can only be exercised at maturity. These options are called *European*. Options which can be exercised anytime up to maturity are called *American*. This is purely terminology and not indicative that either product is predominant on their respective continent. Following this terminology, options which can be exercised on certain dates before

Table 5.8   Adaptive time stepping (Crank–Nicolson)

Underlying	FE tol = 0.01	Difference to analytical solution (%)	FE tol = 0.000001	Difference to analytical solution (%)
30	0.091410571	0.000526	0.091410734	0.000705
33	0.41153699	−0.001161	0.41153765	−0.001001
36	1.2201494	−0.000294	1.2201496	−0.000277
39	2.6800543	0.000023	2.6800538	0.000004
42	4.7594238	0.000030	4.7594233	0.000019
45	7.2878207	0.000001	7.2878206	0.000000
48	10.077493	−0.000003	10.077493	−0.000003
51	12.994325	0.000000	12.994325	0.000000
54	15.964647	0.000003	15.964647	0.000003
Time-steps	101		160	

Table 5.9   Adaptive time stepping (Backward Euler)

Underlying	FE tol = 0.01	Difference to analytical solution (%)	FE tol = 0.000001	Difference to analytical solution (%)
30	0.091415707	0.006145	0.091411908	0.0020
33	0.41155655	0.003591	0.41154237	0.0001
36	1.2201563	0.000272	1.2201513	−0.0001
39	2.6800402	−0.000503	2.6800504	−0.0001
42	4.759409	−0.000281	4.7594197	−0.0001
45	7.2878179	−0.000037	7.2878199	0.0000
48	10.077497	0.000036	10.077494	0.0000
51	12.994329	0.000031	12.994326	0.0000
54	15.964649	0.000016	15.964647	0.0000
	102		1069	

maturity are called *Bermudan*. Another variant of exercise is *early-exercise windows*, which are certain time periods in which the holder of the option can exercise the option anytime. Since early exercise of any type gives an additional right to the holder, such an option cannot be worth less than its European counterpart. The more exercise dates a Bermudan option bears, the closer its value approaches its American counterpart. Another approach for pricing American options will be discussed in Section 10.2.2.

Bermudan exercise can be introduced into our model by inspecting the spatial values at every exercise date. If the value of the option is below its payoff, it is worthwhile to exercise immediately. If the option is worth more than its payoff, the holder should keep the option. Assume there is an exercise date at t_i, then for each node $S_{(i)}$, the following substitution has to be performed:

$$V(S_{(i)}, t_{i+1}) \leftarrow \text{payoff}(S_{(i)}) \ \forall \ S_{(i)} \text{ with } V(S_{(i)}, t_i) \leq \text{payoff}(S_{(i)}) \tag{5.59}$$

If this substitution is applied at every time-step, the numerical solution converges to the price of the American option (Tomas, 1996).

For the example given in Table 5.10, a binomial tree with 500 time-steps was employed as a benchmark. The results are reported in Table 5.11. The exercise dates are distributed in an

Table 5.10   Data for a Bermudan put

Parameter	Value
Asset price	9
Strike price	10
Interest rate	0.12
Volatility	0.5
Maturity	1 year

Table 5.11   Results for Bermudan exercise

Dates of exercise	Binomial tree		FEM	
	Premium	Δ	Premium	Δ
1	1.8164	−0.4373	1.8163	−0.4375
10	1.8905	−0.4638	1.8989	−0.4640
50	1.9049	−0.4659	1.9052	−0.4667

equidistant fashion over time to maturity so that all time-steps are of the same length and no additional binomial steps have to be introduced to catch any exercise date. This is also true for the backward Euler time stepping in the FE algorithm, which uses 500 equidistant time-steps as well. The spatial discretization consists of 50 elements of equal size with $S^{min} = 0.0$ and $S^{max} = 50.0$. A collocation FEM with cubic Hermite basis function has been used. The BCs are

$$V(S^{min}) = 0 \tag{5.60}$$

$$\frac{\partial V(S^{max})}{\partial S} = 1 \tag{5.61}$$

Non-smooth payoffs

Options usually have payoffs that are not differentiable, such as all options involving the *max*-operator in their payoffs. Common exotic options such as digital options even possess discontinuities in their payoffs. The associated pricing problems have discontinuous final conditions or final conditions which are not differentiable. The Crank–Nicolson method, as explained in Section 3.4.3, suffers from oscillating *Deltas* and *Gammas* around the strike. This feature carries over to semidiscretizations with FE for the spatial variable and FD for time to maturity. Solving the resulting system of ODEs with the method presented in Section 3.2.9 leads to oscillating Greeks as well. Crank–Nicolson time stepping, however, is of order two, $O(dt^2)$, in contrast to explicit time stepping which is only, $O(dt)$. Various ways to use the faster Crank–Nicolson time stepping without incurring oscillating Greeks have been explored. In a pure FD setting, only a couple of time-steps are computed with explicit FD. Then, the time stepping is changed to Crank–Nicolson. Usually, around five explicit time-steps are sufficient to avoid unstable Greeks. In the setting of FE semidiscretizations, this approach can be employed accordingly: only a few time-steps of the system of ODEs are computed with the explicit method. This approach has been called *Rannacher time stepping* by Pooley *et al.* (2003b). In Rannacher (1984) this idea was suggested for parabolic PDEs with a nonfinancial background. Another approach is to replace the non-smooth final condition by one not suffering from this drawback. PDE2D takes the following approach: normally, interpolation is

Table 5.12 Modern Greeks

Name	Symbol	Definition
Delta	Δ	V_S
Gamma	Γ	V_{SS}
Speed		V_{SSS}
Rho	ρ	V_r
Rho of dividend	ρ_D	V_D
Vega	υ	V_σ
Theta	Θ	V_t
Charm		V_{St}
Color		V_{SSt}
Leverage/Gearing	λ	$(S/V)V_S$
Vomma/Volga	υ'	$V_{\sigma\sigma}$
Vanna		$V_{S\sigma}$

employed to approximate the initial values using cubic Hermites. Since some derivatives must be interpolated if the initial values are not smooth (i.e. have large or infinite derivatives), the resulting cubic interpolants may have undesired noise or large spikes. Then, a least-squares approximation of the final condition is computed first, which is usually much smoother but requires one extra linear system solution.

A different approach (*mollification*) has been suggested by Jackson and Süli (1997). Assume there is a discontinuity in the payoff or its first derivative at S^*. Then, within the interval $[S^* - \epsilon_{\text{mol}}, S^* + \epsilon_{\text{mol}}]$ (with ϵ_{mol} being a small number) the payoff function is replaced by a cubic Hermite interpolant. This method has the advantage that it is easy to monitor the error incurred through mollification and that experimentally, Jackson and Süli (1997) found it to be extremely small when ϵ_{mol} was chosen sensibly.

5.2.2 Hedging parameters

Hedging parameters are the sensitivities of a security with respect to certain risk factors. The collective label is usually *the (modern) Greeks*, since some of them have Greek or Greek-sounding[2] names (see Table 5.12).

Usually, the Greeks are computed with FD since only a few analytical formulae are available. When the option's premium is computed with FD it is recommended to use central differences for all sensitivities with respect to the underlying and time to maturity, since all necessary computations have already been performed:

$$\Delta(S, t) \approx \frac{V(S + \Delta S, t) - V(S - \Delta S, t)}{2\Delta S} \tag{5.62}$$

$$\Gamma(S, t) \approx \frac{V(S + \Delta S, t) - 2V(S, t) + V(S - \Delta S, t)}{\Delta S^2} \tag{5.63}$$

$$\text{speed}(S, t) \approx \frac{V(S + 2\Delta) - V(S - 2\Delta) - 2V(S + \Delta) + 2V(S - \Delta)}{2\Delta S^3} \tag{5.64}$$

$$\Theta(S, t) \approx \frac{V_{t+\Delta t}(S, t) - V_{t-\Delta t}(S, t)}{2\Delta t} \tag{5.65}$$

---

[2] For instance *Vega* is neither a letter of antiquity, nor of current Greek.

Here, all intervals ΔS are assumed to be of equal length. This is not mandatory. In particular, when the FD net is not equidistant this has to be taken into account. Since the Crank–Nicolson FD scheme has difficulties with ICs which are not differentiable, this method should be used with special care. The (spatial) oscillations arising through this scheme are magnified when taking spatial derivatives. A similar argument applies to trees (Pelsser and Vorst, 1994). Obviously, the above formulae cannot be applied to points very close to the boundary of the domain. This is especially a disadvantage when it comes to computing the Greeks of a knock-out option. Close to the barrier the hedge parameters are most important and difficult to compute with FD. This problem can be cured entirely with FE. For ρ and v, usually only one-sided differences are employed:

$$\rho \approx \frac{V_r(S,t) - V_{r-\Delta r}(S,t)}{\Delta r} \tag{5.66}$$

$$v \approx \frac{V_\sigma(S,t) - V_{\sigma-\Delta\sigma}(S,t)}{\Delta\sigma} \tag{5.67}$$

In an FD setting, this means that in order to compute either of the two hedging parameters, the entire FD net has to be computed twice. Once with the parameter value at which the sensitivity is needed (this one is usually computed anyway since one wants the price of the option *and* its Greeks) and once slightly perturbed. Obviously, central differences could be employed as well; however this would be at the cost of increasing computer time.

Computing the Greeks with finite elements

Approximating the solution of a PDE using finite element methods (Galerkin or collocation), using piecewise polynomials of degree n, the approximation to the solution itself should be of the order $O(h^{n+1})$; the approximation to the mth derivative should be $O(h^{n+1-m})$, i.e. one loses one order of precision for each derivative; see Section 4.7. Why is this? Let us assume a smooth function $f(x)$ in the interval $0 \le x \le h$. We want to approximate this function using a Taylor approximant of degree n:

$$T(x) = f(0) + f'(0)x + \frac{1}{2}f''(0)x^2 + \cdots + f^{(n)}(0)\frac{x^n}{n!} \tag{5.68}$$

By the Taylor series remainder theorem, the error is

$$f(x) - T(x) = \frac{f^{(n+1)}(c)}{(n+1)!}x^{n+1} \tag{5.69}$$

$$0 \le c \le x \tag{5.70}$$

which is $O(h^{n+1})$ when $0 < x < h$. If the first derivative of $T(x)$ is taken, it can be noted that it is exactly the Taylor polynomial approximant of degree $(n-1)$ to $f'(x)$, and thus $f'(x) - T'(x)$ is of the order $O(h^n)$. Similarly, $f''(x) - T''(x)$ is of the order $O(h^{n-1})$, etc. Though it requires a little more work to prove other cases, similar error bounds on $f'(x) - T'(x)$, $f''(x) - T''(x)$, etc. can be derived when $T(x)$ is the Lagrange polynomial interpolant, Hermite polynomial interpolant or spline interpolant of degree n.

The results in Table 5.13 show that Δ is computed in a reliable way, even on a rather coarse net. In practice, the theoretical loss of one order in accuracy often is not a big problem. A more serious problem is the choice of the appropriate basis function, as shown in Figure 5.2: Γ was computed twice with a quadratic basis function. Since the second derivative with respect to

Table 5.13 Results for Δ of a plain vanilla call (data source: Table 5.1)

Underlying	Analytical	FE (100 × 100)	Difference (%)	FE (3200 × 3200)	Difference (%)	FE (25 600 × 25 600)	Difference (%)
30	0.0537	0.0541	−0.745	0.0537	−0.0032	0.0536	0.00319
33	0.1746	0.1742	0.229	0.1746	−0.0011	0.1745	−0.00109
36	0.3742	0.3731	0.294	0.3742	0.0011	0.3742	−0.00109
39	0.5969	0.5964	0.084	0.5969	−0.0003	0.5969	−0.00033
42	0.7791	0.7794	−0.039	0.7791	0.0001	0.7790	0.00011
45	0.8956	0.8961	−0.056	0.8956	0.0002	0.8955	0.00018
48	0.9567	0.9569	−0.021	0.9567	0.0001	0.9567	0.00010
51	0.9839	0.9840	−0.010	0.9839	0.0000	0.9838	0.00002
54	0.9946	0.9945	0.010	0.9946	0.0000	0.9946	−0.00001

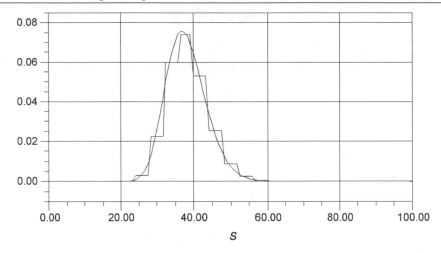

Figure 5.2 Γ of a plain vanilla call: high and low accuracy

S is a horizontal line, either a very fine spatial discretization has to be constructed (as in this example) or a polynomial with a higher degree has to be used.

Computing the Greeks through their governing PDEs

It is possible to construct a PDE for every hedge parameter. The basic idea is to differentiate the pricing PDE with respect to the parameter in question p_i:

$$\frac{\partial}{\partial p_i}\left[\frac{\partial V}{\partial t} + \frac{1}{2}\sigma^2 S^2 \frac{\partial^2 V}{\partial S^2} + (r-D)S\frac{\partial V}{\partial S} - rV\right] = 0 \qquad (5.71)$$

- $p_1 = S$ giving $\Delta \equiv \frac{\partial V}{\partial S}$. We will derive the PDE for both the plain vanilla put and call, including all BCs and FCs. For constant σ, D and r, differentiating the Black–Scholes equation with respect to S gives:

$$\frac{\partial}{\partial S}\left[\frac{\partial V}{\partial t} + \frac{1}{2}\sigma^2 S^2 \frac{\partial^2 V}{\partial S^2} + (r-D)S\frac{\partial V}{\partial S} - rV\right] \qquad (5.72)$$

$$= \frac{\partial}{\partial S}\frac{\partial V}{\partial t} + \frac{1}{2}\sigma^2 \frac{\partial}{\partial S}\left[S^2\frac{\partial^2 V}{\partial S^2}\right] + (r-D)\frac{\partial}{\partial S}\left[S\frac{\partial V}{\partial S}\right] - \frac{\partial}{\partial S}[rV] \qquad (5.73)$$

$$= \frac{\partial \Delta}{\partial t} + \frac{1}{2}\sigma^2\left[2S\frac{\partial \Delta}{\partial S} + S^2\frac{\partial^2 \Delta}{\partial S^2}\right] + (r-D)\left[\Delta + S\frac{\partial \Delta}{\partial S}\right] - r\Delta \qquad (5.74)$$

$$= \underbrace{\frac{\partial \Delta}{\partial t} + \frac{1}{2}\sigma^2 S^2 \frac{\partial^2 \Delta}{\partial S^2} + (r-D)S\frac{\partial \Delta}{\partial S} - r\Delta}_{\text{Black–Scholes operator acting on } \Delta} +(r-D)\Delta + \sigma^2 S\frac{\partial \Delta}{\partial S} = 0 \quad (5.75)$$

In a similar fashion, the FC for the PDE governing *Delta* of a plain vanilla put $P(S,t)$ can be derived:

$$\Delta(S,T) = \frac{\partial}{\partial S}\max(E-S,0) = \begin{cases} -1 & \forall \quad S \leq E \\ 0 & \forall \quad S > E \end{cases} \qquad (5.76)$$

The BC at $S \to \infty$ is derived as follows:

$$P(S, t) \to 0 \tag{5.77}$$

$$\implies \frac{\partial P(S, t)}{\partial S} \to 0 \tag{5.78}$$

$$\implies \frac{\partial \Delta(S, t)}{\partial S} \to 0 \tag{5.79}$$

The left-hand BC is

$$\Delta(0, t) = -1 \tag{5.80}$$

Via put–call parity

$$P + S = E e^{-(T-t)r} + C \tag{5.81}$$

$$\frac{\partial P}{\partial S} + 1 = \frac{\partial C}{\partial S} \tag{5.82}$$

it is possible to deduce the Δ of the call from the Δ of the put. For the more general model with $\sigma = \sigma(S, t)$, $D = D(t)$ and $r = r(t)$, differentiating the Black–Scholes equation with respect to S leads to a slightly more complicated expression. With

$$\frac{\partial}{\partial S} \left(\sigma^2(S, t) S^2 \frac{\partial^2 V}{\partial S^2} \right) = 2 \sigma(S, t) \frac{\partial \sigma(S, t)}{\partial S} S^2 \frac{\partial \Delta}{\partial S} + 2 \sigma^2(S, t) S \frac{\partial \Delta}{\partial S} + \sigma^2(S, t) S^2 \frac{\partial^2 \Delta}{\partial S^2}$$

and

$$\frac{\partial}{\partial S} \left(S \frac{\partial V}{\partial S} \right) = \Delta + S \frac{\partial \Delta}{\partial S}$$

we obtain

$$\frac{\partial}{\partial S} \left[\frac{\partial V}{\partial t} + \frac{1}{2} \sigma^2(S, t) S^2 \frac{\partial^2 V}{\partial S^2} + (r(t) - D(t)) S \frac{\partial V}{\partial S} - r(t) V \right]$$

$$= \frac{\partial}{\partial S} \frac{\partial V}{\partial t} + \frac{1}{2} \frac{\partial}{\partial S} \left(\sigma^2(S, t) S^2 \frac{\partial^2 V}{\partial S^2} \right) + (r(t) - D(t)) \frac{\partial}{\partial S} \left(S \frac{\partial V}{\partial S} \right) - r(t) \frac{\partial V}{\partial S}$$

$$= \frac{\partial \Delta}{\partial t} + \sigma(S, t) \frac{\partial \sigma(S, t)}{\partial S} S^2 \frac{\partial \Delta}{\partial S} + \sigma^2(S, t) S \frac{\partial \Delta}{\partial S} + \frac{1}{2} \sigma^2(S, t) S^2 \frac{\partial^2 \Delta}{\partial S^2}$$

$$+ (r(t) - D(t)) \left(\Delta + S \frac{\partial \Delta}{\partial S} \right) - r(t) \Delta$$

$$= \underbrace{\frac{\partial \Delta}{\partial t} + \frac{1}{2} \sigma^2(S, t) S^2 \frac{\partial^2 \Delta}{\partial S^2} + (r(t) - D(t)) S \frac{\partial \Delta}{\partial S} - r(t) \Delta}_{\text{Black–Scholes operator acting on } \Delta}$$

$$+ \sigma(S, t) \frac{\partial \sigma(S, t)}{\partial S} S^2 \frac{\partial \Delta}{\partial S} + \sigma^2(S, t) S \frac{\partial \Delta}{\partial S} + (r(t) - D(t)) \Delta$$

$$= 0$$

For $\sigma^2(S, t) = \sigma^2 = $ constant, it reduces to

$$\frac{\partial \Delta}{\partial t} + \frac{1}{2}\sigma^2 S^2 \frac{\partial^2 \Delta}{\partial S^2} + (r(t) - D(t))S \frac{\partial \Delta}{\partial S} - r(t)\Delta + \sigma^2 S \frac{\partial \Delta}{\partial S} + (r(t) - D(t))\Delta = 0$$

- $p_2 = S$ applied to Equation (5.75), giving $\Gamma \equiv \frac{\partial^2 V}{\partial S^2}$:

$$\frac{\partial^2}{\partial S^2}\left[\frac{\partial V}{\partial t} + \frac{1}{2}\sigma^2 S^2 \frac{\partial^2 V}{\partial S^2} + (r - D)S\frac{\partial V}{\partial S} - rV\right] \tag{5.83}$$

$$= \frac{\partial}{\partial S}\left[\frac{\partial \Delta}{\partial t} + \frac{1}{2}\sigma^2 S^2 \frac{\partial^2 \Delta}{\partial S^2} + (r - D)S\frac{\partial \Delta}{\partial S} - r\Delta + (r - D)\Delta + \sigma^2 S \frac{\partial \Delta}{\partial S}\right] \tag{5.84}$$

$$= \underbrace{\frac{\partial \Gamma}{\partial t} + \frac{1}{2}\sigma^2 S^2 \frac{\partial^2 \Gamma}{\partial S^2} + (r - D)S\frac{\partial \Gamma}{\partial S} - r\Gamma}_{\text{Black–Scholes operator acting on } \Gamma} + 2(r - D)\Gamma + 2\sigma^2 S \frac{\partial \Gamma}{\partial S} + \sigma^2\Gamma = 0 \tag{5.85}$$

For $\sigma = \sigma(S, t)$, $D = D(t)$ and $r = r(t)$, we get:

$$\frac{\partial}{\partial S}\frac{\partial \Delta}{\partial t} + \frac{1}{2}\underbrace{\frac{\partial}{\partial S}\left(\sigma^2(S, t) S^2 \frac{\partial \Gamma}{\partial S}\right)}_{\text{(i)}} + (r(t) - D(t))\left(\Gamma + S\frac{\partial \Gamma}{\partial S}\right) - r(t)\frac{\partial \Delta}{\partial S}$$

$$+ \underbrace{\frac{\partial}{\partial S}\left(\sigma(S, t)\frac{\partial \sigma(S, t)}{\partial S} S^2 \Gamma\right)}_{\text{(ii)}} + \underbrace{\frac{\partial}{\partial S}\left(\sigma^2(S, t) S \Gamma\right)}_{\text{(iii)}} + (r(t) - D(t))\Gamma = 0 \tag{5.86}$$

with:
(i)

$$\frac{\partial}{\partial S}\left(\sigma^2(S, t) S^2 \frac{\partial \Gamma}{\partial S}\right) = 2\sigma(S, t)\frac{\partial \sigma(S, t)}{\partial S} S^2 \frac{\partial \Gamma}{\partial S}$$

$$+ 2\sigma^2(S, t) S \frac{\partial \Gamma}{\partial S} + \sigma^2(S, t) S^2 \frac{\partial^2 \Gamma}{\partial S^2} \tag{5.87}$$

(ii)

$$\frac{\partial}{\partial S}\left(\sigma(S, t)\frac{\partial \sigma(S, t)}{\partial S} S^2 \Gamma\right) = \left(\frac{\partial \sigma(S, t)}{\partial S}\right)^2 S^2 \Gamma + \sigma(S, t)\frac{\partial^2 \sigma(S, t)}{\partial S^2} S^2 \Gamma$$

$$+ 2\sigma(S, t)\frac{\partial \sigma(S, t)}{\partial S} S \Gamma + \sigma(S, t)\frac{\partial \sigma(S, t)}{\partial S} S^2 \frac{\partial \Gamma}{\partial S} \tag{5.88}$$

(iii)

$$\frac{\partial}{\partial S}\left(\sigma^2(S, t) S \Gamma\right) = 2\sigma(S, t)\frac{\partial \sigma(S, t)}{\partial S} S \Gamma + \sigma^2(S, t)\Gamma + \sigma^2(S, t) S \frac{\partial \Gamma}{\partial S} \tag{5.89}$$

As a whole it follows that:

$$0 = \underbrace{\frac{\partial \Gamma}{\partial t} + \frac{1}{2}\sigma^2(S, t) S^2 \frac{\partial^2 \Gamma}{\partial S^2} + (r(t) - D(t))S\frac{\partial \Gamma}{\partial S} - r(t)\Gamma}_{\text{Black–Scholes operator acting on } \Gamma}$$

$$+ 2\left(r(t) - D(t)\right)\Gamma + 2\sigma^2(S, t) S \frac{\partial \Gamma}{\partial S} + \sigma^2(S, t)\Gamma$$

$$+ 2\sigma(S, t)\frac{\partial \sigma(S, t)}{\partial S} S^2 \frac{\partial \Gamma}{\partial S} + \left(\frac{\partial \sigma(S, t)}{\partial S}\right)^2 S^2\,\Gamma + \sigma(S, t)\frac{\partial^2 \sigma(S, t)}{\partial S^2} S^2\,\Gamma$$

$$+ 4\sigma(S, t)\frac{\partial \sigma(S, t)}{\partial S} S\,\Gamma \tag{5.90}$$

For $\sigma^2(S, t) = \sigma^2 = $ constant, we obtain

$$\frac{\partial \Gamma}{\partial t} + \frac{1}{2}\sigma^2 S \frac{\partial^2 \Gamma}{\partial S^2} + (r(t) - D(t))S\frac{\partial \Gamma}{\partial S} - r(t)\Gamma$$

$$+ 2\left(r(t) - D(t)\right)\Gamma + 2\sigma^2 S\frac{\partial \Gamma}{\partial S} + \sigma^2\Gamma = 0 \tag{5.91}$$

- $p_3 = \sigma$, giving $\upsilon \equiv \frac{\partial V}{\partial \sigma}$:

$$\frac{\partial}{\partial \sigma}\left[\frac{\partial V}{\partial t} + \frac{1}{2}\sigma^2 S^2 \frac{\partial^2 V}{\partial S^2} + (r - D)S\frac{\partial V}{\partial S} - rV\right] \tag{5.92}$$

$$= \underbrace{\frac{\partial \upsilon}{\partial t} + \frac{1}{2}\sigma^2 S^2 \frac{\partial^2 \upsilon}{\partial S^2} + (r - D)S\frac{\partial \upsilon}{\partial S} - r\upsilon}_{\text{Black–Scholes operator acting on } \upsilon} + \sigma S^2\Gamma = 0 \tag{5.93}$$

Since usually $\Gamma = \frac{\partial^2 V}{\partial S^2}$ is not known, both quantities, υ and Γ, have to be computed. As with r and r_D this gives rise to a system of PDEs. For both plain vanilla calls and puts, the FCs and BCs are given by:

$$\upsilon(S, T) = 0 \tag{5.94}$$

$$\upsilon(0, t) = 0 \tag{5.95}$$

$$\lim_{S \to \infty} \upsilon(S, t) = 0 \tag{5.96}$$

For models with local volatility and time-dependent interest rate curves, i.e. $\sigma = \sigma(S, t)$, $r = r(t)$, and $D = D(t)$, this approach has to be modified. Define a one-parameter family of surfaces $\sigma_\lambda(S, t)$ which smoothly depends on λ and satisfies $\sigma_0(S, t) = \sigma(S, t)$. We assume that the solutions to the Black–Scholes model with local volatility $\sigma_\lambda(S, t)$ likewise provide a smooth one-parameter family of prices $V_\lambda(S, t)$ which satisfies $V_0(S, t) = V(S, t)$, i.e.

$$\frac{\partial V_\lambda}{\partial t} + \frac{1}{2}\sigma^2(S, t) S^2 \frac{\partial^2 V_\lambda}{\partial S^2} + (r(t) - D(t)) S\frac{\partial V_\lambda}{\partial S} - r(t) V_\lambda = 0 \tag{5.97}$$

The variation of price with respect to changes in volatility is the tangent to this family at $\lambda = 0$ (Shadwick and Shadwick, 2002):

$$\upsilon = \upsilon(S, t) = \left.\frac{dV_\lambda(S, t)}{d\lambda}\right|_{\lambda = 0} \tag{5.98}$$

Without loss of generality, it is assumed that the family $\sigma_\lambda(S, t)$ can be represented as a series in λ:

$$\sigma_\lambda(S, t) = \sum_{i=0}^{\infty} \lambda^i \sigma^{(i)}(S, t) \tag{5.99}$$

with $\sigma^{(0)}(S, t) = \sigma(S, t)$. Then it follows that

$$
\frac{\partial}{\partial t} \frac{dV_\lambda(S, t)}{d\lambda}\bigg|_{\lambda=0} + \sigma_\lambda(S, t)\frac{d\sigma_\lambda(S, t)}{d\lambda}S^2\frac{\partial^2 V_\lambda}{\partial S^2}\bigg|_{\lambda=0} + \frac{1}{2}\sigma_\lambda^2(S, t)S^2\frac{\partial^2}{\partial S^2}\frac{dV_\lambda}{d\lambda}\bigg|_{\lambda=0}
$$

$$
+ (r(t) - D(t))S\frac{\partial}{\partial S}\frac{dV_\lambda}{d\lambda}\bigg|_{\lambda=0} - r(t)\frac{dV_\lambda}{d\lambda}\bigg|_{\lambda=0} = 0 \tag{5.100}
$$

and therefore

$$
\frac{\partial v}{\partial t} + \frac{1}{2}\sigma^2(S, t)S^2\frac{\partial^2 v}{\partial S^2} + (r(t) - D(t))S\frac{\partial v}{\partial S} - r(t)
$$

$$
= -\sigma(S, t)\sigma^{(1)}(S, t)S^2\frac{\partial^2 V}{\partial S^2} \tag{5.101}
$$

with $v(S, T) = 0$ and $v(S, t) \to 0$ as $S \to \infty$. In the case of constant volatility σ, the family σ_λ is defined as $\sigma_\lambda = \sigma + \lambda + \lambda^2 + \ldots$, i.e. $\sigma^{(1)} = 1$. Thus, we obtain the above-mentioned equation for constant volatilities:

$$
\frac{\partial v}{\partial t} + \frac{1}{2}\sigma^2 S^2\frac{\partial^2 v}{\partial S^2} + (r(t) - D(t))S\frac{\partial v}{\partial S} - r(t)v = -\sigma S^2\Gamma \tag{5.102}
$$

- $p_4 = r$, giving $\rho \equiv \frac{\partial V}{\partial r}$:

$$
\frac{\partial}{\partial r}\left[\frac{\partial V}{\partial t} + \frac{1}{2}\sigma^2 S^2\frac{\partial^2 V}{\partial S^2} + (r - D)S\frac{\partial V}{\partial S} - rV\right] \tag{5.103}
$$

$$
= \underbrace{\frac{\partial \rho}{\partial t} + \frac{1}{2}\sigma^2 S^2\frac{\partial^2 \rho}{\partial S^2} + (r - D)S\frac{\partial \rho}{\partial S} - r\rho}_{\text{Black–Scholes operator acting on } \rho} + S\frac{\partial V}{\partial S} - V = 0 \tag{5.104}
$$

Since usually $\Delta = \frac{\partial V}{\partial S}$ is not known, both quantities, ρ and Δ, have to be computed. This can be done either in steps by first calculating Δ and inserting the results into the second step computing ρ. Alternatively, this problem can be considered as a system of PDEs. However, this is not a 'true' system, in the sense that the equations are coupled.

For time-dependent interest rates $r(t)$ the equation changes. By analogy with the mentioned computation of Vega, we define a one-parameter family $r_\lambda(t)$ which satisfies $r_0(t) = r(t)$. We assume that the solutions of the Black–Scholes model with interest rate $r_\lambda(t)$ provide a one-parameter family of prices V_λ with $V_0(S, t) = V(S, t)$, i.e.

$$
\frac{\partial V_\lambda}{\partial t} + \frac{1}{2}\sigma^2 S^2\frac{\partial^2 V_\lambda}{\partial S^2} + (r_\lambda(t) - D(t))S\frac{\partial V_\lambda}{\partial S} - r_\lambda(t)V_\lambda = 0 \tag{5.105}
$$

Then, ρ is defined as

$$
\rho = \rho(t) = \frac{dV_\lambda(S, t)}{d\lambda}\bigg|_{\lambda=0} \tag{5.106}
$$

We also assume that the family $r_\lambda(t)$ can be represented by

$$
r_\lambda(t) = \sum_{i=0}^{\infty} \lambda^i r^{(i)}(t) \tag{5.107}
$$

with $r^{(0)}(t) = r(t)$. Differentiation of the above equation with respect to λ yields

$$\frac{\partial}{\partial t}\frac{dV_\lambda}{d\lambda}\bigg|_{\lambda=0} + \frac{1}{2}\sigma^2 S^2 \frac{\partial^2}{\partial S^2}\frac{dV_\lambda}{d\lambda}\bigg|_{\lambda=0} + \frac{\partial r_\lambda(t)}{\partial \lambda} S \frac{\partial V_\lambda}{\partial S}\bigg|_{\lambda=0}$$

$$+ (r_\lambda(t) - D(t))S \frac{\partial}{\partial S}\frac{dV_\lambda}{d\lambda}\bigg|_{\lambda=0} - \frac{\partial r_\lambda(t)}{\partial \lambda} V_\lambda\bigg|_{\lambda=0} - r_\lambda(t)\frac{dV_\lambda}{d\lambda}\bigg|_{\lambda=0} = 0 \quad (5.108)$$

Because $\dfrac{\partial r_\lambda(t)}{\partial \lambda}\bigg|_{\lambda=0} = r^{(1)}(t)$, we obtain

$$\frac{\partial \rho}{\partial t} + \frac{1}{2}\sigma^2 S^2 \frac{\partial^2 \rho}{\partial S^2} + (r(t) - D(t))S \frac{\partial \rho}{\partial S} - r(t)\rho = r^{(1)}(t)(V - S\Delta) \quad (5.109)$$

In the case of $r(t) = r = $ constant, it follows that $r^{(1)} = 1$ and we obtain the above-mentioned equation

$$\frac{\partial \rho}{\partial t} + \frac{1}{2}\sigma^2 S^2 \frac{\partial^2 \rho}{\partial S^2} + (r - D(t))S \frac{\partial \rho}{\partial S} - r\rho = V - S\Delta \quad (5.110)$$

- $p_5 = D$, giving $\rho_D \equiv \frac{\partial V}{\partial D}$:

$$\frac{\partial}{\partial D}\left[\frac{\partial V}{\partial t} + \frac{1}{2}\sigma^2 S^2 \frac{\partial^2 V}{\partial S^2} + (r - D)S \frac{\partial V}{\partial S} - rV\right] \quad (5.111)$$

$$= \underbrace{\frac{\partial \rho_D}{\partial t} + \frac{1}{2}\sigma^2 S^2 \frac{\partial^2 \rho_D}{\partial S^2} + (r - D)S \frac{\partial \rho_D}{\partial S} - r\rho_D}_{\text{Black–Scholes operator acting on } \rho_D} - S \frac{\partial V}{\partial S} = 0 \quad (5.112)$$

Similar remarks as for $p_3 = r$ apply here. By analogy with the above-mentioned computations on stochastic volatility and time-depending interest rate, we define a one-parameter family $D_\lambda(t)$ with $D_0(t) = D(t)$ and $D_\lambda(t) = \sum_{i=0}^{\infty} \lambda^i D^{(i)}(t)$. $V_\lambda(S, t)$ denotes the corresponding family of prices to $D_\lambda(t)$ with $V_0(S, t) = V(S, t)$ which satisfies

$$\frac{\partial V_\lambda}{\partial t} + \frac{1}{2}\sigma^2 S^2 \frac{\partial^2 V_\lambda}{\partial S^2} + (r(t) - D_\lambda(t))S \frac{\partial V_\lambda}{\partial S} - r(t)V_\lambda = 0 \quad (5.113)$$

Define

$$\rho_D = \rho_D(t) = \frac{dV_\lambda}{d\lambda}\bigg|_{\lambda=0} \quad (5.114)$$

Differentiation of the partial differential equation with respect to λ yields

$$\frac{\partial}{\partial t}\frac{dV_\lambda}{d\lambda}\bigg|_{\lambda=0} + \frac{1}{2}\sigma^2 S^2 \frac{\partial^2}{\partial S^2}\frac{dV_\lambda}{d\lambda}\bigg|_{\lambda=0} - \frac{\partial D_\lambda(t)}{\partial \lambda} S \frac{\partial V_\lambda}{\partial S}\bigg|_{\lambda=0}$$

$$+ (r(t) - D_\lambda(t))S \frac{\partial}{\partial S}\frac{dV_\lambda}{d\lambda}\bigg|_{\lambda=0} - r(t)\frac{dV_\lambda}{d\lambda}\bigg|_{\lambda=0} = 0 \quad (5.115)$$

and we obtain

$$\frac{\partial \rho_D}{\partial t} + \frac{1}{2}\sigma^2 S^2 \frac{\partial^2 \rho_D}{\partial S^2} + (r(t) - D(t))S \frac{\partial \rho_D}{\partial S} - r(t)\rho_D = D^{(1)}(t)S\Delta \quad (5.116)$$

In the case of $D(t) = D = $ constant, $D^{(1)} = 1$ follows and therefore

$$\frac{\partial \rho_D}{\partial t} + \frac{1}{2}\sigma^2 S^2 \frac{\partial^2 \rho_D}{\partial \rho^2} + (r(t) - D)S\frac{\partial \rho_D}{\partial S} - r(t)\rho_D = S\Delta \qquad (5.117)$$

- $p_6 = t$, giving $\Theta \equiv \frac{\partial V}{\partial t}$. For constant σ, D and r, differentiating the Black–Scholes equation with respect to t gives:

$$\frac{\partial}{\partial t}\left[\frac{\partial V}{\partial t} + \frac{1}{2}\sigma^2 S^2 \frac{\partial^2 V}{\partial S^2} + (r - D)S\frac{\partial V}{\partial S} - rV\right] \qquad (5.118)$$

$$= \frac{\partial \Theta}{\partial t} + \frac{1}{2}\sigma^2 S^2 \frac{\partial^2 \Theta}{\partial S^2} + (r - D)S\frac{\partial \Theta}{\partial S} - r\Theta = 0 \qquad (5.119)$$

Black–Scholes operator acting on Θ

For $\sigma = \sigma(S, t)$, $D = D(t)$ and $r = r(t)$:

$$\frac{\partial}{\partial t}\left[\frac{\partial V}{\partial t} + \frac{1}{2}\sigma^2(S, t)S^2 \frac{\partial^2 V}{\partial S^2} + (r(t) - D(t))S\frac{\partial V}{\partial S} - r(t)V\right]$$

$$= \frac{\partial \Theta}{\partial t} + \sigma(S, t)\frac{\partial \sigma(S, t)}{\partial t}S^2 \frac{\partial^2 V}{\partial S^2} + \frac{1}{2}\sigma^2(S, t)S^2 \frac{\partial}{\partial t}\frac{\partial^2 V}{\partial S^2}$$

$$+ (r'(t) - D'(t))S\frac{\partial V}{\partial S} + (r(t) - D(t))S\frac{\partial}{\partial t}\frac{\partial V}{\partial S} - r'(t)V - r(t)\frac{\partial V}{\partial t}$$

$$= \frac{\partial \Theta}{\partial t} + \frac{1}{2}\sigma^2(S, t)S^2 \frac{\partial^2 \Theta}{\partial S^2} + (r(t) - D(t))S\frac{\partial \Theta}{\partial S} - r(t)\Theta$$

Black–Scholes operator acting on Θ

$$+ \sigma(S, t)\frac{\partial \sigma(S, t)}{\partial t}S^2\Gamma + (r'(t) - D'(t))S\Delta - r'(t)V = 0 \qquad (5.120)$$

In the case of $\sigma^2(S, t) = \sigma^2 = $ constant, it reduces to

$$\frac{\partial \Theta}{\partial t} + \frac{1}{2}\sigma^2 S^2 \frac{\partial^2 \Theta}{\partial S^2} + (r(t) - D(t))S\frac{\partial \Theta}{\partial S} - r(t)\Theta$$

$$+ (r'(t) - D'(t))S\Delta - r'(t)V = 0 \qquad (5.121)$$

This approach of computing the hedge parameters can be generalized to more general pricing equations depending on multiple assets (Randall and Tavella, 2000, p. 58f).

5.2.3   Various exotic options

Barrier options

Barrier options are examples of *path dependency*. This means that the value of the option is dependent on the history of the underlying asset. In the PDE setting, barrier options are even simpler than plain vanilla options because we have more information about the boundaries. Path dependency does not necessarily imply a higher degree of difficulty. Barrier options can be subdivided into *knock-in barrier* options and *knock-out barrier* options. A knock-in barrier option stays worthless until the knock-in barrier is touched. Depending on whether, at start of the contract, the current quotation is below or above the barrier, we distinguish between up-and-in and down-and-in barrier options. Following this line of argument, we also have

Table 5.14 Data for the double barrier option

Parameter	Value
Strike price	100
Down-and-out barrier	75
Up-and-out barrier	130
Rebates	none
Interest rate	0.1
Volatility	0.2
Maturity	1 year

up-and-out and down-and-out options. Additional numerical difficulties are incurred by *reverse barrier options*, i.e. for instance a barrier call with a strike below the knock-out barrier. This option has the nasty feature of possibly becoming worthless while being in the money. The numerical difficulty results from the jump at the corner of the time–space domain where the final condition and the knock-out boundary condition meet. Although the problem is still well-posed, this discontinuity may adversely affect numerical schemes. When there are two barriers of any kind linked to an option, this instrument is called a *double barrier option*. Most double barrier options are of the down-and-out-up-and-out type. Since a plain vanilla call can be decomposed into an up-and-out call and an up-and-in call with the same barrier, it is sufficient to treat the up-and-out case. All other cases can be derived with similar relationships. We have chosen the up-and-out and down-and-out cases since these are the easiest to model in a PDE framework. We consider an up-and-out-down-and-out call option continuously monitored, with the data given in Table 5.14. This leads to the following well-posed backward parabolic PDE problem:

$$\frac{\partial V}{\partial t} + rS\frac{\partial V}{\partial S} + \frac{1}{2}\sigma^2 S^2 \frac{\partial^2 V}{\partial S^2} = rV \tag{5.122}$$

$$V(T, S) = \max(S - E, 0) \tag{5.123}$$

$$V\left(t, S_{\mathrm{u}}^{\mathrm{out}}\right) = 0 \tag{5.124}$$

$$V\left(t, S_{\mathrm{d}}^{\mathrm{out}}\right) = 0 \tag{5.125}$$

The pricing PDE, Equation (5.122), is the Black–Scholes equation (Black and scholes, 1973). Equation (5.123) constitutes the payoff at maturity. Equations (5.124) and (5.125) are the knock-out barriers. The results of Table 5.15 were computed with a Galerkin semidiscretization with quadratic shape functions, as explained in Section 5.1.1.

Sometimes, when an option knocks out a rebate is paid, either to soften the frustration of the holder of the option or for tax purposes. The knock-out BC has to be changed in such a way to represent the rebate $R \geq 0$:

$$V(t, S^{\mathrm{out}}) = R \tag{5.126}$$

When there is no rebate, then $R = 0$ as in Equations (5.124) and (5.125), which represent knock-out BCs without any rebates. A formula which also takes (constant) rebates into account, has been given by Pelsser (2000); some numerical computations from this source are reproduced here by using FE. The data and results are given in Tables 5.16 and 5.17. The FD scheme by Pelsser (2000) is of the Crank–Nicholson type with a mesh of 1000 steps each in time

Table 5.15　Results for the double barrier option

		Fair value				
		Numerical				
		11 elements		44 elements		88 elements
Underlying	Analytical		Error (%)		Error (%)		Error (%)
76	0.27306	0.27376	−0.256	0.27317	−0.040	0.27317	−0.040
80	1.22027	1.22357	−0.270	1.22092	−0.053	1.22087	−0.049
90	2.90287	2.90875	−0.203	2.90378	−0.031	2.90378	−0.031
100	3.52511	3.52456	0.016	3.52395	0.033	3.52533	−0.006
110	2.89967	2.89187	0.269	2.89670	0.102	2.89932	0.012
120	1.47489	1.46833	0.445	1.47269	0.149	1.47458	0.021
129	0.13192	0.13137	0.417	0.13181	0.083	0.13192	0.000

Table 5.16　Data for the double barrier option with rebate

Parameter	Value
Asset price	1000
Strike price	1000
Rebates	Intrinsic value of the option
Interest rate	0.05
Time to maturity	0.5

Table 5.17　Results for the double barrier option with rebate

Volatility σ	Upper barrier	Lower barrier	Analytical solution	Numerical solution		
				FD	FE	Time-steps
0.2	1500	500	68.87	68.26	68.8706	202
	1200	800	66.49	66.42	66.4941	202
	1050	950	26.48	26.45	26.4793	218
0.3	1500	500	95.97	95.38	95.9751	202
	1200	800	86.54	86.47	86.5388	205
	1050	950	25.66	25.63	25.6596	228
0.4	1500	500	122.46	121.87	122.4642	202
	1200	800	97.57	97.50	97.5731	208
	1050	950	25.37	25.34	25.3713	238

and space. The FE solution is a collocation semidiscretization with 99 elements using cubic Hermite basis functions and an adaptive implicit Euler time stepping method.

The barriers discussed so far have a constant barrier level, a constant rebate $R \geq 0$ and are monitored continuously. In real-world applications these assumptions are often violated. One extension in the area of barrier options is to vary the location of the boundaries as time passes. First, we consider the (more common) case of barriers moving in steps. The difference with a time-dependent rebate is that the moving barrier is a function of the underlying *and* time,

Table 5.18 Number of nodes for double barrier option with jumps in the barriers

Volatility	FE1		FE2	
σ	Nodes	Elements	Nodes	Elements
0.05	244	105	733	340
0.10	244	105	519	268
0.15	122	49	454	203
0.20	85	32	358	159

Table 5.19 Data for a double barrier option with jumps in the barriers

Parameter	Value			
Asset price	100			
Strike price	100			
Maturity (years)	0.5	1.0	1.5	2
Down-and-out barrier	95	90	85	80
Up-and-out barrier	142.5	135	127.5	120
Rebates	none			
Interest rate	0.05			
Volatility	0.05, 0.1, 0.15, 0.2			
Maturity	2 years			

Source: SciComp, 1999

while a moving rebate is *fixed with respect to the underlying* and is a function of time only. We assume that a step in the barrier occurs at t_i. The first time interval ranges from time to maturity $\tau = t_0$ to the first step in either one of the boundaries or both boundaries t_1. The second interval, goes from t_1 to t_2 and so on up to the last interval, which goes from t_{n-1} to t_n. To solve this problem with PDE2D we have to divide it into n subproblems which are solved sequentially. The solution at t_i is passed to the next subproblem as part of the initial condition which is

$$V(t_i, S) = \begin{cases} 0, & S_i^l < S < S_{i+1}^l \\ V(t_i, S), & S_{i+1}^l < S < S_{i+1}^r \\ 0, & S_{i+1}^r < S < S_i^r \end{cases} \qquad (5.127)$$

The time-steps for the FE solutions are not of equal length. Because of the adaptive nature of the algorithm, the step size is small at the beginning of each time-step. The number of nodes in Table 5.18 denotes the number of nodes at the end of the computations. This number varies during the computations. The number of elements is only indicative, since triangular elements and a dummy variable are used. The data for the problem are given in Table 5.19. The numerical results, for both FD and FE, are reported in Table 5.20.

Barriers that move continuously can be considered the limit as the time-steps and jump size decrease to zero. However, this type of barrier is not observed in the market because of the trouble it would cause to observe the barrier. In order to demonstrate the complexity of analytical solutions to option pricing problems, the analytical solution is given below. This

Table 5.20   Results for the double barrier option with jumps in the barriers

Volatility	FD1 Premium	FD2 Premium	FE1 Premium	Δ	Γ	FE2 Premium	Δ	Γ
0.05	9.7228	9.7313	9.7564	0.9139	0.0204	9.7423	0.9247	0.0076
0.10	8.1387	8.1336	8.1533	0.2951	−0.0645	8.1225	0.2977	−0.0427
0.15	3.9968	3.9948	3.9896	0.0110	−0.0265	3.9826	0.0110	−0.0266
0.20	1.3876	1.3874	1.3827	−0.0116	−0.0088	1.3828	−0.0115	−0.0088

solution involves a series which goes from $-\infty$ to ∞ (Haug, 1997, p. 73):

$$C = Se^{D(T-t)} \sum_{n=-\infty}^{\infty} \left\{ \left(\frac{S_u^n}{S_d^n}\right)^{\mu_1} \left(\frac{S_d}{S}\right)^{\mu_2} [N(d_1) - N(d_2)] \right.$$

$$\left. - \left(\frac{S_d^{n+1}}{S_d S^n}\right)^{\mu_3} [N(d_3) - N(d_4)] \right\}$$

$$-Ee^{-r(T-t)} \sum_{n=-\infty}^{\infty} \left\{ \left(\frac{S_u^d}{S_d^n}\right)^{\mu_1-2} \left(\frac{S_d}{S}\right)^{\mu_2} \right.$$

$$\left[N(d_1 - \sigma\sqrt{T-t}) - N(d_2 - \sigma\sqrt{T-t}) \right]$$

$$\left. - \left(\frac{S_d^{n+1}}{S_u^n S}\right)^{\mu_3-2} \left[N(d_3 - \sigma\sqrt{T-t}) - N(d_4 - \sigma\sqrt{T-t}) \right] \right\} \qquad (5.128)$$

where

$$d_1 = \frac{\ln\left(\frac{SS_u^{2n}}{ES_d^{2n}}\right) + (T-t)\left((r-D) + \frac{\sigma^2}{2}\right)}{\sigma\sqrt{T-t}}$$

$$d_2 = \frac{\ln\left(\frac{SS_u^{2n}}{FS_d^{2n}}\right) + (T-t)\left((r-D) + \frac{\sigma^2}{2}\right)}{\sigma\sqrt{T-t}}$$

$$d_3 = \frac{\ln\left(\frac{S_d^{2n+2}}{ESS_u^{2n}}\right) + (T-t)\left((r-D) + \frac{\sigma^2}{2}\right)}{\sigma\sqrt{T-t}}$$

$$d_4 = \frac{\ln\left(\frac{S_d^{2n+2}}{FSS_u^{2n}}\right) + (T-t)\left((r-D) + \frac{\sigma^2}{2}\right)}{\sigma\sqrt{T-t}}$$

$$\mu_1 = \frac{2[(r-D) - \delta_2 - n(\delta_1 - \delta_2)]}{\sigma^2} + 1$$

$$\mu_2 = \frac{2n(\delta_1 - \delta_2)}{\sigma^2}$$

$$\mu_3 = \frac{2[(r-D) - \delta_2 + n(\delta_1 - \delta_2)]}{\sigma^2} + 1$$

$$F = S_u e^{\delta_1(T-t)}$$

This formula involves two infinite sums and eight indefinite integrals of the cumulative normal distributions $N(\cdot)$. For both infinite sums their behavior with respect to both of their limits ∞ and $-\infty$ has to be examined. Furthermore, the formula remains messy and error-prone. For numerical purposes this series has to be cut off after some finite number of terms. With all these questions in mind, the reader might want to employ a PDE solver instead of implementing this complicated formula. There are (at least) two ways to do this. Either, the problem is transformed into a problem with a fixed geometry, or the continuously moving barriers are discretized so that Equation (5.127) can be employed.

The discussion up to this point has dealt solely with *continuous monitoring*, which means that the price of the underlying is observed all the time. No matter at what time the barrier is breached or touched, this event causes the option to start or to cease to live. This kind of monitoring is typical for the FX market. In the equity market, usually only the (semi)official daily fixing is used to determine whether a knock-in or a knock-out has taken place. This gives an underlying the chance to cross the barrier and move back to the earlier position without triggering the barrier event. A knock-out option with discrete monitoring is worth less than its counterpart with continuous monitoring because the likelihood of being knocked out is smaller. The smaller the number of fixings, the higher the price of the knock-out option and the smaller the price of the knock-in option. A barrier option with no fixings left for the rest of its time to maturity is the same as a plain vanilla option.

For n monitoring instants of the barrier, the analytical solution in the Black–Scholes framework involves the computation of an $n + 1$-dimensional cumulative normal distribution (Gürtler, 1998) which becomes time consuming for growing n. Discrete monitoring does not make the FE solution more expensive: the time axis is subdivided into n intervals, which are not necessarily of equal length: $T < t_1 < \cdots < t_n < t$. To find the price of the option in t, you have to go backwards from the future point at time T. In terms of time to maturity τ you move forward, starting from T, where you apply the payoff of the option as initial condition. You keep on going forward until you reach t_1. The result $V(S, t_1)$ is manipulated to integrate the barrier fixing at t_1:

$$V(S, t_1) \leftarrow \begin{cases} 0 & \text{for} \quad S \leq H \\ V(S, t_1) & \text{for} \quad S > H \end{cases} \qquad (5.129)$$

The calculated $V(S, t_1)$ is then used as initial condition for the next diffusive problem, running until t_2. The process of manipulating the solution at the monitoring instants and feeding them back to the FE program can be automated. Because of the solution losing its continuity every time it is monitored, some extra elements should be added around H. Also, adaptive time stepping is recommended so that right after a fixing, the time-steps become small, to grow larger when the solution has become smooth again. It has to be emphasized that it is not sufficient to put a time-step on every monitoring instant. This only leads to a poor numerical solution of the problem with continuous monitoring. Between each monitoring instant there have to be enough time-steps to catch the diffusive character of the PDE. The process for integrating discrete monitoring of the barrier can easily be extended to problems with a time-dependent barrier, i.e. when the price of the underlying causing the option to knock in or out is a function of τ. Then, each monitoring instant t_i is associated with a barrier H_i:

$$V(S, t_i) \leftarrow \begin{cases} 0 & \text{for} \quad S \leq H_i \\ V(S, t_i) & \text{for} \quad S > H_i \end{cases} \qquad (5.130)$$

Also, rebates can be included in a similar fashion. For the sake of generality, assume rebates and barrier to be functions of time:

$$V(S, t_i) \leftarrow \begin{cases} R_i & \text{for} \quad S \leq H_i \\ V(S, t_i) & \text{for} \quad S > H_i \end{cases} \qquad (5.131)$$

Table 5.21   Barrier options: discrete monitoring

Parameter	Example 1	Example 2	Example 3
Strike price	40	varies	varies
Type of option	Call	Call	Call
Spot price	40	100	100
Barrier level	35	varies	varies
Type of barrier	down-and-out	down-and-out	down-and-out
Interest rate	0.1	0.1	0.1
Volatility	0.4	0.2	0.2
Maturity	1	$\frac{1}{4}$	$\frac{1}{2}$
No. of fixings	1	2	5

Sources: Ahn *et al.* (1999a)[a], Gürtler (1998)

[a]A typo in the original paper has been corrected

Table 5.22   Results for barrier options: discrete monitoring

Parameter		Example 2		Example 3	
Strike	Barrier	Analytical	FE	Analytical	FE
95.0	85.0	8.58	8.5783	11.4628	11.4623
100.0	90.0	5.28	5.2820	8.1199	8.1201
105.0	95.0	2.89	2.8865	5.1930	5.1932

This way of integrating discrete monitoring is the most obvious to implement and probably the most common approach. However, there are several other possibilities to achieve the same in a PDE framework (Tavella, 2001).

To demonstrate the effectiveness of the method we will rework several examples from the literature. The data for these examples are given in Table 5.21. The first example has only one fixing. The analytical value is reported to be 7.7766 (Ahn *et al.*, 1999a); the value computed with FE is 7.7688, which is reasonably close. This result has been achieved with 209 elements (with local refinement around the barrier and the strike) and 1000 adaptive time-steps. The results for the second and third examples are given in Table 5.22.

Another modification of the basic barrier option contract is partial-time barriers. There, the barrier only applies for fractions of the life of the option. A common distinction is between *partial-time-start options*, which have a knock-out barrier at the beginning of the life of the option that is dropped at some prespecified point in time, and *partial-time-end options*, also called *protected barrier options* (Carr, 1995), which cannot be knocked out at the beginning of time to expiration. Also, it is possible to construct options with different periods of the life of option covered with barriers, these two types seem to be the most common. Here we will give an example for a partial-time-start call. The data are shown in Table 5.23.

There is a closed-form solution to this contract (see Table 5.24). Solutions for various other partial barrier options can be found in Haug (1997) and Carr (1995).

The approach outlined above can easily be adapted to deal with time windows, regardless of whether the monitoring of the barriers is discrete or continuous.

Table 5.23 Data for partial-time-start call

Parameter	Value
Strike price	90
Up-and-out barrier	100
Rebate	0
Interest rate	0.1
Dividend yield	0.05
Volatility	0.25
Maturity T_2	1 year
Barrier monitoring T_1	0.5 year

Table 5.24 Results for partial-time-start call

Underlying	Analytical	FE	Δ_{FE}	Γ_{FE}
90	1.77208	1.77512	−0.15983	−0.00936
95	0.89902	0.90033	−0.18377	−0.00160
99	0.17351	0.17379	−0.17583	0.00513

Asian options

Večeř (2001) introduced a model which can be used to price Asian options in the Black–Scholes framework by solving a simple PDE in time and just one spatial variable. Večeř shows that Asian options can be considered to be a special case of options on a trading account, as introduced in Section 10.2.3. The starting point is the PDE for options on a trading account corresponding to PDE (10.64):

$$-V_t = -rV + (r - \gamma)SV_S \tag{5.132}$$
$$+ \max_{\mu \in [a,b]} \left\{ [r_w w + (r - \gamma - \nu)\mu S] V_w + \frac{\sigma^2 S^2}{2} \left[V_{SS} + 2\mu V_{Sw} + \mu^2 V_{ww} \right] \right\}$$

Setting $\gamma = 0$ and $\nu = r_w$ the usual transformation $x = w/S$ leads to:

$$-v_t = \max_{\mu \in [a,b]} \left\{ (\mu - x)(r - r_w)v_x + \frac{\sigma^2}{2}(\mu - x)v_{xx} \right\} \tag{5.133}$$

Večeř (2001) shows the strategies $q = \mu$ that need to be taken in order to price Asian options, with $\bar{S}$ denoting the average of S up to T. The strategies are given in Table 5.25. For the case of discrete fixings, the control q has to be changed. In the case of the fixed strike call this leads to

$$q = 1 - \frac{1}{n} \left[n\frac{t}{T} \right] \tag{5.134}$$

with [·] denoting the integer part function. For the other types of Asian option the changes have to be made accordingly. The relationship between the price of an Asian option V and v is given by:

$$V(0, S_0, w_0) = S_0 v \left(0, \frac{w_0}{S_0} \right) \tag{5.135}$$

with $t = 0$ indicating the date of pricing. The data for the example are given in Table 5.26. It

Table 5.25   Various types of Asian option

Asian option type	Payoff	Stock position q	Initial wealth w_0
Fixed strike call	$(\bar{S} - K)^+$	$1 - t/T$	$S_0 - K$
Fixed strike put	$(K - \bar{S})^+$	$t/T - 1$	$K - S_0$
Floating strike call	$(KS - \bar{S})^+$	$t/T - 1 + K$	$S_0(K - 1)$
Floating strike put	$(\bar{S} - KS)^+$	$-t/T + 1 - K$	$S_0(1 - K)$

Table 5.26   Data for Asian options

Parameter	Symbol	Value
Interest rate	r	0.15
Price of underlying in $t = 0$	S_0	100
Time to maturity	T	1

is not clear how to integrate dividends into this framework. In a later paper, Večeř developed a second approach which cures this problem (Večeř, 2002).

For an Asian call with continuous sampling, Equation (5.133) can be rewritten as:

$$v_t + r(q - x)v_x + \frac{1}{2}(q - x)^2 \sigma^2 v_{xx} = 0 \tag{5.136}$$

with the final condition

$$v(T, x) = x^+ \tag{5.137}$$

The function q depends on the contract at hand. For continuous sampling, q is a continuous function:

$$q = 1 - \frac{t}{T} \tag{5.138}$$

In order to be able to tackle the final value problem, Equations (5.136) and (5.137), numerically, it has to be converted into a final boundary value problem, i.e. the infinite domain has to be cut off by replacing $-\infty$ and ∞ by finite values for z. Also, these points have to be prescribed with values for q or its derivative. Following Večeř (2001) we choose

$$v(x_{\min} = -1) = 0 \tag{5.139}$$

$$\frac{\partial v(x_{\max} = 1)}{\partial x} = 1 \tag{5.140}$$

Additionally, to be comparable to Večeř (2001) and Zvan et al. (1998b) we take the same number of space and time points. The spatial variable is discretized with 200 elements of equal length with cubic Hermite basis functions. Time is integrated by a Crank–Nicholson scheme with 400 steps. The numerical results are given in Table 5.27.

Discrete sampling calls for a different q:

$$q = 1 - \frac{1}{n}\left[n\frac{t}{T} \right] \tag{5.141}$$

Table 5.27 Results for Asian options (1)

Parameter				Method				
σ	K	FE (Coll.)	Difference to MC	FD Večeř (2001)	Zvar et al. (1998b)	Monte Carlo Fu et al. (1999)	Lower bound	Upper bound Rogers and Stapleton (1998)
0.05	95	11.094	0.000 %	11.094	11.094	11.094	11.094	11.114
	100	6.794	−0.015 %	6.795	6.793	6.795	6.794	6.810
	105	2.746	0.036 %	2.744	2.744	2.745	2.744	2.761
0.10	90	15.399	0.000 %	15.399	15.399	15.399	15.399	15.445
	100	7.028	0.000 %	7.029	7.030	7.028	7.028	7.066
	110	1.415	−0.212 %	1.415	1.410	1.418	1.413	1.451
0.20	90	15.642	0.000 %	15.643	15.643	15.642	15.641	15.748
	100	8.409	0.000 %	8.412	8.409	8.409	8.408	8.515
	110	3.556	0.000 %	3.560	3.554	3.556	3.554	3.661
0.30	90	16.513	−0.018 %	16.516	16.514	16.516	16.512	16.732
	100	10.210	0.000 %	10.215	10.210	10.210	10.208	10.429
	110	5.731	0.000 %	5.736	5.729	5.731	5.728	5.948

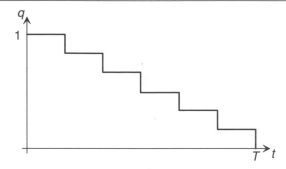

Figure 5.3 Discrete sampling: the control function q

with $[\cdot]$ denoting the integer part function. Note that this formula allows for non equidistant time-steps between the fixings. In terms of Figure 5.3, each time interval is the length of each step, while the weight of the fixing is the drop in q associated with each step. By manipulating the length and/or drop of each step, arbitrary fixing intervals, and also arbitrary weights of the fixings in the final payoff, can be integrated in a simple fashion. The results are shown in Table 5.28.

The details of the numerical implementation are the same as in the previous section except for time integration, which is performed with an adaptive second order Adams–Moulton scheme. This offers the advantage that steps in the coefficients of the PDE, as introduced by Equation (5.141), are detected automatically. After a jump, the step size decreases, as indicated by the time-step history in Figure 5.4. The Monte Carlo results are achieved in a simple fashion, as described by Haug (1997). We use 100 000 sample paths for all examples so that these results are rather crude estimates. Obviously, the computational burden of solving the PDE is much smaller than generating the Monte Carlo paths.

With another example from the literature (Turnbull and Wakemann, 1991), given in Table 5.29, we will show that finite elements can be used to compute Δ *directly* from the numerical solution for the option premium. For the only run necessary for the finite element computation we use 199 elements of equal length; time integration is performed with an adaptive second order Adams–Moulton scheme. For details of this time integration technique, see Sewell (1985), Burden and Faires (1998), the numerical results are given in Table 5.30.

The above approach is much more efficient than earlier work on the pricing of Asian options with the help of PDEs (Bakshi and Madan, 1999; Forsyth *et al.*, 2002; Zhang, 1999). This is mainly based on the use of a transformation reducing the number of spatial variables from two to one. Earlier approaches using this idea are given by Kwok (1998) and Rogers and Shi (1995).

5.2.4   The CEV model

An option pricing model that allows volatility to change with the price level of the underlying has been suggested by Cox and Ross (1976). The dynamics of the underlying are described by:

$$dS = \mu S dt + \sigma S^{\Psi} dX \tag{5.142}$$

Although suggested much earlier, this model can be regarded as a special case of the *local volatility* approach to be discussed briefly in Section 10.2.4. Obviously, this includes the

Table 5.28 Results for Asian options (2)

Parameter		3 fixings			10 fixings			50 fixings		
					Method					
σ	K	FE	MC	Difference	FE	MC	Difference	FE	MC	Difference
0.05	95	13.435	13.450	−0.112 %	11.792	11.780	0.102 %	11.233	11.222	0.098 %
	100	9.134	9.145	−0.120 %	7.492	7.495	−0.040 %	6.934	6.933	0.014 %
	105	4.943	4.937	0.121 %	3.384	3.388	−0.118 %	2.872	2.861	0.383 %
0.10	90	17.742	17.767	−0.141 %	16.097	16.081	0.099 %	15.538	15.538	0.000 %
	100	9.377	9.385	−0.085 %	7.729	7.737	−0.104 %	7.168	7.150	0.251 %
	110	3.026	3.009	0.562 %	1.852	1.853	−0.054 %	1.499	1.487	0.801 %
0.20	90	18.166	18.146	0.110 %	16.393	16.364	0.177 %	15.791	15.748	0.272 %
	100	11.019	10.976	0.390 %	9.189	9.202	−0.141 %	8.564	8.557	0.082 %
	110	5.772	5.762	0.173 %	4.196	4.199	−0.071 %	3.682	3.673	0.244 %
0.30	90	19.429	19.522	−0.479 %	17.383	17.392	−0.052 %	16.687	16.622	0.390 %
	100	13.237	13.325	−0.869 %	11.116	11.081	0.315 %	10.391	10.327	0.616 %
	110	8.512	8.502	0.117 %	6.547	6.540	0.107 %	5.892	5.851	0.696 %

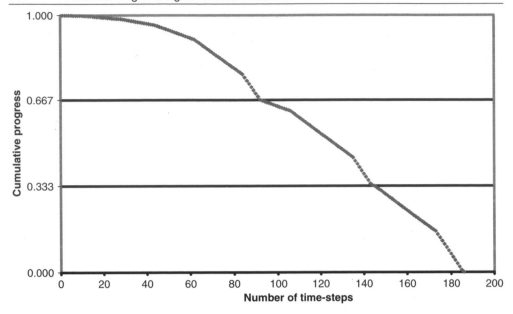

Figure 5.4 Adaptive time stepping (three fixings, $\sigma = 0.2$, $K = 100$)

Table 5.29 Data for Asian options (2)

Parameter	Symbol	Value
Interest rate	r	0.09
Price of underlying in $t = 0$	S_0	100
Time to maturity	T	15 days
Equidistant samplings		14

Black–Scholes model with $\psi = 1$.[3] Further special cases are the *square-root model* with $\psi = 0.5$ and the *absolute diffusion model* with $\psi = 0$. Some authors doubt the existence of solutions for $\psi > 1$ (Jarrow and Rudd, 1983, p. 154), although there are econometric studies suggesting that ψ may be as large as 2.85 for certain underlyings (Gemmill, 1993, p. 234). Other empirical studies find values of ψ to be less than one (Beckers, 1980; Macbeth and Merville, 1980). With $0 < \psi < 1$ the volatility of the underlying $\sigma S^{\psi-1}$ increases as S decreases.[4] With an equity as underlying, the economic rationale behind this process is described nicely by

---

[3] Other common representations of the SDE (5.142) from the literature are:

$$dS = \mu S + \delta S^{\frac{\beta}{2}} \qquad \text{(Zhang, 1998b, p. 64)}$$
$$dS = \mu S + \sigma S^{1-\alpha} \qquad \text{(Hull, 2000, p. 444)}$$

The author recommends that the reader keep this issue in mind because it regularly leads to confusion.

[4] For $\psi > 1$, the volatility increases as the underlying increases.

Table 5.30 Results for Asian options (3)

Strike	Premium				Delta			
	Approximation	MC	FE	Difference	Approximation	MC	FE	Difference
95	5.202	5.194	5.203	−0.019 %	0.979	0.974	0.975	0.409 %
100	1.139	1.135	1.144	−0.439 %	0.536	0.534	0.523	2.425 %
105	0.038	0.038	0.039	−2.632 %	0.037	0.030	0.036	2.703 %

Jarrow and Rudd (1983):

> One justification for this effect results from changes in the firm's financial leverage. As the stock price falls, the market value of the firm's liabilities will also fall because of an increased perception of bankruptcy. Of the two liabilities (debt and equity), the decrease in the market value of the equity will be larger, and this produces an increase in the firm's debt-to-equity ratio. This increase in financial leverage causes an increase in the equity's risk, leading to a rise in stock volatility.

The name of this approach stems from the (economic) definition of elasticity ε:

$$\varepsilon = \frac{dy}{y} : \frac{dx}{x} \equiv \frac{dy}{dx} \cdot \frac{x}{y} \tag{5.143}$$

with x and y being some quantities. Here, elasticity indicates the change in variance for an infinitesimal change in the price of the underlying. Employing the following property

$$\text{Var}\left(\frac{dS}{S}\right) = [\sigma(S,t)]^2 dt \tag{5.144}$$

in Equation (5.142) leads to:

$$\text{Var}\left(\frac{dS}{S}\right) = \sigma^2 S^{2\psi-2} dt \tag{5.145}$$

Applying the definition of elasticity:

$$\varepsilon_{CEV} = \frac{d \, \text{Var}\left(\frac{dS}{S}\right)}{dS} \cdot \frac{S}{\text{Var}\left(\frac{dS}{S}\right)} \tag{5.146}$$

$$= \left[(2\psi - 2)\sigma^2 S^{2\psi-3} dt\right] \cdot \frac{S}{\sigma^2 S^{2\psi-2} dt} \tag{5.147}$$

$$= 2\psi - 2 = \text{constant} \tag{5.148}$$

For $0 < \psi < 1$ the above expression is always less than zero. The same is valid for the expression

$$\frac{d \, \text{Var}\left(\frac{dS}{S}\right)}{dS} = (2\psi - 2)\sigma^2 S^{2\psi-3} dt < 0 \tag{5.149}$$

indicating that an increase in S results in a decrease of volatility. The CEV model can be employed to produce a volatility smile, as we will see below. Practitioners usually prefer the following representation of Equation (5.142) (Lipton, 2001, p. 359):

$$dS = \mu S dt + \sigma(t)^S \left(\frac{S}{S_0}\right)^{-\beta} dX \tag{5.150}$$

S_0 is the current spot price. This representation offers the advantage that σ from Equation (5.150) is usually well approximated by the implied ATM Black–Scholes volatility. The associated pricing PDE reads:

$$\frac{\partial V}{\partial t} + \frac{[\sigma(t)]^2}{2} S^2 \left(\frac{S}{S_0}\right)^{-2\beta} \frac{\partial^2 V}{\partial S^2} + rS \frac{\partial V}{\partial S} - rV = 0 \tag{5.151}$$

As always, FCs and BCs depend on the type of product at hand. Two examples will be discussed below.

For European puts and calls rather complicated formulae have been given by Cox and Ross (1976). These pricing formulae involve the Γ density function and the Γ distribution

Table 5.31 Data for a double barrier call in the CEV model (Linetsky, 1999, p. 88)

Parameter	Symbol	Value
Asset price	S_0	100.0
Interest rate	r	0.1
Volatility	σ	0.25
Lower barrier	S_d	90.0
Upper barrier	S_u	120.0
Strike	E	0.0

Table 5.32 Results for the double barrier call in the CEV model

Method	$\beta = 0$	$\beta = 0.5$	$\beta = 1$	$\beta = 2$	$\beta = 3$	$\beta = 4$
			$T = 0.08333$ (1 month)			
Eigenfunction expansion	3.0154	3.0820	3.1376	3.2171	3.2607	3.2794
Laplace inversion	3.0154	3.0820	3.1376	3.2171	3.2606	3.2793
FE	3.0154	3.0820	3.1377	3.2172	3.2607	3.2794
Time-steps (FE)	191	190	192	176	177	161
			$T = 1$ (1 year)			
Eigenfunction expansion	0.1410	0.1672	0.1994	0.2859	0.4103	0.5822
Laplace inversion	0.1410	0.1673	0.1994	0.2860	0.4104	0.5823
FE	0.1411	0.1673	0.1995	0.2861	0.4105	0.5824
Time-steps (FE)	293	290	278	265	244	243

function; see Section B.1.5. These authors also derived simplified formulae for the square-root and absolute models. The integration of both discrete and continuous dividends is discussed in Jarrow and Rudd (1983). Binomial tree implementations of the process Equation (5.142) are given in Jarrow and Rudd (1983) and Gemmill (1993). These can easily be adapted for American options. An alternative way to integrate early exercise in this setting was suggested by Meyer and van der Hoek (1979) who employed a method of lines. The complicated pricing formulae derived by Cox and Ross (1976) have been simplified somewhat by Schroder (1989) who put his versions in terms of the noncentral χ^2 distribution; see Section B.1.7. A derivation of a representation of these formulae in terms of the noncentral χ^2 distribution can be found in Lipton (2001). This source also provides complicated analytical formulae for various barrier options (Lipton, 2001, Section 12.5.3).

As a numerical example we will reproduce the results for double barrier options computed by Linetsky (2002). Together with the following BCs and FCs and the data from Table 5.31, the PDE (5.151) has to be solved:

$$V(S_d, t) = 0 \tag{5.152}$$

$$V(S_u, t) = 0 \tag{5.153}$$

$$V(S, T) = \max(S - E, 0) \tag{5.154}$$

Since the BCs and the FC are not compatible in (S_u, T), an adaptive version of the backward Euler scheme is employed. With incompatible BCs/FCs there is a sudden jump at $T = 0$ from the FC conditions to satisfying the BC, so one needs a small step size here, but once one moves away from $T = 0$, the solution changes slowly and one does not need a small step size anymore. The results and the number of time-steps are reported in Table 5.32. The time-step

Table 5.33 Data for a plain vanilla call in the CEV model (Gemmill, 1993, Table 13.6)

Parameter	Symbol	Value
Asset price	S_0	40.0
Interest rate	r	0.05
Volatility	σ	0.2
Maturity	T	0.333 year

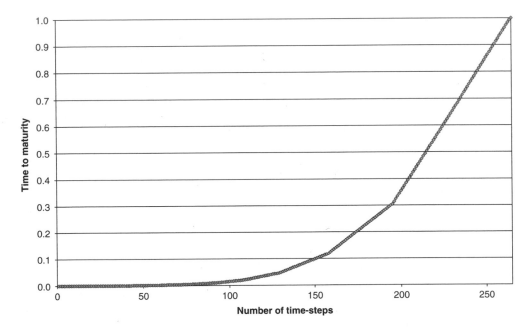

Figure 5.5 Progress of an adaptive backward Euler method ($\beta = 2$)

history is shown in Figure 5.5. The spatial variable S is discretized in 59 elements of equal length. A collocation finite element method with cubic Hermite basis functions is used.

As a second numerical example we will show how a smirk can be produced with the CEV model. We consider a European call. Together with the following BCs and FCs and the data from Table 5.33, the PDE Equation (5.151) has to be solved:

$$\frac{\partial V(S^{\max} = 100.0, t)}{\partial S} = 1 \tag{5.155}$$

$$V(S^{\min} = 0, t) = 0 \tag{5.156}$$

$$V(S, T) = \max(S - E, 0) \tag{5.157}$$

The results in Table 5.34 have been computed with 250 equidistant time-steps using a backward Euler scheme and 99 elements of equal size in the spatial variable S. A collocation finite element method with cubic Hermite basis functions has been used.

Table 5.34 Results for a plain vanilla call in the CEV model

Strike	Black–Scholes $\beta = 0$			Square-root $\beta = 0.25$			Absolute diffusion $\beta = 0.5$		
	Analytic	FE	Implied Volatility	Analytic	FE	Implied Volatility	Analytic	FE	Implied Volatility
35	5.76	5.7724	0.200	5.79	5.7979	0.207	5.81	5.8255	0.214
40	2.17	2.5750	0.200	2.17	2.1752	0.200	2.17	2.1760	0.200
45	0.51	0.5099	0.200	0.47	0.4722	0.194	0.43	0.4365	0.189

5.2.5 Some practicalities: Dividends and settlement

Integrating dividends

Besides volatility, dividends are also often unknown and need to be estimated. For the rest of this section we assume that both the height and date of payments are known. Some hints on how to approach uncertain dividend models are given in Chapter 5 of Bakstein and Wilmott (2002). The modeling of dividend payments can be subdivided into two approaches:

1. Discrete dividends in the form of a cash payment to shareholders.
2. A continuous dividend yield, as exemplified by an option on a foreign currency. The dividend yield is equivalent to the foreign risk free interest rate r_f but much smaller.

For simplicity, we start the exposition with dividend yield. The integration of dividend yield D into the Black–Scholes framework and its extensions is achieved by simply subtracting it from the convection term:

$$\frac{\partial V}{\partial t} + \frac{1}{2}\sigma^2 S^2 \frac{\partial^2 V}{\partial S^2} + (r - D)S \frac{\partial V}{\partial S} - rV = 0 \qquad (5.158)$$

The sign of the dividend yield D is not restricted to being positive. In the case of currency options, i.e. $D = r_f$, the foreign interest rate r_f might be higher or lower than the domestic rate, so that not even the sign of the convection term is known in advance. It might even be a function of time when the two term structures $r(t)$ and $r_f(t)$ cross. Commodity options often have a negative dividend yield resulting from the fact that the storage of commodities incurs a cost of carry q. In the case of options on futures, the convection terms drops out since $D = r$. When the underlying of the option is an index or large basket of equities, the dividend payments resulting from the ingredients of the index or basket are often modeled as a continuous stream of dividends, rather than many discrete payments.

Discrete dividends need a more sophisticated treatment. The methods to integrate discrete dividend payments mainly differ in the way volatility is interpreted. After a dividend payment D has been announced, the underlying S can be separated into a purely stochastic part S^* and a deterministic part $e^{-r(T-t)}$. The full price process for S is usually referred to as the *dirty process*, while the adjusted price process S^* is called the *clean process*. The volatilities associated with these processes are called the *clean volatility* σ^* and the *dirty volatility* σ, respectively. Usually, data providers only compute dirty quotes for both historical and implied volatilities.

When working with a clean process, the standard method to integrate discrete dividends is to compute the NPV of all dividends up to maturity and to subtract this number from the spot price. For n dividend payments between t_0 (today) and T (maturity) this reads:

$$S^*(t_0) = S(t_0) - \sum_{i=1}^{n} D_i \exp\left(-\int_{t_0}^{T} r(s)\mathrm{d}s\right) \qquad (5.159)$$

This approach incurs discontinuities when considering options with different maturities, as we will see shortly. Also, this approach cannot be applied directly to options with early exercise.

Next, we discuss the dirty process. This approach assumes a stochastic process with a drop at each dividend payment. This usually enforces the use of numerical methods. The moment just before the dividend is paid is denoted by t^-, while the instant right after the underlying goes ex dividend is denoted by t^+. Due to there being no arbitrage considerations, across a dividend date the underlying equity drops by the amount of the dividend:

$$S(t_i^+) = S(t_i^-) - D_i \qquad (5.160)$$

Table 5.35   Data for a call with one discrete dividend

Parameter	Symbol	Value
Asset price	S	100
Interest rate	r	0.05
Volatility	σ	0.2
Dividend date	t_1	1 year
Dividend	D_1	10

Source: Berger and Klein, 1998

Table 5.36   Results for a call with one discrete dividend

Maturity	Strike	Premium with volatility on full price (numerical)	Premium with volatility on adjusted price (analytical)
0.92	100	9.9290	9.9305
0.94	100	10.0604	10.0618
0.96	100	10.1908	10.1923
0.98	100	10.3203	10.3219
0.9999	100	10.4485	10.4499
1.0001	90	10.4511	9.7192
1.02	90	10.5592	9.8343
1.04	90	10.6671	9.9492
1.06	90	10.7744	10.0634

Methods based on dirty processes share the common feature that the *continuity condition* is satisfied:

$$\lim_{\epsilon \to 0} V(S, t_i - \epsilon) = \lim_{\epsilon \to 0}(S - D_i, t_i + \epsilon), \quad \epsilon > 0 \qquad (5.161)$$

In short:

$$V(S, t_i^-) = V(S - D_i, t_i^+) \qquad (5.162)$$

Assuming that the dividend payment D_i is independent of S, this shifts V to the left by D_i. Supposing that the dividend payment is proportional to the asset value, $D_i = DS$, Equation (5.160) reduces to:

$$S(t_i^+) = S(t_i^-) - DS(t_i^-) \qquad (5.163)$$
$$= (1 - D)S(t_i^-) \qquad (5.164)$$

This stretches the curve of V by a factor of $\frac{1}{1-D}$. Working with the dirty process avoids the discontinuity with options with different maturities, as the following examples show. Assume a scenario as given by Table 5.35. Writing a call with a strike of 100, expiring shortly before the dividend payment, and buying a call with a strike of 90, maturing just after the dividend payment, constitutes an almost perfect hedge. Consequently, option premiums should be roughly equal. With the time gap vanishing, both premiums converge. This behavior is reflected only when the computations are based on the full price volatility. When working with the volatility on the adjusted price there is a jump in the option premium when it crosses the dividend date, as shown in Table 5.36.

Table 5.37   Data for a call with three discrete dividends

Parameter	Symbol	Value
Asset price	S	100
Strike price	E	100
Interest rate	r	0.05
Volatility	σ	0.2
Maturity	T	3.5 years
1. Dividend date	t_1	1 year
1. Dividend	D_1	5
2. Dividend date	t_2	2 years
2. Dividend	D_2	5
3. Dividend date	t_3	3 years
3. Dividend	D_3	5

Source: Berger and Klein, 1998

Table 5.38   Premiums of a call with three discrete dividends

Strike	Clean process	Dirty process	Absolute difference in premium
100	13.9139	15.0595	1.1456
105	12.1071	13.2626	1.1555
110	10.5073	11.6556	1.1483
115	9.0977	10.2244	1.1267
120	7.8613	8.9544	1.0931
125	6.7810	7.8310	1.0500
130	5.8403	6.8401	0.9998

The difference in value between the clean and the dirty process can be striking, as the next example shows. A numerical value of 20% is used for both σ and σ^*. The results associated with the clean process were computed analytically using Equation (5.159), while the results for the dirty process were computed with an FE routine, as outlined in Section 5.2.1, with 2000 equidistant time-steps and 299 elements of equal length. Data and numerical results for the example are given in Table 5.37 and 5.38.

Sometimes it can happen that the dirty process needs to be the basis for computations, but only clean volatilities are given. This is the case with barrier options, since the monitoring of the barrier is done on the basis of the dirty process. Working with the clean process would lead to boundary conditions for the barrier being a function of time in the spatial variable S^*. The Black–Scholes PDE for the clean process is given by:

$$\frac{\partial V}{\partial t} + [r(t) - q(t)] S^* \frac{\partial V}{\partial S^*} + \frac{(\sigma^* S^*)^2}{2} \frac{\partial^2 V}{\partial (S^*)^2} - r(t)V = 0 \tag{5.165}$$

with $q(t)$ being an additional dividend yield as above. The task is to eliminate S^* from the above PDE based on:

$$S = S^* + \sum_{i, t_i > t}^{n} D_i \exp\left(-\int_t^T r(s)\, ds\right) \tag{5.166}$$

We get:

$$\frac{\partial V(S^*, t)}{\partial S^*} = \frac{\partial S}{\partial S^*} \frac{\partial V(S, t)}{\partial S} = \frac{\partial V(S, t)}{\partial S} \tag{5.167}$$

$$\frac{\partial V(S^*, t)^2}{\partial (S^*)^2} = \frac{\partial}{\partial S^*} \frac{\partial}{\partial S^*} V(S, t) = \frac{\partial^2 V(S, t)}{\partial S^2} \tag{5.168}$$

$$\frac{\partial V(S^*, t)}{\partial t} = \frac{\partial V(S, t)}{\partial t} + \frac{\partial S}{\partial t} \frac{\partial V(S, t)}{\partial S} \tag{5.169}$$

$$\frac{\partial S}{\partial t} = \frac{\partial}{\partial t} \sum_{i, t_i > t}^{n} D_i \exp\left(-\int_t^T r(s)\,ds\right) \tag{5.170}$$

$$= r(t) \underbrace{\sum_{i, t_i > t}^{n} D_i \exp\left(-\int_t^T r(s)\,ds\right)}_{\equiv \tilde{D}} = r(t)\tilde{D} \tag{5.171}$$

Inserting the above expressions into PDE Equation (5.165) gives:

$$\frac{\partial V}{\partial t} + \left([r(t) - q(t)]S + q(t)\tilde{D}(t)\right)\frac{\partial V}{\partial S} + \frac{(\sigma^*)^2}{2}\left(S - \tilde{D}(t)\right)^2 \frac{\partial^2 V}{\partial S^2} - r(t)V = 0 \tag{5.172}$$

Now we have a PDE based on the dirty process depending on the volatility of the clean process. In the case of $q(t) = 0$ (which is the standard case) the dirty volatility can be expressed as:

$$\sigma(t) = \frac{S(t) - \tilde{D}(t)}{S(t)} \sigma^* \tag{5.173}$$

Note that, even for $\sigma^*(t) = $ constant, dirty volatility $\sigma(t)$ is a function of time!

For long-dated options, a situation can occur in which some discrete dividends are known, but for the later future only a dividend yield can be given. In such situations, usually called *hybrid dividends*, both the methods for discrete dividends and dividend yield can be combined. Up to the last discrete dividend payment, the dividend yield is set to zero. So far it has been presupposed that both the date of payment and its height are currently known. Besides this, it has been assumed that the dividend payout occurs on the ex-dividend date. Also, Equation (5.161) is based purely on a simple no-arbitrage argument, while there is some empirical evidence that the drop in the stock price does not exactly equal the dividend payment due to tax reasons. The reader interested in a more sophisticated treatment of dividends is referred to Chapter 5 of Bakstein and Wilmott (2002); Section 4.6 of Overhaus *et al.* (2002) and Haug *et al.* (2003).

Settlement

In the equity and commodity markets, cash settlement is the predominant way of settling derivatives. There are some exceptions to this quite general rule, such as equity options embedded in reverse convertibles, which are a very popular retail product in some European countries such as Italy and Germany. These products are settled physically, i.e. the holder of the option actually receives equities. The different ways of settlement are summarized in Table 5.39.

Usually, some days pass between the fixing of an option and its final settlement. How does this influence the price? In the case of cash settlement, the NPV of the option has to be discounted

Table 5.39   Different ways of settlement

Settlement	$t_{\text{fixing}} = T$	$t_{\text{fixing}} < T$
physical	no	true optionality
	adjustment	false optionality
cash	necessary	discounting

with the forward rate from the fixing date to the settlement date. The details are discussed by Hakala (2002). In the case of physical settlement, the situation is more complicated. We first discuss the case of false optionality, i.e. the amount of the underlying which has to be delivered is fixed at the same time as the price of the underlying itself. The price of a plain vanilla put being fixed at t_{fixing} and being settled at T is given by:

$$P_{\text{pD}} = e^{-r(T-t)} \mathbf{E}_Q \left\{ (E - S_T) \mathbf{1}_{S_{t_{\text{fixing}}} < E} | \mathcal{F}_t \right\} \tag{5.174}$$

$$= e^{-r(T-t)} \mathbf{E}_Q \left\{ (E - \frac{S_{t_{\text{fixing}}}}{e^{-r(T-t_{\text{fixing}})}}) \mathbf{1}_{S_{t_{\text{fixing}}} < E} | \mathcal{F}_t \right\} \tag{5.175}$$

$$= \frac{e^{-r(T-t)}}{e^{-r(T-t_{\text{fixing}})}} \mathbf{E}_Q \left\{ (E e^{-r(T-t_{\text{fixing}})} - S_{t_{\text{fixing}}}) \mathbf{1}_{S_{t_{\text{fixing}}} < E} | \mathcal{F}_t \right\} \tag{5.176}$$

$$= e^{-r(t_{\text{fixing}}-t)} \mathbf{E}_Q \left\{ (E e^{-r(T-t_{\text{fixing}})} - E + E - S_{t_{\text{fixing}}}) \mathbf{1}_{S_{t_{\text{fixing}}} < E} | \mathcal{F}_t \right\} \tag{5.177}$$

$$= e^{-r(t_{\text{fixing}}-t)} \mathbf{E}_Q \left\{ (E - S_{t_{\text{fixing}}}) \mathbf{1}_{S_{t_{\text{fixing}}} < E} | \mathcal{F}_t \right\}$$
$$+ e^{-r(t_{\text{fixing}}-t)} \mathbf{E}_Q \left\{ (E e^{-r(T-t_{\text{fixing}})} - E) \mathbf{1}_{S_{t_{\text{fixing}}} < E} | \mathcal{F}_t \right\} \tag{5.178}$$

$$= \underbrace{e^{-r(t_{\text{fixing}}-t)} \mathbf{E}_Q \left\{ (E - S_{t_{\text{fixing}}})^+ | \mathcal{F}_t \right\}}_{P_{\text{plain vanilla}} = \text{plain vanilla put expiring at } t_{\text{fixing}}}$$

$$+ e^{-r(t_{\text{fixing}}-t)} E \, \mathbf{E}_Q \left\{ (e^{-r(T-t_{\text{fixing}})} - 1) \mathbf{1}_{S_{t_{\text{fixing}}} < E} | \mathcal{F}_t \right\} \tag{5.179}$$

$$= P_{\text{plain vanilla}} - \left(1 - e^{-r(T-t_{\text{fixing}})}\right) \underbrace{E e^{-r(t_{\text{fixing}}-t)} \mathbf{E}_Q \left\{ \mathbf{1}_{S_{t_{\text{fixing}}} < E} | \mathcal{F}_t \right\}}_{P_{\text{cash-or-nothing}} = \text{cash-or-nothing put expiring at } t_{\text{fixing}}} \tag{5.180}$$

The result for a call is analogous:

$$C_{\text{pD}} = e^{-r(T-t)} \mathbf{E}_Q \left\{ (S_T - E) \mathbf{1}_{S_{t_{\text{fixing}}} > E} | \mathcal{F}_t \right\} \tag{5.181}$$

$$= e^{-r(t_{\text{fixing}}-t)} \mathbf{E}_Q \left\{ (E - E e^{-r(T-t_{\text{fixing}})} + S_{t_{\text{fixing}}} - E) \mathbf{1}_{S_{t_{\text{fixing}}} > E} | \mathcal{F}_t \right\} \tag{5.182}$$

$$= \underbrace{e^{-r(t_{\text{fixing}}-t)} \mathbf{E}_Q \left\{ (S_{t_{\text{fixing}}} - E)^+ | \mathcal{F}_t \right\}}_{C_{\text{plain vanilla}} = \text{plain vanilla call expiring at } t_{\text{fixing}}}$$

$$+ \left(1 - e^{-r(T-t_{\text{fixing}})}\right) \underbrace{E e^{-r(t_{\text{fixing}}-t)} \mathbf{E}_Q \left\{ \mathbf{1}_{S_{t_{\text{fixing}}} > E} | \mathcal{F}_t \right\}}_{C_{\text{cash-or-nothing}} = \text{cash-or-nothing call expiring at } t_{\text{fixing}}} \tag{5.183}$$

$$= C_{\text{plain vanilla}} + \left(1 - e^{-r(T-t_{\text{fixing}})}\right) C_{\text{cash-or-nothing}} \tag{5.184}$$

A put with delayed payment and physical delivery is worth less than one with immediate payment. The holder of the put delivers the underlying and receives the money some time later. For this period of time he does not receive any interest. For a call the opposite is true: The owner of the call is delivered the underlying before he has to pay, without being charged any interest payments. Consequently, for $r = 0$ delayed payment has no influence on the price of the option just as in equations (5.180) and (5.184).

Since the above derivation is model independent, it can be applied to any Itô process governing the dynamics of the underlying. In the Black–Scholes framework, it is particularly easy to compute the differences to options with cash settlement and options with $T - t_{\text{fixing}} = 0$. The value of a European put without dividends in the Black–Scholes model is given by:

$$P^{\text{BS}}_{\text{plain vanilla}} = Ee^{-r(t_{\text{fixing}}-t)}N(-d_2) - SN(-d_1) \tag{5.185}$$

$$C^{\text{BS}}_{\text{plain vanilla}} = SN(d_1) - Ee^{-r(t_{\text{fixing}}-t)}N(d_2) \tag{5.186}$$

$$P^{\text{BS}}_{\text{cash-or-nothing}} = Ee^{-r(t_{\text{fixing}}-t)}N(-d_2) \tag{5.187}$$

$$C^{\text{BS}}_{\text{cash-or-nothing}} = Ee^{-r(t_{\text{fixing}}-t)}N(d_2) \tag{5.188}$$

with

$$d_{1,2} = \frac{\ln(S/E) + (r \pm \sigma^2/2)(t_{\text{fixing}} - t)}{\sigma\sqrt{t_{\text{fixing}} - t}} \tag{5.189}$$

Consequently,

$$P^{\text{BS}}_{\text{pD}} = P^{\text{BS}}_{\text{plain vanilla}} - \left(1 - e^{-r(T-t_{\text{fixing}})}\right)P^{\text{BS}}_{\text{cash-or-nothing}} \tag{5.190}$$

$$= Ee^{-r(t_{\text{fixing}}-t)}N(-d_2) - SN(-d_1) - \left(1 - e^{-r(T-t_{\text{fixing}})}\right)Ee^{-r(t_{\text{fixing}}-t)}N(-d_2) \tag{5.191}$$

$$= E\left[e^{-r(t_{\text{fixing}}-t)} - \left(1 - e^{-r(T-t_{\text{fixing}})}\right)e^{-r(t_{\text{fixing}}-t)}\right]N(-d_2) - SN(-d_1) \tag{5.192}$$

$$= Ee^{-r(T-t)}N(-d_2) - SN(-d_1) \tag{5.193}$$

and

$$C^{\text{BS}}_{\text{pD}} = C^{\text{BS}}_{\text{plain vanilla}} + \left(1 - e^{-r(T-t_{\text{fixing}})}\right)C^{\text{BS}}_{\text{cash-or-nothing}} \tag{5.194}$$

$$= SN(d_1) - Ee^{-r(t_{\text{fixing}}-t)}N(d_2) + \left(1 - e^{-r(T-t_{\text{fixing}})}\right)Ee^{-r(t_{\text{fixing}}-t)}N(d_2) \tag{5.195}$$

$$= SN(d_1) - E\left[e^{-r(t_{\text{fixing}}-t)} - \left(1 - e^{-r(T-t_{\text{fixing}})}\right)e^{-r(t_{\text{fixing}}-t)}\right]N(d_2) \tag{5.196}$$

$$= SN(d_1) - Ee^{-r(T-t)}N(d_2) \tag{5.197}$$

Before the above results are generalized, we will rederive Equation (5.197) based on economic reasoning, which might be a little more intuitive. The contract under inspection is characterized by the following mechanics. In the case where the price of the share S is quoted at no less than E at t_{fixing}, the holder of the contingent claim has to deliver E monetary units in cash and receives one share. Otherwise, i.e. if $S < E$, nothing happens. In the limit of $t_{\text{fixing}} \to T$ this contract is equivalent to a plain vanilla call (with physical delivery). For $t_{\text{fixing}} \leq T$ the following decomposition holds:

- V_1: the holder of the option possesses one share.
- V_2: in the case $S < E$ at t_{fixing}, the holder of the option delivers the equity at t_{fixing}. Otherwise, i.e. $S \geq E$, the holder of the option delivers an amount E of cash at t_{fixing}.
- V_3: if $S < E$ at t_{fixing}, the holder of the option receives nothing. Otherwise, if $S \geq E$, the holder of the option delivers an amount $E\left(1 - e^{-r(T-t_{\text{fixing}})}\right)$ of cash at t_{fixing}.

At t_{fixing}, the payoffs of these contracts are given by:

- $V_1(S, t_{\text{fixing}}) = S$.
- $V_2(S, t_{\text{fixing}}) = \max(E - S, 0) - E$.
- $V_3(S, t_{\text{fixing}}) = \begin{cases} 0 & \text{if} \quad S < E \\ E\left(1 - e^{-r(T - t_{\text{fixing}})}\right) & \text{if} \quad S \geq E \end{cases}$.

Consequently, the contract is the same as:

- one share;
- one plain vanilla put option with strike E and expiry t_{fixing} minus a cash amount E due at t_{fixing};
- one cash-or-nothing option with strike E and expiry t_{fixing} on a nominal $E\left(1 - e^{-r(T - t_{\text{fixing}})}\right)$.

The Black–Scholes prices of these contracts are given by:

$$V(S, t) = V_1(S, t) + V_2(S, t) + V_3(S, t) \tag{5.198}$$

$$= S + Ee^{-r(t_{\text{fixing}} - t)}N(-d_2) - SN(-d_1) - Ee^{-r(t_{\text{fixing}} - t)}N(d_2)$$
$$+ E\left(1 - e^{-r(T - t_{\text{fixing}})}\right)e^{-r(t_{\text{fixing}} - t)}N(d_2) \tag{5.199}$$

$$= SN(d_1) - Ee^{-r(T - t)}N(d_2) \tag{5.200}$$

with $d_{1,2}$ as given by Equation (5.189).

The separations given by Equations (5.180) and (5.184) work for all linear payoffs (but not, e.g., for power options) with arbitrary additional conditions. For instance, for a down-and-out call CDO_{pD}:

$$CDO_{pD} = e^{-r(T - t)}\mathbf{E}_Q\left\{(S_T - E)\,\mathbf{1}_{S_{t_{\text{fixing}}} > E, S_{\min} > B}|\mathcal{F}_t\right\} \tag{5.201}$$

$$= \underbrace{e^{-r(t_{\text{fixing}} - t)}\mathbf{E}_Q\left\{(S_{t_{\text{fixing}}} - E)^+\,\mathbf{1}_{S_{\min} > B}|\mathcal{F}_t\right\}}_{=CDO_{\text{plain vanilla}} = \text{down-and-out call expiring at } t_{\text{fixing}}}$$

$$+ \left(1 - e^{-r(T - t_{\text{fixing}})}\right)$$

$$\underbrace{e^{-r(t_{\text{fixing}} - t)}\mathbf{E}_Q\left\{\mathbf{1}_{S_{t_{\text{fixing}}} > E, S_{\min} > B}|\mathcal{F}_t\right\}}_{=CDO_{\text{cash-or-nothing}} = \text{cash-or-nothing down-and-out call expiring at } t_{\text{fixing}}} \tag{5.202}$$

$$= CDO_{\text{plain vanilla}} + \left(1 - e^{-r(T - t_{\text{fixing}})}\right)CDO_{\text{cash-or-nothing}} \tag{5.203}$$

To get an impression of the size of the effect discussed above we slightly modify a previous example (Table 5.1). The data shown in Table 5.40 yield the results in Table 5.41. Some of the results in Table 5.41 are plotted in Figure 5.6.

Next, we discuss the case of true optionality. This is not a very common way of settlement and only occurs as an embedded option in reverse convertibles. Because of this, we will first consider the payoff of a simple reverse convertible paying no coupon. At T, the investor receives E. However, if $S_{t_{\text{fixing}}} < E$, the issuer has the right to deliver S_T instead of E at T. Since, if possible, the issuer will deliver $\min(E, S_T)$, this amounts to the following payoff:

$$A_T = \begin{cases} \min(E, S_T) = E - \max(E - S_T, 0) & \text{if} \quad S_{t_{\text{fixing}}} < E \\ E & \text{otherwise} \end{cases} \tag{5.204}$$

Table 5.40   Data for a put with delivery delay

Parameter	Symbol	Value
Strike price	E	40
Interest rate	r	0.1
Volatility	σ	0.2
Maturity	T	0.5 year
Fixing	t_{fixing}	0.45 year

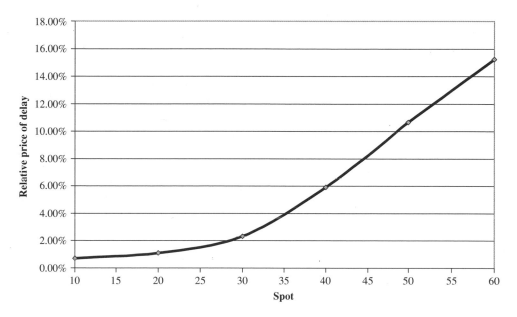

Figure 5.6   Premium of a put with delivery delay

Hence

$$A_t = e^{-r(T-t)} \mathbf{E}_Q \{ A_T \,|\, \mathcal{F}_t \} \tag{5.205}$$

$$= e^{-r(T-t)} E - e^{-r(T-t)} \underbrace{\mathbf{E}_Q \left\{ (E - S_T)^+ \, \mathbf{1}_{S_{t_{\text{fixing}}} < E} \,|\, \mathcal{F}_t \right\}}_{\text{down-and-out put}} \tag{5.206}$$

The down-and-out put has one discrete barrier fixing in t_{fixing}. If the option is below the strike price E it expires worthless. Otherwise, it becomes a plain vanilla put. Since knock-out barrier and strike are the same, and the time interval between t_{fixing} and T is only small, the effect is small. It can be computed with the techniques from Section 5.2.3. The example given in Tables 5.42 and 5.43 gives an impression of the influence of the volatility on this means of settlement for a call.

Some care has to be taken when these models are applied to real-life option pricing problems with physical delivery. Volatilities are usually computed from options with cash settlement and some days between settlement and fixings. For equities, the settlement period is five days in

Table 5.41 Results for the put with delivery delay

Spot	Put expiring at t_{fixing}	Put expiring at T	Discount factor from t_{fixing} to T	Put with cash settlement	Cash-or-nothing put expiring at T	Put with physical delivery	Price of delay	Relative price of delay
10	28.2399	28.0492	0.9950	28.0991	40.0000	28.0404	0.1995	0.71%
20	18.2399	18.0492	0.9950	18.1489	40.0000	18.0404	0.1995	1.09%
30	8.3032	8.1406	0.9950	8.2618	38.7867	8.1098	0.1934	2.33%
40	1.3286	1.3603	0.9950	1.3220	15.7689	1.2500	0.0786	5.92%
50	0.0499	0.0627	0.9950	0.0497	1.0683	0.0446	0.0053	10.68%
60	0.0007	0.0012	0.9950	0.0007	0.0200	0.0006	0.0001	15.25%

Table 5.42   Data for a physical settlement with optionality

Parameter	Symbol	Value
Asset price	S_0	100.0
Interest rate	r	0.1
Down-and-out barrier	S_d	100.0
Strike	E	100.0
Maturity	T	0.5

Table 5.43   Results for the physical settlement with optionality

Volatility	$t_{fixing} = 25/26$	$t_{fixing} = T$
0.05	4.9929	4.9996
0.10	5.8261	5.8503
0.15	6.9749	7.0145
0.20	8.2339	8.2778
0.25	9.5144	9.5822
0.30	10.8250	10.9065

the UK and it is two days in Germany. So, strictly speaking, an option on an equity is an option on a five-day forward or two-day forward, respectively. Usually, the settlement periods for options (with cash settlement) on these equities are set accordingly, so that a possible hedge is perfect.

Static 2D Problems

6.1 INTRODUCTION AND OVERVIEW

We now present the Galerkin method for solving the partial differential equation

$$-\nabla \cdot (\mathbf{D}\nabla u(x, y)) + \mathbf{b} \cdot \nabla u(x, y) + cu(x, y) = f(x, y), \quad (x, y) \in \Omega, \tag{6.1}$$

$$u(x, y) = u_D(x, y), \quad (x, y) \in \Gamma_D, \tag{6.2}$$

$$\mathbf{n} \cdot (\mathbf{D}\nabla u(x, y)) + \gamma u(x, y) = u_N(x, y), \quad (x, y) \in \Gamma_N \tag{6.3}$$

on a two-dimensional domain Ω, where Γ_N and Γ_D form a decomposition of the boundary $\partial\Omega$. The parameters $\mathbf{D}$, $\mathbf{b}$, c and f are functions on the domain. $\mathbf{D}(x, y)$ is a positive-definite matrix, $\mathbf{b}(x, y)$ is a vector and $c(x, y)$ and $f(x, y)$ are scalars. The PDE is given in divergence form (see Definition 9 in Appendix A) so that integration by parts can be performed. Dirichlet conditions are imposed on Γ_D in Equation (6.2), and mixed or Neumann boundary conditions are imposed on Γ_N in Equation (6.3). Here, $\gamma(x, y) \geq 0$ and the unit normal on Γ_N is denoted by $\mathbf{n}(x, y)$. Note that the PDE (6.1) is given in divergence form. Problems in finance are usually not stated in this form, but can be transformed into divergence easily; compare section A.3.6.

In this chapter, the seven-step procedure for solving the above problem introduced in Chapter 4 will be adapted for 2D problems. Since the numerical computations become much more complex, they will be explained with the help of two examples:

Example 9
The first example is the partial differential equation

$$-\frac{1}{2}\left(\frac{\partial^2 u}{\partial x^2} + \frac{\partial^2 u}{\partial y^2}\right) = 1 \tag{6.4}$$

on the domain $\Omega = [1, 2] \times [1, 2]$ with the boundary condition $u(x, y) = 0$ on the entire boundary. This is an instance of Equations (6.1)–(6.3), where

$$\mathbf{D}(x, y) = \frac{1}{2}\begin{pmatrix} 1 & 0 \\ 0 & 1 \end{pmatrix}, \quad f(x, y) = 1 \tag{6.5}$$

and where the parameters $\mathbf{b}$, c and u_D vanish. The Dirichlet boundary Γ_D is the whole boundary here, and the Neumann boundary Γ_N is empty. □

Example 10
The second example is the partial differential equation

$$-\frac{1}{2}\left(\sigma_1^2 x^2 \frac{\partial^2 u}{\partial x^2} + \sigma_2^2 y^2 \frac{\partial^2 u}{\partial y^2}\right) - (r - D_1)x\frac{\partial u}{\partial x} - (r - D_2)y\frac{\partial u}{\partial y} = 1 \tag{6.6}$$

on the domain $\Omega = [10, 30] \times [10, 30]$ with the boundary condition $u(x, y) = 0$ on the entire *boundary. This is an instance of Equations (6.1)–(6.3), where*

$$\mathbf{D}(x, y) = \frac{1}{2} \begin{pmatrix} \sigma_1^2 x^2 & 0 \\ 0 & \sigma_2^2 y^2 \end{pmatrix}, \quad \mathbf{b}(x, y) = - \begin{pmatrix} (r - D_1 - \sigma_1^2)x \\ (r - D_2 - \sigma_2^2)y \end{pmatrix}, \quad f(x, y) = 1$$

and where the parameters c and u_D vanish. Again, the Dirichlet boundary Γ_D is the whole boundary and the Neumann boundary Γ_N is empty.   □

These examples will be solved while working through the seven-step procedure. Because of the complexity of the computations this seems to bear advantages in contrast to first explaining the seven-step procedure and applying it to the examples.

6.2   CONSTRUCTION OF A MESH

For 1D problems, the construction of the mesh reduces to dividing a straight line into nonoverlapping intervals. Segments of the line where high curvature is anticipated are divided into smaller intervals than segments with low curvature. One should avoid having segments with many small elements directly neighboring segments with rather large elements. Areas with a coarse mesh should be refined only gradually to areas with a fine mesh. Before we turn our attention to the practical problems of constructing a mesh, we will define the following expression.

Definition 1 (based on Borouchaki and George, 1998)
T is a mesh *of Ω if:*

1. It holds that: $\Omega = \cup_{e=1}^{n} \Omega_e$
2. Every element Ω_e is nonempty.
3. The intersection of the interiors of any two elements is empty.
4. The intersection of any two elements in T is one of:
   - *the empty set;*
   - *a vertex;*
   - *an edge;*
   - *a face (only for 3D problems).*   □

A mesh is a collection of nonoverlapping finite elements covering the entire domain, with its vertices obeying certain rules. Compared to 1D problems, the task of constructing 2D meshes turns into a problem because of the following facts:

- The domain can be subdivided into triangles, rectangles and more general geometric figures. Usually, either triangles or rectangles are chosen, although it is possible to construct a mesh using different geometric figures. Triangles allow greater geometric flexibility, so triangles are chosen either for the entire domain or the area close to the edge when the domain is rather irregular.
- The domain might become rather complicated. Actually, the ability of FE methods to cope with irregular domains is one of the key factors of their success. In finance, however, currently the domains cannot be considered to be particularly irregular compared to engineering applications. To visualize this compare the domain of a double-barrier rainbow option to a car door. The car door has a much more complicated geometry. It is easy to think of much more complicated geometries from engineering, while the author is unaware of any more

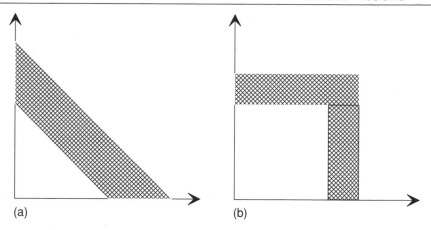

(a)   (b)

Figure 6.1   Knock-out conditions. (a) The spatial domain of an up-and-out-down-and-out rainbow option; (b) an option with barriers applied to each asset individually. Compare this with Pooley *et al.* (2000).

complicated domains arising in finance than an up-and-out-down-and-out multi-asset option. In option pricing, more complicated domains *could* arise when you have barrier features applied to the individual underlyings in certain combinations; see Figure 6.1.

- The problem might exhibit bad scaling. This can happen if the spatial variables have different dimensions. Consider a two-factor model for pricing convertible bonds. The first factor supposedly emulates the interest rate, while the second represents the underlying equity. The interest rate will move somewhere between 0.0 and 0.2 ($= 0$ and 20 %), while the equity might currently trade at \$ 1000. Obviously, problems with much larger differences in the dimensions occur in natural science, and have been solved successfully.
- The construction of a mesh might interfere with the approximation of the unknown boundary conditions. To explain this, we consider a two-asset basket call (with equal weights on the underlyings). When the basket itself is deep in the money, any further growth in the value of the basket leads to an equal growth in the value of the call, so the boundary condition is

$$\frac{\partial V(S_1, S_2)}{\partial n} = 1 \tag{6.7}$$

as also shown in Figure 6.2. This approach requires the computations to be performed on a triangular domain. The alternative is to assume that *each individual asset* being deep in the money moves the price of the option by the same factor

$$\frac{\partial V(S_1, S_2)}{\partial n} = \frac{\partial V(S_1, S_2)}{\partial S_1} = 1 \tag{6.8}$$

and

$$\frac{\partial V(S_1, S_2)}{\partial n} = \frac{\partial V(S_1, S_2)}{\partial S_2} = 1 \tag{6.9}$$

This example will be treated in more detail in Section 7.2.1.

Besides the aforementioned distinction following the geometric form of the individual element, the collection of all elements in a problem is determined to be either a *structured grid* or a

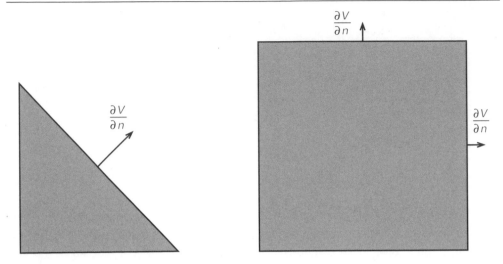

Figure 6.2   Different geometries

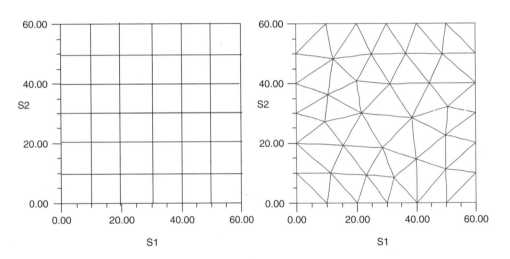

Figure 6.3   Structured vs. unstructured grids

nonstructured grid. Structured grids possess a regular pattern of the nodes, possibly somewhat distorted at the boundaries of the domain. Nonstructured grids do not possess a scheme repeating the location of the nodes. Even a combination of different types of elements is possible. For an example of a structured grid with rectangular elements and a nonstructured grid with triangular elements, see Figure 6.3. In most FE software packages, the generation of the grid is part of a preprocessor. In PDE2D, the preprocessor provides a structured grid for simple geometries in 1D, 2D and 3D. Arbitrary 2D geometries have to be provided with a mesh by the user. We will not dig any deeper into the subject of mesh generation since geometries in financial applications are usually simple. For an introduction to various techniques of grid generation, see Section 4.1 in Angermann and Knabner (2000).

6.3 THE GALERKIN METHOD

6.3.1 The Galerkin method with linear elements (triangles)

1. *Discretization:* This is shown in Figure 6.4 for the nodes in Table 6.1 and the elements in Table 6.2.

   The finite element method provides a convenient way of constructing such shape functions for general domains Ω. The domain Ω is partitioned into nonoverlapping smaller regions $\Omega_1, \ldots, \Omega_n$, called elements. The shape functions are constructed such that they are continuous on the domain and such that they are polynomials when restricted to any element (thus guaranteeing that they are piecewise differentiable). The elements are chosen such that they cover the domain and that their interiors are disjoint, i.e.

$$\Omega = \bigcup_{e=1}^{n} \Omega_e$$

$$\int_{\Omega_e \cap \Omega_f} 1 \, \mathrm{d}x \, \mathrm{d}y = 0, \quad \text{if } e \neq f. \tag{6.10}$$

Finite elements in one dimension were constructed in Section 4.5. Every element was an interval $\Omega_e = [x_e, x_{e-1}]$, where $e = 1, \ldots, n$ denotes the element number, and where $x_0, \ldots, x_n$ denote the node positions.

In two dimensions, the decomposition of the domain into elements is more complex, as there are more ways in which a decomposition of the domain into smaller regions can be constructed. Two examples are shown in Figure 6.5: the left part of the figure shows a decomposition of the region by quadrilateral elements and the right part shows a decomposition of the region by triangular elements.

We will assume here that a decomposition of the domain is given (the process of decomposing a given domain is called mesh generation and is quite a complex topic) and that each element Ω_e is given by a set of n_e nodes $\mathbf{r}_1^e, \ldots, \mathbf{r}_{n_e}^e$, where neighboring elements share some of their nodes. We first consider the case where the elements are polygons. In this case, the nodes are the vertices of the elements (as in Figure 6.5). Then, some of the element nodes will specify the vertices of the element and the other nodes will be nodes on the element boundaries determining the curvature of the boundaries.

Table 6.1 Nodes of exemplary FE discretization

Nodes
$r_1 = (1.0, 1.0)$
$r_2 = (2.0, 1.0)$
$r_3 = (2.0, 2.0)$
$r_4 = (1.0, 2.0)$
$r_5 = (1.5, 1.5)$

Table 6.2   Elements of exemplary FE discretization

Elements	r_1^e r_2^e r_3^e	Boundary elements	r_1^b r_2^b
Ω_1	[1 2 5]	b_1	[1 2]
Ω_2	[2 3 5]	b_2	[2 3]
Ω_3	[3 4 5]	b_3	[3 4]
Ω_4	[4 1 5]	b_4	[4 1]

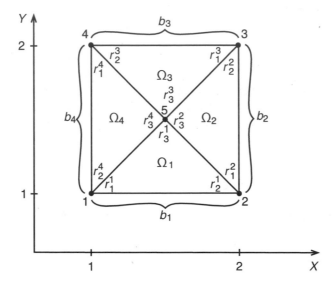

Figure 6.4   Exemplary FE discretization

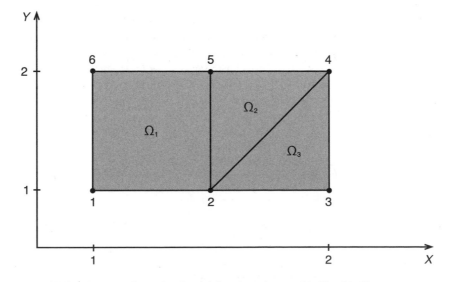

Figure 6.5   Finite element discretization of the domain $\Omega = [1, 2] \times [1, 2]$

Table 6.3 Nodes and elements of FE discretization in Figure 6.5

Nodes		Elements
$r_1 = (1.0, 1.0)$	$r_4 = (2.0, 2.0)$	$\Omega_1 = [1\ 2\ 5\ 6]$
$r_2 = (1.5, 1.0)$	$r_5 = (1.5, 2.0)$	$\Omega_2 = [2\ 4\ 5]$
$r_3 = (2.0, 1.0)$	$r_6 = (1.0, 2.0)$	$\Omega_3 = [2\ 3\ 4]$

Altogether there are m nodes in the finite element mesh, which we will denote by $\mathbf{r}_1 = (x_1, y_1), \ldots, \mathbf{r}_m = (x_m, y_m)$. We will define the element Ω_e, $e = 1, \ldots, n$, by stating the subset of nodes of which it is composed. Hence, the eth element is identified by a list $\mathbf{p}^e = (p_1^e, \ldots, p_{n_e}^e)$, where n_e is the number of nodes in the eth element, and where $p_1^e, \ldots, p_{n_e}^e \in \{1, \ldots, m\}$ denote the indices of the nodes.

Hence, there are two ways of naming the nodes. In total, there are m nodes in the finite element mesh and these are called $\mathbf{r}_1, \ldots, \mathbf{r}_m$. This is the *global* ordering of the nodes. The nodes in the eth element are denoted by $\mathbf{r}_1^e, \ldots, \mathbf{r}_{n_e}^e$. This is the *local* ordering of the nodes (in the eth element). The lists of node indices $\mathbf{p}^e$ define mappings of the local element node orderings onto the global node orderings.

Example 11 (Example 9 revisited)
Figure 6.5 and Tables 6.1 and 6.2 show a decomposition of the domain $\Omega = [1, 2] \times [1, 2]$ into one quadrilateral element Ω_1 and into two triangular elements Ω_2 and Ω_3. Six nodes $\mathbf{r}_1, \ldots, \mathbf{r}_6$ are used to construct the elements and each element Ω_e is specified by stating the list p^e of its node indices ($e = 1, \ldots, 3$). Figure 6.4 and Table 6.3 show a triangulation that will be used for the exemplary calculations. □

2. *Interpolation:* As outlined in Section 4.5, the first step in discretizing a partial differential equation by the Galerkin method is an expansion of the solution in a series of ansatz functions $\phi_1(x, y), \ldots, \phi_{n_{\text{dof}}}(x, y)$ (the so-called 'shape functions' or 'interpolation functions'):

$$\tilde{u}(x, y, \mathbf{a}) = \tilde{u}_D(x, y) + \sum_{i=1}^{n_{\text{dof}}} a_j \phi_j(x, y) \tag{6.11}$$

Here, $a_1, \ldots, a_{n_{\text{dof}}}$ are unknown parameters (the degrees of freedom, abbreviated as dof) that have to be determined, and $\tilde{u}_D(x, y)$ is a function that accounts for the Dirichlet boundary conditions. We will assume here that $\tilde{u}_D(x, y) = u_D(x, y)$ for all $(x, y) \in \Gamma_D$ (in practice, this will only hold approximately).

We will discuss the treatment of Dirichlet conditions in detail in step 5. For now, let us mention that adding $\tilde{u}_D$ to the linear combination of the shape functions $\phi_1, \ldots, \phi_{n_{\text{dof}}}$ leads to the requirement

$$\phi_j(x, y) = 0, \qquad (x, y) \in \Gamma_D \tag{6.12}$$

To obtain the approximation $\tilde{u}(x, y, \mathbf{a})$, we need to obtain and solve n_{dof} equations for the unknowns $a_1, \ldots, a_{n_{\text{dof}}}$, which we collect in a vector $\mathbf{a} = (a_1, \ldots, a_{n_{\text{dof}}})$. We will use the Galerkin method to obtain these equations. The expansion Equation (6.11) is inserted into the PDE (6.1)–(6.3). Upon doing this, a residual

$$R(x, y, \mathbf{a}) = f(x, y) + \nabla \cdot (\mathbf{D} \nabla \tilde{u}(x, y, \mathbf{a})) - \mathbf{b} \cdot \nabla \tilde{u}(x, y, \mathbf{a}) - c\tilde{u}(x, y, \mathbf{a}) \tag{6.13}$$

remains. If $\tilde{u}(x, y, \mathbf{a})$ were the solution of the PDE, the residual would vanish. However, usually the solution cannot be written in the form of Equation (6.11) and it is thus usually not possible to choose the $a_1, \ldots, a_{n_{\text{dof}}}$ such that the residual vanishes.

Example 12 (Examples 9 and 10 revisited)
For the first example, Equation (6.4), the resulting residual is

$$R(x, y, \mathbf{a}) = 1 + \frac{1}{2}\left(\frac{\partial^2 \tilde{u}(x, y, \mathbf{a})}{\partial x^2} + \frac{\partial^2 \tilde{u}(x, y, \mathbf{a})}{\partial y^2}\right)$$

$$= 1 + \frac{1}{2}\Delta\tilde{u}_D(x, y) + \frac{1}{2}\sum_{j=1}^{n_{\text{dof}}} a_j \, \Delta\phi_j(x, y)$$

The residual for the second example, Equation (6.6), is computed in the same manner. $\square$

As already mentioned, Galerkin methods are a special case of weighted residual methods: a set of n_{dof} equations is obtained for the unknowns $\mathbf{a} = (a_1, \ldots, a_{n_{\text{dof}}})$ by demanding that the result of weighting the residual by any of the ansatz functions vanishes. This condition yields the set of equations

$$\int_\Omega \phi_1(x, y)R(x, y, \mathbf{a})\,\mathrm{d}x\,\mathrm{d}y = 0$$
$$\vdots \qquad\qquad\qquad (6.14)$$
$$\int_\Omega \phi_{n_{\text{dof}}}(x, y)R(x, y, \mathbf{a})\,\mathrm{d}x\,\mathrm{d}y = 0$$

As Equations (6.1)–(6.3) contain second order derivatives, the residual Equation (6.13) can only be computed if the shape functions are differentiable twice. However, as we will see, the integral in Equation (6.14) relaxes this requirement. In fact, we will find that we may choose the shape functions such that they are continuous and piecewise differentiable (but not necessary twice differentiable).

Inserting the residual from Equation (6.13) and the ansatz from Equation (6.11) into Equation (6.14) and omitting the dependence on (x, y), we may rewrite any single Equation (6.14) for $i = 1, \ldots, n_{\text{dof}}$ as

$$\sum_{j=1}^{n_{\text{dof}}} \int_\Omega \left(-\phi_i \, \nabla \cdot \left(\mathbf{D}\nabla\phi_j\right) + \phi_i \mathbf{b} \cdot \nabla\phi_j + c\phi_i\phi_j\right)\mathrm{d}x\,\mathrm{d}y \, a_j = \int_\Omega \left(f - R(x, y, 0)\right)\phi_i \, \mathrm{d}x\,\mathrm{d}y$$
$$(6.15)$$

Note that the contribution of $\tilde{u}_D(x, y)$ to the residual is here accounted for on the right-hand side by the term $R(x, y, 0)$. This is the residual obtained for a zero vector of unknowns and is hence the residual obtained by inserting the function $\tilde{u}_D(x, y)$ into the PDE.

Our set of equations given by Equation (6.15) may be written as a single matrix equation: every single Equation (6.15) corresponds to a row of a matrix equation and the sum over j is the result of multiplying a matrix row with the vector $\mathbf{a}$. We may write the system of equations (6.15) as

$$(\mathbf{M} + \mathbf{B} + \mathbf{C})\mathbf{a} = \mathbf{f} \qquad\qquad (6.16)$$

where $\mathbf{M}$, $\mathbf{B}$ and $\mathbf{C}$ are $n_{\text{dof}} \times n_{\text{dof}}$ matrices with components $M_{i,j}$, $B_{i,j}$ and $C_{i,j}$ ($i, j = 1, \ldots, n_{\text{dof}}$) and where $\mathbf{f}$ is a vector with n_{dof} components f_i ($i = 1, \ldots, n_{\text{dof}}$). The terms in Equation (6.16) are defined as

$$M_{i,j} = -\int_\Omega \phi_i(x, y)\nabla \cdot \left(\mathbf{D}(x, y)\nabla\phi_j(x, y)\right) \, dx \, dy$$

$$B_{i,j} = \int_\Omega \phi_i(x, y)\,\mathbf{b}(x, y) \cdot \nabla\phi_j(x, y) \, dx \, dy$$

$$C_{i,j} = \int_\Omega \phi_i(x, y)c(x, y)\phi_j(x, y) \, dx \, dy$$

$$f_i = \int_\Omega \phi_i(x, y)\left(f(x, y) - R(x, y, 0)\right) \, dx \, dy$$

The integrands occurring in the matrix $\mathbf{M}$ involve second order derivatives, but these can be avoided by performing an integration by parts. It follows from Gauss's theorem that

$$M_{i,j} = \underbrace{\int_\Omega \nabla\phi_i \cdot \left(\mathbf{D}\nabla\phi_j\right) \, dx \, dy}_{=:K_{i,j}} - \int_{\partial\Omega} \phi_i\, \mathbf{n} \cdot \left(\mathbf{D}\nabla\phi_j\right)ds \qquad (6.17)$$

where $\partial\Omega$ is the boundary of Ω, $\mathbf{n}$ is the unit normal on $\partial\Omega$ and ds indicates that the integration is performed along the boundary.

The boundary integral in Equation (6.17) can be simplified further. Because of Equation (6.12),

$$\int_{\partial\Omega} \phi_i\, \mathbf{n} \cdot \left(\mathbf{D}\nabla\phi_j\right)ds = \int_{\Gamma_N} \phi_i\, \mathbf{n} \cdot \left(\mathbf{D}\nabla\phi_j\right)ds$$

and due to the boundary conditions on Γ_N in Equations (6.1)–(6.3), this is

$$= \int_{\Gamma_N} \phi_i \left(u_N - \gamma\tilde{u}(x, y, \mathbf{a})\right)ds$$

$$= \underbrace{\int_{\Gamma_N} \phi_i(u_N - \gamma\tilde{u}_D)ds}_{=:g_i} - \sum_{j=1}^{n_{\text{dof}}} a_j \underbrace{\int_{\Gamma_N} \phi_i\gamma\phi_j ds}_{=:L_{i,j}}$$

Altogether, the resulting system of equations may be written as the set of linear equations

$$\mathbf{T}\mathbf{a} = (\mathbf{K} + \mathbf{B} + \mathbf{C} + \mathbf{L})\,\mathbf{a} = \mathbf{f} + \mathbf{g} \qquad (6.18)$$

which has to be solved for the vector of unknowns $\mathbf{a} \in \mathbb{R}^{n_{\text{dof}}}$.

For convenience, the definitions of the individual terms are summarized below ($i, j = 1, \ldots, n_{\text{dof}}$):

$$K_{i,j} = \int_\Omega \nabla\phi_i(x, y) \cdot \mathbf{D}(x, y)\nabla\phi_j(x, y) \, dx \, dy \qquad (6.19)$$

$$B_{i,j} = \int_\Omega \phi_i(x, y)\,\mathbf{b}(x, y) \cdot \nabla\phi_j(x, y) \, dx \, dy \qquad (6.20)$$

$$C_{i,j} = \int_\Omega \phi_i(x, y)\,c(x, y)\,\phi_j(x, y) \, dx \, dy \qquad (6.21)$$

$$L_{i,j} = \int_{\Gamma_N} \phi_i(x, y)\, \gamma(x, y)\, \phi_j(x, y)\, \mathrm{d}s \tag{6.22}$$

$$f_i = \int_{\Omega} \phi_i(x, y)\big(f(x, y) - R(x, y, 0)\big)\, \mathrm{d}x\, \mathrm{d}y \tag{6.23}$$

$$g_i = \int_{\Gamma_N} \phi_i(x, y)(u_N(x, y) - \gamma(x, y)\tilde{u}_D(x, y))\, \mathrm{d}s \tag{6.24}$$

Apparently, the computation of these terms requires that the gradient of the shape functions can be computed. It is not necessary though that the shape functions are differentiable twice.

What regularity conditions should the shape functions satisfy? A discussion of these topics is beyond the scope of this book and the interested reader is referred to Oden and Reddy (1982) and Strang and Fix (1973).

Let us briefly note that the partial differential equation (6.1)–(6.3) is usually not required to hold pointwise for each $(x, y) \in \Omega$. Instead, one requires a partial differential equation to hold in an averaged manner: a (so-called weak) solution of the PDE (6.1)–(6.3) must satisfy

$$\int_{\Omega} \phi\left(-\nabla \cdot (\mathbf{D}\nabla u) + \mathbf{b} \cdot \nabla u + cu\right) \mathrm{d}x\, \mathrm{d}y = \int_{\Omega} \phi f \mathrm{d}x\, \mathrm{d}y \tag{6.25}$$

for every infinitely often differentiable function ϕ vanishing outside a closed disc. By applying Gauss's theorem, as we have done in Equation (6.17), the second order derivatives may be changed into a product of first order derivatives. Hence, weak solutions need not be second order differentiable, but need only have first order differentials.

The shape functions $\phi_1, \ldots, \phi_{n_{\mathrm{dof}}}$ must be chosen such that they satisfy certain regularity conditions (depending on the PDE being solved). For our PDE (6.1)–(6.3), we may choose the shape functions such that they are continuous and piecewise differentiable (shape functions are piecewise differentiable if there is a partition of the domain Ω into a finite number of smaller regions on each of which the shape functions are differentiable). Additionally, it is important that the solution of the PDE can be approximated well by the linear combination of the shape functions in order to obtain small approximation errors.

Our goal is now to construct shape functions for interpolating a given continuous function u on the domain Ω. For every node $\mathbf{r}_1, \ldots, \mathbf{r}_m$ we construct one shape function and in a similar way to before, we represent the interpolant of a given function u on the domain Ω by a function

$$\tilde{u}(x, y, \mathbf{a})(x, y) = \sum_{i=1}^{m} a_i \phi_i(x, y) \tag{6.26}$$

where $a_1, \ldots, a_m$ are real parameters chosen such that a 'good' interpolation of u is obtained and which are collected in the vector $\mathbf{a} = (a_1, \ldots, a_m)$.

The interpolation Equation (6.26) differs from the ansatz Equation (6.11) for the PDE solution in that the function u_D does not appear here. The reason for this is that, in contrast to the ansatz for the PDE, here we consider the interpolation of arbitrary functions on the domain. Therefore, we must not demand a priori that u satisfies the Dirichlet boundary conditions. As Equation (6.26) allows one to interpolate functions with arbitrary values on the boundary, it contains more parameters (we have m parameters $a_1, \ldots, a_m$) than the ansatz Equation (6.11) for the PDE solution (there, the values on the Dirichlet boundary are fixed and there are only $a_1, \ldots, a_{n_{\text{dof}}}$ parameters).

Later, we will use Equation (6.26) as an ansatz for the PDE, but then we will prescribe some of the parameters $a_1, \ldots, a_m$ such that the Dirichlet boundary conditions are satisfied. Assuming that the coefficients $a_{n_{\text{dof}}+1}, \ldots, a_m$ determine the values of $\tilde{u}(x, y, \mathbf{a})$ on the Dirichlet boundary, we will fix $a_{n_{\text{dof}}+1}, \ldots, a_m$ such that $\tilde{u}(x, y, \mathbf{a})$ matches the Dirichlet boundary conditions as well as possible. Then we will only need to compute the remaining coefficients $a_1, \ldots, a_{n_{\text{dof}}}$. Hence, we will see Equation (6.11) as a special case of Equation (6.26). We will discuss this in more detail in step 5.

The shape functions are chosen such that ϕ_i is nonzero only on the elements that contain the node $\mathbf{r}^i$. Hence, the restriction of a shape function ϕ_i to an element Ω_e is nonzero only if the node index i is contained in the list of node indices $\mathbf{p}^e$. Further, we will consider here only so-called nodal shape functions. These vanish on all nodes except one and satisfy

$$\phi_j(\mathbf{r}_i) = \phi_j(x_i, y_i) = \begin{cases} 1, & \text{if } i = j \\ 0, & \text{if } i \neq j \end{cases} \qquad (6.27)$$

Because of Equation (6.27), $\tilde{u}(x_i, y_i, \mathbf{a}) = a_i$ for $i = 1, \ldots, m$. The parameters a_i are therefore the values of the interpolant at the nodal positions, and a given function $u(x, y)$ on the domain Ω is interpolated by $\tilde{u}(x, y, \mathbf{a})$ if $a_i = u(x_i, y_i), i = 1, \ldots, m$.

For the one-dimensional case, piecewise linear shape functions were defined in Equation (4.151) as piecewise linear hat functions on the elements. In two dimensions, we proceed similarly and use continuous shape functions that satisfy Equation (6.27) and are piecewise linear or bilinear hat functions on the elements. Figure 6.6 (a) shows an example of a piecewise bilinear polynomial on quadrilateral elements and Figure 6.6 (b) shows an example of a piecewise linear polynomial on triangular elements.

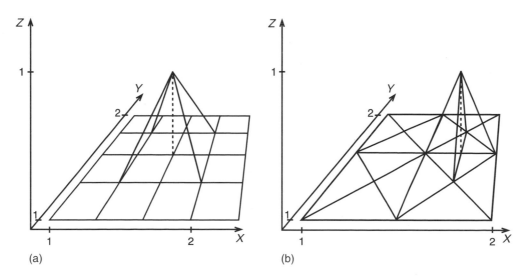

Figure 6.6   Shape functions in two dimensions

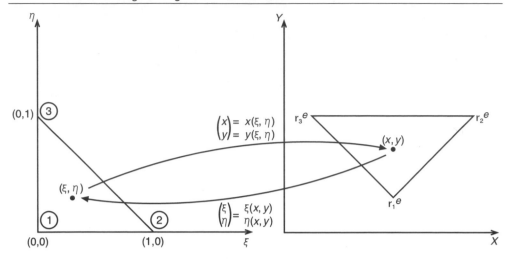

Figure 6.7 Construction of a triangular element with linear shape functions

We construct the shape functions by specifying their restrictions to the individual elements: a local coordinate system is set up in each element and the definitions of the shape functions and all further computations are performed in these local element coordinate systems.

Figure 6.7 shows the construction of a triangular element Ω_e at the points $\mathbf{r}_1^e$, $\mathbf{r}_2^e$, $\mathbf{r}_3^e$. Every position $\mathbf{r}^e(x, y)$ in the element Ω_e is mapped to a coordinate (ξ, η) in the reference element with the vertices $(0, 0)$, $(1, 0)$ and $(0, 1)$. Obviously, a linear map from the reference element to the element Ω_e is given by

$$\mathbf{r}^e(\xi, \eta) = (1 - \xi - \eta)\mathbf{r}_1^e + \xi\mathbf{r}_2^e + \eta\mathbf{r}_3^e \qquad (6.28)$$

$$\Leftrightarrow \begin{pmatrix} x^e(\xi, \eta) \\ y^e(\xi, \eta) \end{pmatrix} = \begin{pmatrix} x_1^e \\ y_1^e \end{pmatrix} + \begin{pmatrix} x_2^e - x_1^e & x_3^e - x_1^e \\ y_2^e - y_1^e & y_3^e - y_1^e \end{pmatrix} \begin{pmatrix} \xi \\ \eta \end{pmatrix} \qquad (6.29)$$

We may write this as

$$\mathbf{r}^e(\xi, \eta) = \sum_{i=1}^{3} N_i(\xi, \eta)\mathbf{r}_i^e \qquad (6.30)$$

where the local shape functions N_1, N_2, N_3 are the linear polynomials

$$\begin{aligned} N_1(\xi, \eta) &= 1 - \xi - \eta \\ N_2(\xi, \eta) &= \xi \\ N_3(\xi, \eta) &= \eta \end{aligned} \qquad (6.31)$$

These local shape functions are used to define the element shape functions. The element shape functions have the same shape as the local shape functions, but they are not defined on the reference element, rather on the actual element Ω_e as

$$N_i^e(x, y) := N_i(\xi(x, y), \eta(x, y)), \qquad i = 1, \ldots, n_e \qquad (6.32)$$

where $\xi(x, y)$ and $\eta(x, y)$ denote the inverse mapping to Equation (6.28). The thus-defined element shape functions $N_1^e(x, y), \ldots, N_{n_e}^e(x, y)$ on a triangular element Ω_e are shown in Figure 6.8.

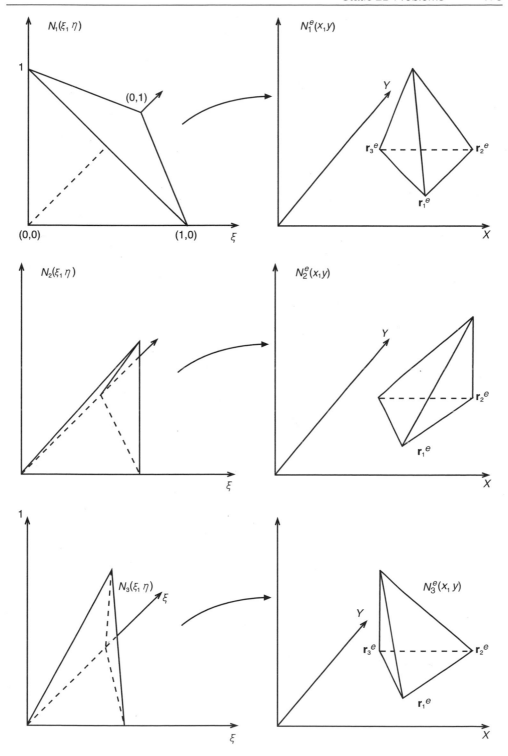

Figure 6.8 Construction of the element shape functions $N_1^e(x, y)$, $N_2^e(x, y)$ and $N_3^e(x, y)$ from the local element shape functions $N_1(\xi, \eta)$, $N_2(\xi, \eta)$ and $N_3(\xi, \eta)$

The global shape functions $\phi_1(x, y), \ldots, \phi_m(x, y)$ are now constructed using the element shape functions. If the node $\mathbf{r}_i$ is a vertex of the element Ω_e and if its local node number in the element is j (i.e. if $p_j^e = i$), then the restriction of the shape function to the eth element is given by N_j^e, i.e. $\phi_i(x, y) := N_j^e(x, y)$ for all $(x, y) \in \Omega_e$.

We will not need to evaluate the global shape functions $\phi_1, \ldots, \phi_m$ for a given (x, y) when using the finite element method. Hence, we will not require the explicit definition of the inverse maps $\xi(x, y)$ and $\eta(x, y)$ when implementing the FE method. However, if the approximation $\tilde{u}(x, y, \mathbf{a})$ is to be evaluated at a given position (x, y), then we need to evaluate the shape functions and then we need the inverse maps. Obviously, the inverse to Equation (6.29), yielding the local coordinates for a point $(x, y) \in \Omega_e$, is

$$\begin{pmatrix} \xi \\ \eta \end{pmatrix} = \begin{pmatrix} x_2 - x_1 & x_3 - x_1 \\ y_2 - y_1 & y_3 - y_1 \end{pmatrix}^{-1} \begin{pmatrix} x - x_1 \\ y - y_1 \end{pmatrix} \tag{6.33}$$

Hence, the shape function $\phi_i(x, y)$ may be evaluated by the following algorithm:

(i) Find the index e of an element Ω_e with $(x, y) \in \Omega_e$.

(ii) If the node $\mathbf{r}_i$ is not a node of the element Ω_e, i.e. if $\mathbf{p}^e$ does not contain the node index i, then $\phi_i(x, y) = 0$.

(iii) If the node $\mathbf{r}_i$ is a node of the element Ω_e, then find out its local element number, i.e. find the index l with $p_l^e = i$. Evaluate $\phi_i(x, y) = N_l(\xi(x, y), \eta(x, y))$.

The construction of the shape functions for the quadrilateral element is analogous to the construction for the triangular elements. Figure 6.11 in Section 6.3.2 shows the construction of the quadrilateral element from its reference element. The quadrilateral reference element $[-1, 1] \times [-1, 1]$ is mapped to the actual element, just as in Equation (6.30), but now four local bilinear element shape functions

$$N_1(\xi, \eta) = \frac{1}{4}(1 - \xi)(1 - \eta) \tag{6.34}$$

$$N_2(\xi, \eta) = \frac{1}{4}(1 + \xi)(1 - \eta) \tag{6.35}$$

$$N_3(\xi, \eta) = \frac{1}{4}(1 + \xi)(1 + \eta) \tag{6.36}$$

$$N_4(\xi, \eta) = \frac{1}{4}(1 - \xi)(1 + \eta) \tag{6.37}$$

are used.

3. *Elemental formulation:* The previous section was concerned with interpolation and with the construction of shape functions. We will now use the finite elements constructed there to discretize the PDE (6.1)–(6.3). For this, we need to compute the terms in the linear system (6.18), i.e. we need to evaluate Equations (6.19)–(6.24) for the shape functions $\phi_1, \ldots, \phi_m$ defined in the previous section. For now, we will ignore the restrictions arising from the Dirichlet boundary conditions and we will compute the terms in Equations (6.19)–(6.24) for all shape functions defined in the previous section. In step 5 we will see how to incorporate the Dirichlet conditions into the system of equations computed here.

All computations performed here will be restricted to the individual elements. This way, working only with the restrictions of the shape functions to the individual elements, is a chief advantage of the finite element method. All computations on the domain are broken up into simpler computations on the individual elements and later, the contributions of the elements are assembled to perform computations on the whole domain (this will be done in step 4).

We will first present the formulation for triangular elements with linear shape functions. The formulation for quadrilateral elements with bilinear shape functions will be shown, in an analogous manner, in Section 6.3.2.

All expressions in Equations (6.19)–(6.24) involve integrals over the domain or over the boundary. We will first treat the integrals over the domain and then treat the integrals over the boundary.

As the domain is a union of the elements and as the elements are nonoverlapping (see Equation (6.10)), integrals over the domain can be computed by adding up the integrals over the individual elements.

Let us demonstrate this for the matrix $\mathbf{C}$ defined in Equation (6.21):

$$C_{i,j} = \int_{\Omega} \phi_i(x, y)c(x, y)\phi_j(x, y) \, dx \, dy \tag{6.38}$$

$$= \sum_{e=1}^{n} \underbrace{\int_{\Omega_e} \phi_i(x, y)c(x, y)\phi_j(x, y) \, dx \, dy}_{=:C_{i,j}^e} \tag{6.39}$$

$$= \sum_{e=1}^{n} C_{i,j}^e \tag{6.40}$$

The matrices $\mathbf{K}$ and $\mathbf{B}$ and the vector $\mathbf{f}$ in Equations (6.19)–(6.24) are also defined as integrals over the domain. Hence, they may also be written as sums of element contributions $\mathbf{K}^e$, $\mathbf{B}^e$ and $\mathbf{f}^e$. The matrices $\mathbf{K}^e$, $\mathbf{B}^e$ and $\mathbf{C}^e$ are called (global) element matrices and the vectors $\mathbf{f}^e$ are called (global) element vectors.

Denoting the contribution of the eth element to the integral by the superscript e, we may write the matrix equation (6.18) as

$$\left[\left(\sum_{e=1}^{n} \mathbf{K}^e + \mathbf{B}^e + \mathbf{C}^e\right) + \mathbf{L}\right]\mathbf{a} = \mathbf{g} + \sum_{e=1}^{n} \mathbf{f}^e \tag{6.41}$$

where $\mathbf{C}^e$ is defined by Equation (6.39) and where the other terms are defined in an analogous manner.

The global element matrices and vectors have sparse nonzero patterns, as any element shape function ϕ_i is nonzero only inside those elements that have the node $\mathbf{r}_i$ as a vertex. As a consequence, the elements of the global element matrices and of the global element vectors are nonzero only in the rows and columns that correspond to a node of the element. For example, $C_{i,j}^e \neq 0$ only if both i and j are contained in the list of element nodes $\mathbf{p}^e$.

Example 13 (Example 11 revisited)

For the element mesh shown in Figure 6.5, the following nonzero pattern results for all global element matrices $\mathbf{K}^e$, $\mathbf{B}^e$ and $\mathbf{C}^e$ (an X stands for a nonzero number):

$$
\mathbf{C}^1 \simeq \begin{pmatrix} X & X & 0 & 0 & X & X \\ X & X & 0 & 0 & X & X \\ 0 & 0 & 0 & 0 & 0 & 0 \\ 0 & 0 & 0 & 0 & 0 & 0 \\ X & X & 0 & 0 & X & X \\ X & X & 0 & 0 & X & X \end{pmatrix} \qquad \mathbf{C}^3 \simeq \begin{pmatrix} 0 & 0 & 0 & 0 & 0 & 0 \\ 0 & X & X & X & 0 & 0 \\ 0 & X & X & X & 0 & 0 \\ 0 & X & X & X & 0 & 0 \\ 0 & 0 & 0 & 0 & 0 & 0 \\ 0 & 0 & 0 & 0 & 0 & 0 \end{pmatrix}
$$

$$
\mathbf{C}^2 \simeq \begin{pmatrix} 0 & 0 & 0 & 0 & 0 & 0 \\ 0 & X & 0 & X & X & 0 \\ 0 & 0 & 0 & 0 & 0 & 0 \\ 0 & X & 0 & X & X & 0 \\ 0 & X & 0 & X & X & 0 \\ 0 & 0 & 0 & 0 & 0 & 0 \end{pmatrix}
$$

□

We can exploit the known zero-pattern of the element matrices in the computation. The global element matrix $\mathbf{C}^e \in \mathbb{R}^{m \times m}$ may be represented by a smaller matrix (the so-called local element matrix) $\mathbf{C}^{\mathrm{loc},e} \in \mathbb{R}^{n_e \times n_e}$ containing only those terms of the global element matrix which correspond to element nodes. The elements of the local element matrix are assembled into the global element matrix by setting $C^e_{p_i,p_j} := C^{\mathrm{loc},e}_{i,j}$ for $i, j = 1, \ldots, n_e$ and by setting the remaining entries of $\mathbf{C}^e$ to zero.

The restriction of each shape function to an element is a local element shape function, as defined in Equation (6.32). Hence, every local element matrix $\mathbf{C}^{\mathrm{loc},e}$ can be computed as

$$
C^{\mathrm{loc},e}_{i,j} = \int_{\Omega_e} N_i(\xi(x, y), \eta(x, y)) c(x, y) N_j(\xi(x, y), \eta(x, y)) \, \mathrm{d}x \, \mathrm{d}y \tag{6.42}
$$

Transforming the integral to the reference element, we obtain

$$
C^{\mathrm{loc},e}_{i,j} = \int_\xi \int_\eta N_i(\xi, \eta) c(x(\xi, \eta), y(\xi, \eta)) N_j(\xi, \eta) \det(\mathbf{J}^e) \, \mathrm{d}\xi \, \mathrm{d}\eta \tag{6.43}
$$

where the integrals over ξ and η are performed over the reference element and where $\mathbf{J}^e = \left(\frac{\partial(x,y)}{\partial(\xi,\eta)} \right)$ is the Jacobian of the coordinate transformation for the eth element.

Example 14

To illustrate the FEM procedure, we will apply the findings of this section to the PDE examples Equation (6.4) and Equation (6.6) discretized by the triangular finite elements shown in Figure 6.4 with linear shape functions. For a triangular element with vertices $\mathbf{r}^e_1 = (x^e_1, y^e_1), \mathbf{r}^e_2 = (x^e_2, y^e_2)$ and $\mathbf{r}^e_3 = (x^e_3, y^e_3)$ the Jacobian appearing in Equation (6.43) is given by

$$
\mathbf{J}^e = \frac{\partial(x, y)}{\partial(\xi, \eta)} = \begin{pmatrix} x^e_2 - x^e_1 & x^e_3 - x^e_1 \\ y^e_2 - y^e_1 & y^e_3 - y^e_1 \end{pmatrix} \tag{6.44}
$$

due to Equation (6.29). Hence,

$$
\det(\mathbf{J}^e) = (x^e_2 - x^e_1)(y^e_3 - y^e_1) - (y^e_2 - y^e_1)(x^e_3 - x^e_1) = 2A_{\mathbf{r}_1,\mathbf{r}_2,\mathbf{r}_2} \tag{6.45}
$$

where $A_{\mathbf{r}_1,\mathbf{r}_2,\mathbf{r}_3}$ is the size of the triangle with nodes $\mathbf{r}_1$, $\mathbf{r}_2$, and $\mathbf{r}_3$. The local element matrix $\mathbf{C}^e$ is

$$C_{i,j}^{\text{loc},e} = 2A_{\mathbf{r}_1,\mathbf{r}_2,\mathbf{r}_3} \int_{\xi=0}^{1} \int_{\eta=0}^{1-\xi} N_i(\xi,\eta) c(x(\xi,\eta), y(\xi,\eta)) N_j(\xi,\eta) \, d\xi \, d\eta$$

and hence

$$\mathbf{C}^{\text{loc},e} = 2A_{\mathbf{r}_1,\mathbf{r}_2,\mathbf{r}_3} \int_{\xi=0}^{1} \int_{\eta=0}^{1-\xi} \begin{pmatrix} N_1(\xi,\eta) \\ N_2(\xi,\eta) \\ N_3(\xi,\eta)\eta \end{pmatrix} c(x,y) \begin{pmatrix} N_1(\xi,\eta) \\ N_2(\xi,\eta) \\ N_3(\xi,\eta) \end{pmatrix}^{\text{T}} d\xi \, d\eta$$

$$= 2A_{\mathbf{r}_1,\mathbf{r}_2,\mathbf{r}_3} \int_{\xi=0}^{1} \int_{\eta=0}^{1-\xi} c(x,y) \begin{pmatrix} 1-\xi-\eta \\ \xi \\ \eta \end{pmatrix} \begin{pmatrix} 1-\xi-\eta \\ \xi \\ \eta \end{pmatrix}^{\text{T}} d\xi \, d\eta$$

If $c = c(x,y)$ is constant over the element, we obtain

$$\mathbf{C}^e = c A_{\mathbf{r}_1,\mathbf{r}_2,\mathbf{r}_3} \frac{1}{12} \begin{pmatrix} 2 & 1 & 1 \\ 1 & 2 & 1 \\ 1 & 1 & 2 \end{pmatrix} \tag{6.46}$$

If $c(x,y)$ is not constant, the element matrix is usually computed by quadrature (this will be discussed later). Alternatively, Equation (6.46) may be used with $c = c(x(1/4, 1/4), y(1/4, 1/4))$, which is the value of c in the element midpoint.

For the finite element discretization shown in Figure 6.4, and for $c(x,y) = 1$, we obtain the Jacobian

$$\mathbf{J}^1 = \begin{pmatrix} x_2^e - x_1^e & x_3^e - x_1^e \\ y_2^e - y_1^e & y_3^e - y_1^e \end{pmatrix} = \begin{pmatrix} 1 & \frac{1}{2} \\ 0 & \frac{1}{2} \end{pmatrix} \Rightarrow \det(\mathbf{J}^1) = \frac{1}{2} \tag{6.47}$$

As $\det(\mathbf{J}^e)$ is twice the area of the triangle through the points $\mathbf{r}^e, \ldots, \mathbf{r}^e$, this shows that $A_{\mathbf{r}_1^1, \mathbf{r}_2^1, \mathbf{r}_3^1} = \frac{1}{4}$, but of course, this is also obvious from geometric considerations. All triangular elements in Figure 6.5 have the same area, $\frac{1}{4}$. As we assume $c(x,y) = 1$, we obtain from Equation (6.46) the local element matrices

$$\mathbf{C}^1 = \mathbf{C}^2 = \mathbf{C}^3 = \mathbf{C}^4 = \frac{1}{48} \begin{pmatrix} 2 & 1 & 1 \\ 1 & 2 & 1 \\ 1 & 1 & 2 \end{pmatrix} \tag{6.48}$$

□

The remaining element matrices in Equation (6.41) are computed similarly. For computing the so-called local element stiffness matrix

$$K_{i,j}^{\text{loc},e} = \int_{\Omega_e} \nabla N_i(\xi(x,y), \eta(x,y)) \cdot \mathbf{D}(x,y) \nabla N_j(\xi(x,y), \eta(x,y)) \, dx \, dy \tag{6.49}$$

Note that the gradient operator $\nabla = \nabla_{(x,y)}$ takes derivatives with respect to x and y. To transform the integral to the reference element, we have to replace $\nabla = \nabla_{(x,y)}$ by a gradient operator $\nabla = \nabla_{(\xi,\eta)}$ with respect to (ξ, η). Note that

$$\nabla_{(x,y)} = \begin{pmatrix} \frac{d}{dx} \\ \frac{d}{dy} \end{pmatrix} = \begin{pmatrix} \frac{\partial x}{\partial \xi} & \frac{\partial x}{\partial \eta} \\ \frac{\partial y}{\partial \xi} & \frac{\partial y}{\partial \eta} \end{pmatrix} \begin{pmatrix} \frac{d}{d\xi} \\ \frac{d}{d\eta} \end{pmatrix} \Rightarrow \nabla_{(\xi,\eta)} = \begin{pmatrix} \frac{d}{d\xi} \\ \frac{d}{d\eta} \end{pmatrix} = \begin{pmatrix} \frac{\partial x}{\partial \xi} & \frac{\partial x}{\partial \eta} \\ \frac{\partial y}{\partial \xi} & \frac{\partial y}{\partial \eta} \end{pmatrix}^{-1} \begin{pmatrix} \frac{d}{dx} \\ \frac{d}{dy} \end{pmatrix}$$

$$\tag{6.50}$$

Therefore $\nabla_{(\xi,\eta)} = (\mathbf{J}^e)^{-1}\nabla_{(x,y)}$, where $\mathbf{J}^e$ is the Jacobian of the coordinate transformation for the eth element. Transforming Equation (6.49) to the reference element, we obtain

$$K_{i,j}^{\text{loc},e} = \int_{\xi}\int_{\eta}\left(\nabla_{(\xi,\eta)}N_i\right)^{\text{T}}(\mathbf{J}^e)^{-\text{T}}\mathbf{D}(\mathbf{J}^e)^{-1}\left(\nabla_{(\xi,\eta)}N_j\right)\det(\mathbf{J}^e)\,d\xi\,d\eta \tag{6.51}$$

where $(\mathbf{J}^e)^{-\text{T}}$ denotes the transpose of $(\mathbf{J}^e)^{-1}$, and where the integrals with respect to ξ and η are again performed on the reference element.

Example 15 (Examples 9 and 10 revisited)
For a triangular element with vertices $\mathbf{r}_1^e = (x_1^e, y_1^e)$, $\mathbf{r}_2^e = (x_2^e, y_2^e)$ and $\mathbf{r}_3^e = (x_3^e, y_3^e)$ the determinant of the Jacobian is given by Equation (6.45) and hence

$$\mathbf{K}^{\text{loc},e} = 2A_{\mathbf{r}_1,\mathbf{r}_2,\mathbf{r}_3}\int_{\xi=0}^{1}\int_{\eta=0}^{1-\xi}\nabla\begin{pmatrix}1-\xi-\eta\\\xi\\\eta\end{pmatrix}(\mathbf{J}^e)^{-\text{T}}\mathbf{D}(\mathbf{J}^e)^{-1}\nabla\begin{pmatrix}1-\xi-\eta\\\xi\\\eta\end{pmatrix}^{\text{T}}d\xi\,d\eta$$

$$= 2A_{\mathbf{r}_1,\mathbf{r}_2,\mathbf{r}_3}\int_{\xi=0}^{1}\int_{\eta=0}^{1-\xi}\begin{pmatrix}-1 & -1\\1 & 0\\0 & 1\end{pmatrix}(\mathbf{J}^e)^{-\text{T}}\mathbf{D}(\mathbf{J}^e)^{-1}\begin{pmatrix}-1 & 1 & 0\\-1 & 0 & 1\end{pmatrix}d\xi\,d\eta$$

$$= 2A_{\mathbf{r}_1,\mathbf{r}_2,\mathbf{r}_3}\tilde{\mathbf{K}}^{\text{T}}\left(\int_{\xi=0}^{1}\int_{\eta=0}^{1-\xi}\mathbf{D}(x(\xi,\eta),y(\xi,\eta))\,d\xi\,d\eta\right)\tilde{\mathbf{K}}$$

where (using Equation (6.44) for the the Jacobian)

$$\tilde{\mathbf{K}} = \begin{pmatrix}-1 & -1\\1 & 0\\0 & 1\end{pmatrix}\begin{pmatrix}x_2^e-x_1^e & y_2^e-y_1^e\\x_3^e-x_1^e & y_3^e-y_1^e\end{pmatrix}^{-1} \tag{6.52}$$

The remaining integral may be evaluated by quadrature and usually a one-point quadrature rule is used, yielding

$$\int_{\xi=0}^{1}\int_{\eta=0}^{1-\xi}\mathbf{D}(x^e(\xi,\eta),y^e(\xi,\eta))\,d\xi\,d\eta \approx \frac{1}{2}\mathbf{D}(x^e(1/4,1/4),y^e(1/4,1/4)) =: \mathbf{D}^e$$

Hence,

$$\mathbf{K}^{\text{loc},e} = A_{\mathbf{r}_1,\mathbf{r}_2,\mathbf{r}_3}\tilde{\mathbf{K}}\,\mathbf{D}^e\,\tilde{\mathbf{K}}^{\text{T}} \tag{6.53}$$

□

Example 16 (Example 9 revisited)
For the finite element discretization shown in Figure 6.4, it is easy to see that the Jacobians of the individual elements are

$$\mathbf{J}^1 = \begin{pmatrix}1 & \frac{1}{2}\\0 & \frac{1}{2}\end{pmatrix}\qquad \mathbf{J}^2 = \begin{pmatrix}0 & -\frac{1}{2}\\1 & \frac{1}{2}\end{pmatrix} \tag{6.54}$$

$$\mathbf{J}^3 = \begin{pmatrix}-1 & -\frac{1}{2}\\0 & -\frac{1}{2}\end{pmatrix}\qquad \mathbf{J}^4 = \begin{pmatrix}0 & \frac{1}{2}\\-1 & -\frac{1}{2}\end{pmatrix} \tag{6.55}$$

For Equation (6.6), the matrix in the diffusive term is

$$\mathbf{D}(x,y) = \begin{pmatrix}\frac{1}{2} & 0\\0 & \frac{1}{2}\end{pmatrix}$$

Evaluating Equations (6.53) and (6.52) with these Jacobians, we obtain the following matrices $\mathbf{K}^{loc,e}$:

$$\mathbf{K}^{loc,1} = \mathbf{K}^{loc,2} = \mathbf{K}^{loc,3} = \mathbf{K}^{loc,4} = \frac{1}{4}\begin{pmatrix} 1 & 0 & -1 \\ 0 & 1 & -1 \\ -1 & -1 & 2 \end{pmatrix} \tag{6.56}$$

All local element stiffness matrices are the same in this example, as all elements have the same size and as the matrix $\mathbf{D}$ *is constant.*   □

Example 17 (Example 10 revisited)

For Equation (6.6) and the finite element discretization shown in Figure 6.4, the Jacobians are the same as in Equations (6.54) and (6.55). For demonstration purposes, we evaluate the local element stiffness matrices both exactly and by quadrature. If we use a one-point quadrature rule, as in Equation (6.53), then we obtain the following matrices:

$$\mathbf{K}^{loc,1} = \begin{pmatrix} 0.236\sigma_1^2 + 0.158\sigma_2^2 & -0.236\sigma_1^2 + 0.158\sigma_2^2 & -0.316\sigma_2^2 \\ -0.236\sigma_1^2 + 0.158\sigma_2^2 & 0.236328\sigma_1^2 + 0.158\sigma_2^2 & -0.316\sigma_2^2 \\ -0.316\sigma_2^2 & -0.316\sigma_2^2 & 0.632812\sigma_2^2 \end{pmatrix} \tag{6.57}$$

$$\mathbf{K}^{loc,2} = \begin{pmatrix} 0.439\sigma_1^2 + 0.236\sigma_2^2 & 0.439\sigma_1^2 - 0.236\sigma_2^2 & -0.879\sigma_1^2 \\ 0.439\sigma_1^2 - 0.236\sigma_2^2 & 0.439\sigma_1^2 + 0.236\sigma_2^2 & -0.879\sigma_1^2 \\ -0.879\sigma_1^2 & -0.879\sigma_1 & 1.758\sigma_1^2 \end{pmatrix} \tag{6.58}$$

$$\mathbf{K}^{loc,3} = \begin{pmatrix} 0.330\sigma_1^2 + 0.439\sigma_2^2 & -0.330\sigma_1^2 + 0.439\sigma_2^2 & -0.879\sigma_2^2 \\ -0.330\sigma_1^2 + 0.439\sigma_2^2 & 0.330\sigma_1^2 + 0.439\sigma_2^2 & -0.879\sigma_2^2 \\ -0.879\sigma_2^2 & -0.879\sigma_2^2 & 1.758\sigma_2^2 \end{pmatrix} \tag{6.59}$$

$$\mathbf{K}^{loc,4} = \begin{pmatrix} 0.158\sigma_1^2 + 0.330\sigma_2^2 & 0.158\sigma_1^2 - 0.330\sigma_2^2 & -0.316\sigma_1^2 \\ 0.158\sigma_1^2 - 0.330\sigma_2^2 & 0.158\sigma_1^2 + 0.330\sigma_2^2 & -0.316\sigma_1^2 \\ -0.316\sigma_1^2 & -0.316\sigma_1^2 & 0.632\sigma_1^2 \end{pmatrix} \tag{6.60}$$

whereas the exact evaluation of Equation (6.53) yields the following matrices:

$$\mathbf{K}^{loc,1} = \begin{pmatrix} 0.286\sigma_1^2 + 0.172\sigma_2^2 & -0.286\sigma_1^2 + 0.172\sigma_2^2 & -0.344\sigma_2^2 \\ -0.286\sigma_1^2 + 0.172\sigma_2^2 & 0.286\sigma_1^2 + 0.172\sigma_2^2 & -0.344\sigma_2^2 \\ -0.344\sigma_2^2 & -0.344\sigma_2^2 & 0.688\sigma_2^2 \end{pmatrix} \tag{6.61}$$

$$\mathbf{K}^{loc,2} = \begin{pmatrix} 0.422\sigma_1^2 + 0.286\sigma_2^2 & 0.422\sigma_1^2 - 0.286\sigma_2^2 & -0.844\sigma_1^2 \\ 0.422\sigma_1^2 - 0.286\sigma_2^2 & 0.422\sigma_1^2 + 0.286\sigma_2^2 & -0.844\sigma_1^2 \\ -0.844\sigma_1^2 & -0.844\sigma_1^2 & 1.688\sigma_1^2 \end{pmatrix} \tag{6.62}$$

$$\mathbf{K}^{loc,3} = \begin{pmatrix} 0.286\sigma_1^2 + 0.422\sigma_2^2 & -0.286\sigma_1^2 + 0.422\sigma_2^2 & -0.844\sigma_2^2 \\ -0.286\sigma_1^2 + 0.422\sigma_2^2 & 0.286\sigma_1^2 + 0.422\sigma_2^2 & -0.844\sigma_2^2 \\ -0.844\sigma_2^2 & -0.844\sigma_2^2 & 1.688\sigma_2^2 \end{pmatrix} \tag{6.63}$$

$$\mathbf{K}^{loc,4} = \begin{pmatrix} 0.172\sigma_1^2 + 0.286\sigma_2^2 & 0.172\sigma_1^2 - 0.286\sigma_2^2 & -0.344\sigma_1^2 \\ 0.172\sigma_1^2 - 0.286\sigma_2^2 & 0.172\sigma_1^2 + 0.286\sigma_2^2 & -0.344\sigma_1^2 \\ -0.344\sigma_1^2 & -0.344\sigma_1^2 & 0.688\sigma_1^2 \end{pmatrix} \tag{6.64}$$

Apparently, there are slight differences between the matrices that were computed by the one-point quadrature rule and the matrices that were computed by exact integration. If the error introduced by quadrature is larger than tolerable, then a quadrature rule with more integration points may be used.   □

We now treat the matrices that account for the convective terms. Proceeding as above, we can compute

$$B_{i,j}^{\text{loc},e} = \int_{\xi} \int_{\eta} N_i(\xi, \eta)\, \mathbf{b}(x(\xi, \eta), y(\xi, \eta)) \cdot (\mathbf{J}^e)^{-1} \left(\nabla_{(\xi,\eta)} N_j\right) \det(\mathbf{J}^e)\, d\xi\, d\eta \qquad (6.65)$$

Example 18
For a triangular element in the nodes $\mathbf{r}_1, \mathbf{r}_2, \mathbf{r}_3$, we obtain

$$\mathbf{B}^{\text{loc},e} = \int_{\xi=0}^{1} \int_{\eta=0}^{1-\xi} \begin{pmatrix} N_1 \\ N_2 \\ N_3 \end{pmatrix} \mathbf{b} \cdot (\mathbf{J}^e)^{-1} \left(\nabla_{(\xi,\eta)} \begin{pmatrix} N_1 \\ N_2 \\ N_3 \end{pmatrix}^{\mathsf{T}} \right) \det(\mathbf{J}^e)\, d\xi\, d\eta$$

$$= 2A_{\mathbf{r}_1,\mathbf{r}_2,\mathbf{r}_3} \int_{\xi=0}^{1} \int_{\eta=0}^{1-\xi} \begin{pmatrix} 1-\xi-\eta \\ \xi \\ \eta \end{pmatrix} \begin{pmatrix} \mathbf{b}^{\mathsf{T}} \\ \mathbf{b}^{\mathsf{T}} \\ \mathbf{b}^{\mathsf{T}} \end{pmatrix} \cdot (\mathbf{J}^e)^{-1} \begin{pmatrix} -1 & 1 & 0 \\ -1 & 0 & 1 \end{pmatrix} d\xi\, d\eta$$

As before, if $\mathbf{b}$ is constant, then we can evaluate the integrals. This yields

$$\mathbf{B}^{\text{loc},e} = 2A_{\mathbf{r}_1,\mathbf{r}_2,\mathbf{r}_3} \begin{pmatrix} 1/6 \\ 1/6 \\ 1/6 \end{pmatrix} \begin{pmatrix} \mathbf{b}^{\mathsf{T}} \\ \mathbf{b}^{\mathsf{T}} \\ \mathbf{b}^{\mathsf{T}} \end{pmatrix} \cdot (\mathbf{J}^e)^{-1} \begin{pmatrix} -1 & 1 & 0 \\ -1 & 0 & 1 \end{pmatrix} \qquad (6.66)$$

If $\mathbf{b}$ is not constant, then we may again use a quadrature rule. Usually, a one-point quadrature rule is used for this element type, which yields the same expression as Equation (6.66), where $\mathbf{b}$ is replaced by its value at the element midpoint.   □

Setting $\tilde{f}(x, y) := f(x, y) - R(x, y, 0)$, we obtain

$$f_i^{\text{loc},e} = \int_{\xi} \int_{\eta} N_i(\xi, \eta)\tilde{f}(x(\xi, \eta), y(\xi, \eta)) \det(\mathbf{J}^e)\, d\xi\, d\eta \qquad (6.67)$$

In a finite element setting, functions on the domain are usually represented by their interpolation in the finite element shape function basis: the function $\tilde{f}(x, y)$ is thus represented as $\tilde{f}(x, y) \approx \sum_{i=1}^{m} \phi_i(x, y)\tilde{f}(x_i, y_i)$. Recall that the restriction of the functions $\phi_1, \ldots, \phi_m$ to the eth element is equal to the local shape functions. Hence, the restriction of the interpolant of $\tilde{f}$ to the eth element is given by $\tilde{f}^e(x, y) = \sum_{j=1}^{n_e} N_j^e(x, y)\tilde{f}_j^e$, where $\tilde{f}_j^e = \tilde{f}(x_{p_j^e}, y_{p_j^e})$ denotes the values of $\tilde{f}$ at the nodes of the eth element. This yields

$$f_i^{\text{loc},e} \approx \sum_{j=1}^{n_e} \int_{\xi} \int_{\eta} N_i(\xi, \eta)N_j(\xi, \eta)\tilde{f}_j^e \det(\mathbf{J}^e)\, d\xi\, d\eta$$

$$\Rightarrow \mathbf{f}^{\text{loc},e} = \mathbf{M}^e \tilde{\mathbf{f}}^e$$

where $M_{i,j}^e = \int_{\xi} \int_{\eta} N_i N_j\, d\xi\, d\eta$ may be obtained by evaluating the matrix $\mathbf{C}^e$ for a unit material parameter c.

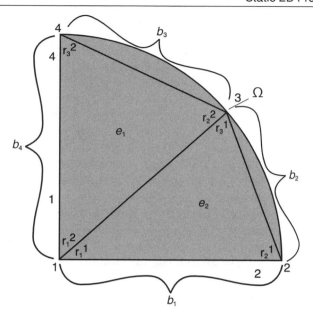

Figure 6.9 Boundary approximation for a curved boundary

Example 19
For a triangular element in the nodes $\mathbf{r}_1^e$, $\mathbf{r}_2^e$, $\mathbf{r}_3^e$ *with linear shape functions, we obtain, due to Equation (6.46),*

$$\mathbf{f}^{\text{loc},e} = A_{\mathbf{r}_1,\mathbf{r}_2,\mathbf{r}_3} \frac{1}{12} \begin{pmatrix} 2 & 1 & 1 \\ 1 & 2 & 1 \\ 1 & 1 & 2 \end{pmatrix} \tilde{\mathbf{f}}^e \tag{6.68}$$

where $A_{\mathbf{r}_1,\mathbf{r}_2,\mathbf{r}_3}$ *is the area of the triangle in the element nodes given by Equation (6.45).* □

We have now computed all terms in Equations (6.19)–(6.24) that involve integrals over the domain by stating their local element matrices $\mathbf{K}^e$, $\mathbf{B}^e$ and $\mathbf{C}^e$ and the local element vectors $\tilde{\mathbf{f}}^e$.

The integrals over the boundary require some additional attention: we need to parameterize the boundary in order to compute the boundary integrals. We will do this in an analogous manner to the handling of the integrals over the domain by decomposing the boundary into boundary elements: if the domain is discretized by finite elements, then the discretized boundary consists of the outer boundaries of the exterior elements. This is demonstrated by Figure 6.9.

Figure 6.9 demonstrates how the finite element discretization may introduce an error caused by approximation of the boundary. The boundary of the domain shown in Figure 6.9. is not approximated well. A curved boundary may be interpolated better by using a finer mesh (more finite elements) or by using finite elements with curved boundaries (e.g. triangular elements with quadratic shape functions). For now, we will stick to linear elements with straight boundaries.

As Figure 6.9 demonstrates, the discretization of the geometry by polygons yields a discretization of the boundary by line segments. These line segments will now be interpreted as finite elements (called boundary elements here), where the shape functions on the boundary elements are given by the restrictions of the shape functions on the original elements to the boundary.

Before we discuss the shape functions, let us consider how the discretized boundary may be described. In total, there are n_B line segments in the discretized boundary and every line segment connects two nodes $\mathbf{r}_1^b$ and $\mathbf{r}_2^b$, $b = 1, \ldots, n_B$.

Figure 6.9 demonstrates the complete description of the finite element mesh: the list of nodes $\mathbf{r}_1, \ldots, \mathbf{r}_m$ stores the positions of the nodes. Every element Ω^e, $e = 1, \ldots, n$ is constructed in a subset $\mathbf{r}_1^e, \ldots, \mathbf{r}_{n_e}^e$ of the nodes and is identified by the indices $\mathbf{p}^e$ of its nodes ($\mathbf{r}_i = \mathbf{r}_{p_i}$, $i = 1, \ldots, n_e$). Similarly, every boundary element is constructed in a subset $\mathbf{r}_1^b, \ldots, \mathbf{r}_{n_b}^b$ of the nodes (for the linear triangular and quadrilateral elements $n_b = 2$) and is identified by its node indices $\mathbf{p}^b$.

Hence, the boundary is approximated by n_B line segments, where every line segment connects two nodes $\mathbf{r}_1^b$ and $\mathbf{r}_2^b$, $b = 1, \ldots, n_B$ (we will denote boundary nodes by $\mathbf{r}_i^b$ and domain element nodes by $\mathbf{r}_i^e$; this is a slight abuse of notation, but what is meant will always be clear from the context). The points on the bth boundary segment are parameterized by

$$\mathbf{r}(s) = (x(s), y(s)) = \frac{1-s}{2}\mathbf{r}_1^b + \frac{1+s}{2}\mathbf{r}_2^b \qquad s \in [-1, 1]$$

$$= \sum_{i=1}^{2} \mathbf{r}_i^b N_1^{1D}(s)$$

where the boundary element shape functions $N_1^{1D}(s)$ and $N_2^{1D}(s)$ are given by

$$\mathbf{N}^{1D}(s) = \begin{pmatrix} N_1^{1D}(s) \\ N_2^{1D}(s) \end{pmatrix} = \frac{1}{2}\begin{pmatrix} 1-s \\ 1+s \end{pmatrix}, \qquad s \in [-1, 1] \tag{6.69}$$

Both for the triangular and for the rectilinear elements with linear shape functions, the boundaries of the elements are straight lines. Each line is given by the list of its node indices $\mathbf{p}^b = (p_1^b, p_2^b)$ and connects the nodes $\mathbf{r}_{p_1^b} = \mathbf{r}_1^b$ and $\mathbf{r}_{p_2^b} = \mathbf{r}_2^b$. The restrictions of the shape functions $\phi_{p_1^b}$ and $\phi_{p_2^b}$ to the boundary element b are linear functions with $\phi_{p_j^b}(\mathbf{r}_i^b) = \delta_{ij}$ (for $i, j = 1, 2$). This is demonstrated in Figure 6.10 for triangular elements.

Every boundary segment b ($b = 1, \ldots, n_B$) is, hence, a line element with linear shape functions given by

$$\mathbf{N}^b(x, y) = \begin{pmatrix} N_1^{1D}(s^b(x, y)) \\ N_2^{1D}(s^b(x, y)) \end{pmatrix} \tag{6.70}$$

where N_1^{1D} and N_2^{1D} are given by Equation (6.69) and where $s^e(x, y)$ is the inverse mapping of Equation (6.69). We will not require an explicit expression for $s(x, y)$; for completeness, note that $s^b(x, y) = 2(x - x_1^b)/(x_2^b - x_1^b) - 1$.

We have now decomposed the boundary into boundary elements and can compute the boundary integrals in the same manner as the domain integrals. Let us demonstrate this for the matrix $\mathbf{L}$ that was defined in Equation (6.22) as

$$L_{i,j} = \int_{\Gamma_N} \phi_i(x, y)\, \gamma(x, y)\phi_j(x, y)\, ds$$

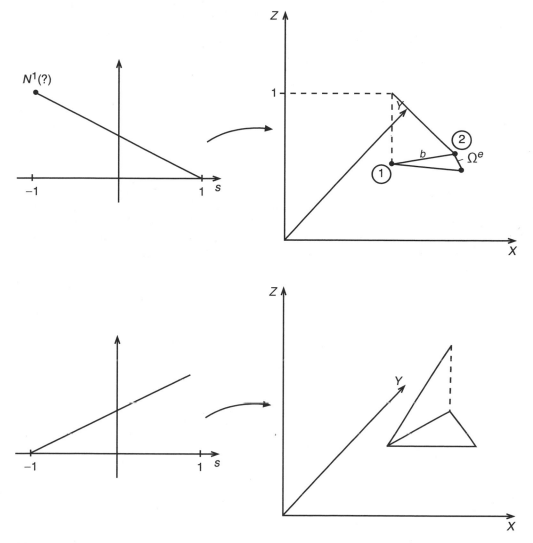

Figure 6.10 Boundary elements and triangular elements

We can write this integral as a sum of integrals over the boundary elements inside the Neumann boundary Γ_N (for simplicity, we assume that the first n_{Γ_N} boundary elements $b = 1, \ldots, n_{\Gamma_N}$ decompose the Neumann boundary, while the remaining boundary elements $b = n_{\Gamma_N} + 1, \ldots, n_B$ decompose the Dirichlet boundary). We obtain

$$
L_{i,j} = \sum_{b=1}^{n_{\Gamma_N}} \int_{\Gamma_b} \phi_i(x, y)\, \gamma(x, y)\, \phi_j(x, y)\, \mathrm{d}s
$$

$$
= \sum_{b=1}^{n_{\Gamma_N}} L_{i,j}^b
$$

where the $L_{i,j}^b$ $(i, j = 1, \ldots, m)$ define the (global) boundary element matrices $\mathbf{L}^b$. Performing a similar decomposition for the vector $\mathbf{g}$ defined in Equation (6.24), we obtain

$$\mathbf{L} = \sum_{b=1}^{n_{\Gamma_N}} \mathbf{L}^b, \qquad \mathbf{g} = \sum_{b=1}^{n_{\Gamma_N}} \mathbf{g}^b \tag{6.71}$$

Just as for the global element matrices $\mathbf{K}^e$, $\mathbf{B}^e$ and $\mathbf{C}^e$, the global boundary element matrices $\mathbf{L}^e$ and the global boundary element vectors $\mathbf{g}^b$ are sparse. The reason is again that any element shape function ϕ_i, $i = 1, \ldots, m$ is nonzero only in those boundary elements that have the node $\mathbf{r}_i$ as vertex. Hence, $\mathbf{L}_{i,j}^b \neq 0$ only if both i and j are contained in the list $\mathbf{p}^b$ of element node indices.

Hence, just as the global element matrices were represented by local element matrices $\mathbf{K}^{\text{loc},e}$, $\mathbf{B}^{\text{loc},e}$ and $\mathbf{C}^{\text{loc},e}$, the global boundary element matrices may be obtained and stored in compressed form by local boundary element matrices and vectors $\mathbf{L}^{\text{loc},e}$ and $\mathbf{g}^{\text{loc},e}$.

As the restrictions of the shape functions to the boundary elements are given by the boundary shape functions $N_1^{1D}(s)$ and $N_2^{1D}(s)$, the local boundary element matrices may be computed as

$$L_{i,j}^{\text{loc},b} = \int_{\Omega} \phi_{p_i^b}(x, y)\, \gamma(x, y)\, \phi_{p_j^b}(x, y)\, \mathrm{d}s, \qquad i, j = 1, 2$$

$$= \int_{\Omega} N_i^{1D}(s(x, y))\, \gamma(x, y)\, N_j^{1D}(s(x, y))\, \mathrm{d}s$$

$$= \int_{s=-1}^{1} N_i^{1D}(s)\, \gamma(x(s), y(s))\, N_j^{1D}(s)\, |\mathbf{J}^b|\, \mathrm{d}s$$

where $\mathbf{J}^b$ is the derivative of the coordinate mapping Equation (6.69). It is given by

$$\mathbf{J}^b = \begin{pmatrix} x'(s) \\ y'(s) \end{pmatrix} = \frac{1}{2}\begin{pmatrix} x_2^b - x_1^b \\ y_2^b - y_1^b \end{pmatrix} \quad \Rightarrow \quad |\mathbf{J}^b| = \frac{1}{2}\sqrt{(x_2^b - x_1^b)^2 + (y_2^b - y_1^b)^2} = \frac{1}{2}A_{\mathbf{r}_1^b, \mathbf{r}_2^b}$$

where $A_{\mathbf{r}_1^b, \mathbf{r}_2^b}$ denotes the length of the boundary element b.

Therefore,

$$\mathbf{L}^{\text{loc},b} = \int_{s=-1}^{1} \mathbf{N}^{1D}(s)\, \gamma(x(s), y(s))\, \mathbf{N}^{1D}(s)^{\mathrm{T}}\, |\mathbf{J}^b|\, \mathrm{d}s$$

$$= \frac{1}{2}A_{\mathbf{r}_1^b, \mathbf{r}_2^b} \int_{s=-1}^{1} \gamma(x(s), y(s)) \begin{pmatrix} (1-s) \\ s+1 \end{pmatrix}\begin{pmatrix} (1-s) \\ s+1 \end{pmatrix}^{\mathrm{T}} \mathrm{d}s$$

For the special case where $\gamma(x, y) = \gamma$ is constant over the boundary element, this evaluates to

$$\mathbf{L}^{\text{loc},b} = \gamma\, A_{\mathbf{r}_1^b, \mathbf{r}_2^b}\, \frac{1}{6}\begin{pmatrix} 2 & 1 \\ 1 & 2 \end{pmatrix} \tag{6.72}$$

If $\gamma(x, y)$ is not constant over the boundary element, then the matrix $\mathbf{L}^{\text{loc},b}$ is usually evaluated by quadrature. Alternatively, an approximation may be obtained by using the value of γ at the boundary element midpoint, $\gamma = \gamma(x(0), y(0))$ in Equation (6.72).

To compute the boundary element vector **g**, let us define $\tilde{g}(x, y) := u_N(x, y) - \gamma(x, y)\tilde{u}_D(x, y)$. Then the local boundary element vector is

$$g_i^{\text{loc},b} = \int_{\Gamma_b} \phi_{p_i^b}(x, y)\tilde{g}(x, y)\,ds, \quad i = 1, 2. \tag{6.73}$$

In finite element approximations, functions are usually represented by their interpolation in the finite element shape functions. We may compute Equation (6.73) by interpolating $\tilde{g}(x, y)$ in the finite element shape functions as $\tilde{g}(x, y) \approx \sum_{i=1}^{m} \tilde{g}(x_i, y_i)\phi_i(x, y)$. Inserting this into Equation (6.73) and noting that the restriction of the shape function to the boundary equals the local boundary shape functions $N_1^{1D}(x, y)$, $N_2^{1D}(x, y)$, we obtain

$$g_i^{\text{loc},b} \approx \sum_{j=1}^{2} \int_{\Gamma_b} N_i^{1D}(x, y)N_j^{1D}(x, y)\,ds\; \tilde{g}_{p_j^b}, \quad i = 1, 2.$$

Note that the integrand is the same integrand that occurs in the computation of the matrix $\mathbf{L}^{\text{loc},b}$ for $\gamma = 1$. Hence, we can compute Equation (6.74) by using Equation (6.72) as

$$\mathbf{g}^{\text{loc},b} = \begin{pmatrix} g_1^{\text{loc},b} \\ g_2^{\text{loc},b} \end{pmatrix} = A_{\mathbf{r}_1^b,\mathbf{r}_2^b} \frac{1}{6} \begin{pmatrix} 2 & 1 \\ 1 & 2 \end{pmatrix} \begin{pmatrix} \tilde{g}_{p_1^b} \\ \tilde{g}_{p_2^b} \end{pmatrix} \tag{6.74}$$

4. *Assembly:* We have seen in the preceding sections how to compute the contributions of the elements to the system of equations (6.19)–(6.24). The contributions of the elements were computed in compressed form as local element matrices (e.g. as $\mathbf{C}^{\text{loc},e} \in \mathbb{R}^{n_e \times n_e}$) and as local element vectors (e.g. as $\mathbf{f}^{\text{loc},e} \in \mathbb{R}^{n_e}$), where every row or column of an element matrix and every row of an element vector corresponds to a local node of the element.

The next step is to assemble the contributions of the individual elements into a system of equations for every degree of freedom in the finite element discretization. The matrices and vectors arising in this system of equations were computed in Equation (6.39) by summing up the global element matrices computed for each individual element. In Example 13, it was shown how these global element matrices are stored in a compressed manner as local element matrices. The rth or the cth column of a local element matrix corresponds to the rth or to the cth node of the element in the local ordering of the element nodes. This allows us to obtain the global element matrices (e.g. $\mathbf{C}^e \in \mathbb{R}^{m \times m}$) where the rth row or the cth column corresponds to the rth or cth node in the global ordering of all nodes in the finite element by setting

$$C_{p_i,p_j}^e := C_{i,j}^{\text{loc},e}, \qquad i, j = 1, \ldots, n_e$$

$$f_{p_i}^e := f_i^{\text{loc},e}, \qquad i = 1, \ldots, n_e$$

and by setting the remaining entries of $\mathbf{C}^e$ and of $\mathbf{f}^e$ to zero. This is performed in an analogous manner for the other element matrices in Equation (6.41). Also, the boundary element matrices and boundary element vectors in Equation (6.71) are assembled in an analogous fashion.

This yields a practical way of computing the system of equations for the vector of unknowns **a** as

$$\hat{\mathbf{T}}\,\mathbf{a} := \left[\sum_{e=1}^{m}(\mathbf{K}^e + \mathbf{B}^e + \mathbf{C}^e) + \sum_{b=1}^{n_{\Gamma_N}}\mathbf{L}^b\right]\mathbf{a} = \sum_{e=1}^{m}\mathbf{f}^e + \sum_{b=1}^{n_{\Gamma_N}}\mathbf{g}^b = \hat{\mathbf{f}} + \hat{\mathbf{g}} \tag{6.75}$$

However, Equation (6.75) is not the system of equations (6.18) that we need to solve. We have not incorporated the Dirichlet boundary conditions in constructing the finite element discretization. Hence, the system (6.75) is indeterminate and the matrix on the left-hand side is noninvertible.

Example 20 (Example 9 revisited)
Equation (6.71) reduces to:

$$\left[\sum_{e=1}^{m} K^e\right] \mathbf{a} = \sum_{e=1}^{m} \mathbf{f}^e = \hat{\mathbf{f}} \tag{6.76}$$

$\mathbf{K}^{\mathrm{loc},e}$ *is given Equation (6.55) so that the expanded matrices are given by:*

$$\mathbf{K}^1 = \frac{1}{4}\begin{pmatrix} 1 & 0 & 0 & 0 & -1 \\ 0 & 1 & 0 & 0 & -1 \\ 0 & 0 & 0 & 0 & 0 \\ 0 & 0 & 0 & 0 & 0 \\ -1 & -1 & 0 & 0 & 2 \end{pmatrix} \quad \mathbf{K}^2 = \frac{1}{4}\begin{pmatrix} 0 & 0 & 0 & 0 & 0 \\ 0 & 1 & 0 & 0 & -1 \\ 0 & 0 & 1 & 0 & -1 \\ 0 & 0 & 0 & 0 & 0 \\ 0 & -1 & -1 & 0 & 2 \end{pmatrix} \tag{6.77}$$

$$\mathbf{K}^3 = \frac{1}{4}\begin{pmatrix} 0 & 0 & 0 & 0 & 0 \\ 0 & 0 & 0 & 0 & 0 \\ 0 & 0 & 1 & 0 & -1 \\ 0 & 0 & 0 & 1 & -1 \\ 0 & 0 & -1 & -1 & 2 \end{pmatrix} \quad \mathbf{K}^4 = \frac{1}{4}\begin{pmatrix} 1 & 0 & 0 & 0 & -1 \\ 0 & 0 & 0 & 0 & 0 \\ 0 & 0 & 0 & 0 & 0 \\ 0 & 0 & 0 & 1 & -1 \\ -1 & 0 & 0 & -1 & 2 \end{pmatrix} \tag{6.78}$$

Consequently:

$$\hat{\mathbf{T}} = \mathbf{K}^1 + \mathbf{K}^2 + \mathbf{K}^3 + \mathbf{K}^4 = \frac{1}{4}\begin{pmatrix} 2 & 0 & 0 & 0 & -2 \\ 0 & 2 & 0 & 0 & -2 \\ 0 & 0 & 2 & 0 & -2 \\ 0 & 0 & 0 & 2 & -2 \\ -2 & -2 & -2 & -2 & 8 \end{pmatrix} \tag{6.79}$$

For the RHS we have:

$$\hat{\mathbf{f}} = \sum_{e=1}^{4} \mathbf{f}^e = \frac{1}{12}\left[\begin{pmatrix} 1 \\ 1 \\ 0 \\ 0 \\ 1 \end{pmatrix} + \begin{pmatrix} 0 \\ 1 \\ 1 \\ 0 \\ 1 \end{pmatrix} + \begin{pmatrix} 0 \\ 0 \\ 1 \\ 1 \\ 1 \end{pmatrix} + \begin{pmatrix} 1 \\ 0 \\ 0 \\ 1 \\ 1 \end{pmatrix}\right] = \frac{1}{12}\begin{pmatrix} 2 \\ 2 \\ 2 \\ 2 \\ 4 \end{pmatrix} \tag{6.80}$$

□

5. *Boundary conditions:* We now apply the constraints to the assembled system of equations. Again, we assume that the nodes $\mathbf{r}_1, \ldots, \mathbf{r}_{n_{\mathrm{dof}}} \notin \Gamma_D$ lie outside the Dirichlet boundary, whereas the nodes $\mathbf{r}_{n_{\mathrm{dof}}+1}, \ldots, \mathbf{r}_m \in \Gamma_D$ lie on the Dirichlet boundary. Hence, the parameters $a_{n_{\mathrm{dof}}+1}, \ldots, a_m$ are prescribed by the Dirichlet conditions, whereas $a_1, \ldots, a_{n_{\mathrm{dof}}}$ are determined by a linear system of equations.

The system of Equations (6.75) may be rewritten as

$$\hat{\mathbf{T}}\mathbf{a} = \begin{pmatrix} \hat{\mathbf{T}}_{\mathrm{free,free}} & \hat{\mathbf{T}}_{\mathrm{free,fix}} \\ \hat{\mathbf{T}}_{\mathrm{fix,free}} & \hat{\mathbf{T}}_{\mathrm{fix,fix}} \end{pmatrix} \begin{pmatrix} \mathbf{a}_{\mathrm{free}} \\ \mathbf{a}_{\mathrm{fix}} \end{pmatrix} = \begin{pmatrix} \mathbf{f}_{\mathrm{free}} + \mathbf{g}_{\mathrm{free}} \\ \mathbf{f}_{\mathrm{fix}} + \mathbf{g}_{\mathrm{fix}} \end{pmatrix} \tag{6.81}$$

Here, the subscript *free* denotes the indices $1, \ldots, n_{\text{dof}}$ and the subscript *fix* denotes the indices $n_{\text{dof}} + 1, \ldots, m$ (for example, $\hat{\mathbf{T}}_{\text{free,free}} = (\hat{T}_{i,j})_{i,j=1,\ldots,n_{\text{dof}}}$ and $\hat{\mathbf{T}}_{\text{free,fix}} = (\hat{T}_{i,j})_{i=1,\ldots,n_{\text{dof}},\,j=n_{\text{dof}}+1,\ldots,m}$).

The vector $\mathbf{a}_{\text{free}} = (a_{n_{\text{dof}}+1}, \ldots, a_m)^{\text{T}}$ is given by the Dirichlet conditions: $a_i = u_D(\mathbf{r}_i)$ for $i = n_{\text{dof}} + 1, \ldots, m$. Hence, the first rows of Equation (6.81) are

$$\left(\hat{\mathbf{T}}_{\text{free,free}} \quad \hat{\mathbf{T}}_{\text{free,fix}} \right) \begin{pmatrix} \mathbf{a}_{\text{free}} \\ \mathbf{a}_{\text{fix}} \end{pmatrix} = \mathbf{f}_{\text{free}} + \mathbf{g}_{\text{free}} \tag{6.82}$$

and hence, the remaining unknowns $\mathbf{a}_{\text{free}}$ are the solutions of the linear system

$$\hat{\mathbf{T}}_{\text{free,free}} \mathbf{a}_{\text{free}} = \mathbf{f}_{\text{free}} + \mathbf{g}_{\text{free}} - \hat{\mathbf{T}}_{\text{free,fix}} \mathbf{a}_{\text{fix}} \tag{6.83}$$

Example 21 (Example 9 revisited)
The values for the four nodes on the boundary are given by the Dirichlet BC. Only the interior node is free. Equation (6.83) reduces to

$$\hat{\mathbf{T}}_{\text{free,free}} a_5 = \mathbf{f}_{\text{free}} - \hat{\mathbf{T}}_{\text{free,fix}} \mathbf{a}_{\text{fix}} \tag{6.84}$$

$$2a_5 = \frac{4}{12} - \frac{1}{2}(-1, -1, -1, -1) \begin{pmatrix} 0 \\ 0 \\ 0 \\ 0 \end{pmatrix} \tag{6.85}$$

so that $a_5 = \frac{1}{6}$. This numerical value can be interpreted as the height of the numerical solution at node 5. $\square$

6. *Solution:* Here, similar comments hold as for the 1D Galerkin approach.

7. *Computation of derived variables:* Here, similar comments hold as for the 1D Galerkin approach.

6.3.2 The Galerkin method with linear elements (rectangular elements)

Figure 6.11 shows the construction of a quadrilateral element. On the left, the reference element is shown with coordinates ξ and η. The coordinates of the reference element are mapped into the real space by the coordinate transformations $\begin{pmatrix} x(\xi, \eta) \\ y(\xi, \eta) \end{pmatrix}$, which has the inverse mapping $\begin{pmatrix} \xi(x, y) \\ \eta(x, y) \end{pmatrix}$. Just as for the triangular element, the basis functions are defined on the reference element and mapped into the physical space as $N_i^e(x, y) = N_i(\xi(x, y), \eta(x, y))$, where the ith local ansatz function, N_i, is defined as a bilinear function which is zero in all nodes of the reference element apart from the ith node, where it is equal to one.

6.4 CASE STUDIES

6.4.1 Brownian motion leaving a disk

The problem to be solved is given by the elliptic PDE:

$$\frac{1}{2} \left(\frac{\partial^2 u}{\partial x^2} + \frac{\partial^2 u}{\partial y^2} \right) = -1 \tag{6.86}$$

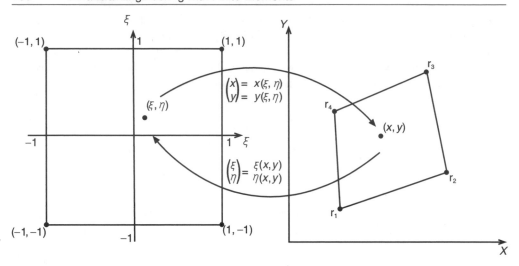

Figure 6.11 Construction of a quadrilateral element with bilinear shape functions

together with the BC $u(x, y) = 0$ on the entire boundary. The domain is a circle of radius $R = 2$. The analytical solution to this problem is given by:

$$u(x, y) = \frac{4 - x^2 - y^2}{2} \tag{6.87}$$

This example shows that FE can easily deal with nonrectangular domains, as shown in Figure 6.12. Here, we use a Galerkin approach with triangular elements of degree 1 and 2. To emphasize the effect of linear elements, in Figure 6.13 less elements than in Figure 6.12 have been employed. In Figure 6.14, numerical and analytical solutions coincide.

6.4.2 Ritz revisited

To demonstrate the Ritz approach in a 2D setting, we consider the first exit time of a 2D Brownian motion on a square given by $[0, 1] \times [0, 1]$. The PDE is given by

$$\frac{u_{xx}}{2} + \frac{u_{yy}}{2} = -1 \tag{6.88}$$

Together with $u(x, y) = 0$ on the entire boundary, this constitutes a well-posed problem. According to Theorem 33 (see Section A.4), the associated variational problem is given by:

$$\max / \min V[f(x, y)] = \int_0^1 \int_0^1 \left(\frac{u_x^2}{4} + \frac{u_y^2}{4} - u \right) dx \, dy \tag{6.89}$$

First we choose $m + 1$ functions ϕ_i $(i = 0, \ldots, m)$ in such a way that the boundary condition is satisfied. In this problem this is simply $\phi_0 = 0$, since $u(x, y)$ is zero on the boundary

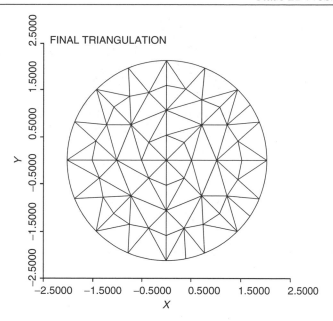

Figure 6.12 Triangulation for Brownian motion on a disk

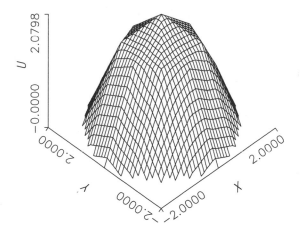

Figure 6.13 Brownian motion on a disk (linear elements)

everywhere. The ansatz is:

$$u(x, y) \approx \tilde{u}(x, y) = \sum_{k=0}^{m} a_k \phi_k(x, y) \qquad (6.90)$$

with the a_i being constants to be determined at a later stage. The variational problem demands that the term in the integral expression be maximized or minimized, i.e. we are looking for a function that maximizes or minimizes the value of the integral. The basic idea of Ritz's

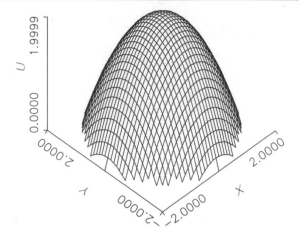

Figure 6.14 Brownian motion on a disk (quadratic elements)

approach is to fix the functional form of $\tilde{u}(x, y, a_1, \ldots, a_m)$ and to find parameters $a_1, \ldots, a_m$ that make $\tilde{u}$ the 'best' approximate solution of all functions of the given functional form. Inserting Equation (6.90) into Equation (6.89) leads to:

$$V(a_1, \ldots, a_m) = \int_0^1 \int_0^1 \left[\frac{1}{4} \left(\sum_{k=1}^m a_k \frac{\partial \phi_k}{\partial x} \right)^2 + \frac{1}{4} \left(\sum_{k=1}^m a_k \frac{\partial \phi_k}{\partial y} \right)^2 - \left(\sum_{k=1}^m a_k \phi_k \right) \right] dx \, dy$$

(6.91)

This way, the *dynamic* variational problem has been reduced to a *static* optimization problem. The first order necessary condition for an (unconstrained) static optimization problem is a vanishing gradient:

$$\frac{\partial V}{\partial a_1} \overset{!}{=} 0, \quad \ldots, \quad \frac{\partial V}{\partial a_m} \overset{!}{=} 0$$

(6.92)

Applied to the above problem, this leads to:

$$\frac{\partial V}{\partial a_j} = \int_0^1 \int_0^1 \left[\sum_{k=1}^m \frac{a_k}{2} \frac{\partial \phi_k}{\partial x} \frac{\partial \phi_j}{\partial x} + \sum_{k=1}^m \frac{a_k}{2} \frac{\partial \phi_k}{\partial y} \frac{\partial \phi_j}{\partial y} - \phi_j \right] dx \, dy \overset{!}{=} 0$$

(6.93)

for $j = 1, \ldots, m$. Choosing $m = 2$ and

$$\phi_1 = xy(1 - x)(1 - y)$$

(6.94)

$$\phi_2 = x^2 y^2 (1 - x)^2 (1 - y)^2 = \phi_1^2$$

(6.95)

we arrive at:

$$\frac{\partial V}{\partial a_1} = \int_0^1 \int_0^1 \left[\frac{a_1}{2} \left(\frac{\partial \phi_1}{\partial x} \frac{\partial \phi_1}{\partial x} + \frac{\partial \phi_1}{\partial y} \frac{\partial \phi_1}{\partial y} \right) + \frac{a_2}{2} \left(\frac{\partial \phi_2}{\partial x} \frac{\partial \phi_1}{\partial x} + \frac{\partial \phi_2}{\partial y} \frac{\partial \phi_1}{\partial y} \right) - \phi_1 \right] dx \, dy$$

(6.96)

$$\frac{\partial V}{\partial a_2} = \int\limits_0^1 \int\limits_0^1 \left[\frac{a_1}{2}\left(\frac{\partial \phi_1}{\partial x}\frac{\partial \phi_2}{\partial x} + \frac{\partial \phi_1}{\partial y}\frac{\partial \phi_2}{\partial y}\right) + \frac{a_2}{2}\left(\frac{\partial \phi_2}{\partial x}\frac{\partial \phi_2}{\partial x} + \frac{\partial \phi_2}{\partial y}\frac{\partial \phi_2}{\partial y}\right) - \phi_2 \right] dx\ dy$$

(6.97)

Applying Equation (6.92) to both of the above expressions, and after some algebraic manipulations leading to $a_1 = \frac{37}{13}$ and $b = -\frac{105}{13}$, the approximate solution is given by:

$$\tilde{u}(x, y) = \frac{37}{13}xy(1 - x)(1 - y) - \frac{105}{13}x^2y^2(1 - x)^2(1 - y)^2$$

(6.98)

Since the PDE at hand is self-adjoint, Galerkin's method results in the same approximate solution. The residual is given by:

$$R(x, y, a_1, a_2) = \frac{1}{2}\left(\tilde{u}_{xx} + \tilde{u}_{yy}\right) + 1$$

(6.99)

$$= \frac{1}{2}\left(a_1\frac{\partial^2 \phi_1}{\partial x^2} + a_2\frac{\partial^2 \phi_2}{\partial x^2} + a_1\frac{\partial^2 \phi_1}{\partial y^2} + a_2\frac{\partial^2 \phi_2}{\partial y^2}\right) + 1$$

(6.100)

Applying the Ritz criterion:

$$\int\limits_0^1 \int\limits_0^1 R\ \phi_1\ dx\ dy \overset{!}{=} 0$$

(6.101)

$$\int\limits_0^1 \int\limits_0^1 R\ \phi_2\ dx\ dy \overset{!}{=} 0$$

(6.102)

leads to a system of two linear equations:

$$-\frac{70a_3 + 3a_4 - 175}{6300} = 0$$

(6.103)

$$-\frac{63a_3 + 4a_4 - 147}{132\,300} = 0$$

(6.104)

which is solved by $a_1 = \frac{37}{13}$ and $b = -\frac{105}{13}$, so that the approximate solution resulting from Galerkin's method is the same as the one computed with Ritz's method. Figure 6.15 shows the difference between this approach and a fine FE discretization.

6.4.3   First exit time in a two-asset pricing problem

The expected first exit time of an Itô process can be computed with the help of Dynkin's equation, see Section B.3.5. In the case of two uncorrelated geometric Brownian motions, Dynkin's equation turns out to be:

$$\frac{1}{2}\left(\sigma_1^2 S_1^2 \frac{\partial^2 u}{\partial S_1^2} + \sigma_2^2 S_2^2 \frac{\partial^2 u}{\partial S_2^2}\right) + (r - D_1)S_1 \frac{\partial u}{\partial S_1} + (r - D_2)S_2 \frac{\partial u}{\partial S_2} = -1$$

(6.105)

with $u(S_1, S_2) = 0$ on the entire boundary. We assume the domain of this PDE to be a square.

Table 6.4   Data for the first exit time of a 2D geometric Brownian motion

Parameter	Symbol	Value
Lower bound of first asset	S_1^{min}	10.0
Upper bound of first asset	S_1^{max}	20.0
Lower bound of second asset	S_2^{min}	10.0
Upper bound of second asset	S_2^{max}	30.0
Volatility of first asset	σ_1	0.2
Volatility of second asset	σ_2	0.3
Dividend yield of first asset	D_1	0.02
Dividend yield of second asset	D_2	0.03
Interest rate	r	0.1

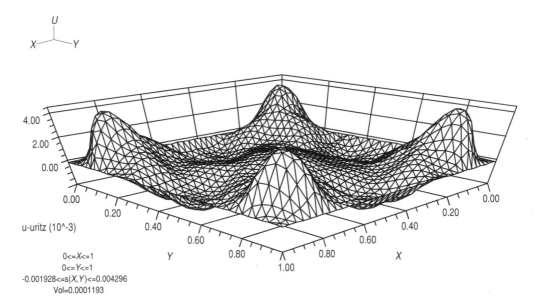

Figure 6.15   Difference between Ritz's method and a fine mesh FE solution

This is a well-posed linear elliptic problem. Together with the data given in Table 6.4, we will produce several numerical solutions with various meshes.

The results shown in Table 6.5 were computed with a collocation method with cubic Hermite basis functions, as discussed in Section 8.1. The initial meshing of one rectangular element was refined up to 2500 elements on a regular grid, i.e. all spacing between the nodes in each direction was equidistant. The last two columns of Table 6.2 show that increasing the number of elements from 100 to 2500 hardly changes the approximate solution. This is also demonstrated in Figure 6.16. The coefficient matrix associated with the 10×10 discretization is of size 484×484; the coefficient matrix associated with the 50×50 discretization is of size $10\,404 \times 10\,404$.

Table 6.5 Results for the first exit time of a 2D geometric Brownian motion ($S_2 = 20$, collocation method)

S_1	Number of elements			
	1×1	3×3	10×10	50×50
11	1.3331	1.0806	1.0372	1.0372
12	2.1937	1.5526	1.5054	1.5055
13	2.6481	1.6745	1.6725	1.6726
14	2.7620	1.6659	1.6660	1.6650
15	2.6018	1.5518	1.5481	1.5481
16	2.2335	1.3515	1.3504	1.3504
17	1.7231	1.0888	1.0892	1.0892
18	1.1366	0.7737	0.7735	0.7735
19	0.5403	0.4092	0.4088	0.4088

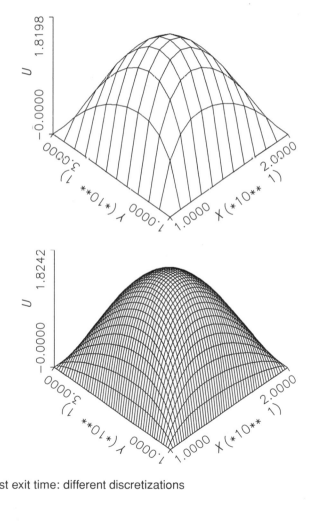

Figure 6.16   First exit time: different discretizations

6.5 NOTES

The automatic construction of a mesh is – in general, and especially for 3D problems – a nontrivial problem. Many techniques exist and a significant amount of research is currently conducted in this area. However, for the purpose of option pricing, this is of little importance since contracts demanding really complicated geometries have not appeared yet. Even the geometry of a multi-asset option with several barriers is far less complicated than a car door. The interested reader who wants to learn more about grid generation methods is referred to Liseikin (1999). More examples for nonstandard domains, as arising in the pricing and hedging of convertible bonds, can be found in Ouachani and Zhang (2004).

Dynamic 2D Problems

7.1 DERIVATION OF ELEMENT EQUATIONS

We now present a semidiscretization for solving the parabolic partial differential equation

$$-\nabla \cdot (\mathbf{D}\nabla u(x, y, t)) + \mathbf{b} \cdot \nabla u(x, y, t) \tag{7.1}$$

$$+ cu(x, y, t) - f(x, y, t) = u_t(x, y) \in \Omega$$

$$u(x, y, t) = u_D(x, y, t), \quad (x, y) \in \Gamma_D \tag{7.2}$$

$$\mathbf{n} \cdot (\mathbf{D}\nabla u(x, y, t)) + \gamma u(x, y, t) = u_N(x, y, t), \quad (x, y) \in \Gamma_N \tag{7.3}$$

$$u(x, y, t_0) = u_0(x, y) \tag{7.4}$$

on a three-dimensional domain $\Omega \times [t_0, t^{\max}]$, where Γ_N and Γ_D are a decomposition of the boundary $\partial\Omega$. This is the most general form of a linear parabolic PDE of second order with two spatial variables. The domain is either bounded in the x, y-plane, as in Figure 7.1, or the problem at hand has to be modified by techniques described in Section 3.1.2 in a way that results in a bounded domain. The time domain is usually a closed interval, since one is interested in the price of a contingent claim as of some specific point in time $t^{\max}$. The parameters $\mathbf{D}, \mathbf{b}, c$ and f are functions on the domain, possibly also depending on t. $\mathbf{D}(x, y, t)$ is a positive-definite matrix, $\mathbf{b}(x, y, t)$ is a vector and $c(x, y, t)$ and $f(x, y, t)$ are scalars. Dirichlet conditions are imposed on Γ_D in Equation (7.2) and mixed or Neumann boundary conditions are imposed on Γ_N in Equation (7.3). Here, $\gamma(x, y) \geq 0$ and the unit normal on Γ_N is denoted by $\mathbf{n}(x, y)$. The BCs do not move in space as time evolves. The semidiscretization employs a Galerkin FE approach, as in Chapter 6, for the spatial variables, and FD for time. Since this semidiscretization is very similar to the case with only one spatial variable, as discussed in Chapter 5, the exposition will be succinct.

1. *Discretization:* Since only the spatial variables are discretized with FE, this step can be performed in just the same manner as in the 2D static problem. In cases of an initial condition (Equation (7.4)) with kinks, as is often the case in option pricing, it is usually recommendable to construct the mesh in such a way that these kinks are located at the intersection of individual elements.

2. *Interpolation:* Employing the usual separation of variables technique, the interpolation function is given by:

$$\tilde{u}(x, y, t) = \sum_{i=1}^{m} \tilde{u}_i(t) n_i(x, y) \tag{7.5}$$

Just as in the 1D dynamic case, the nodal values are now functions of time, instead of constants as in the 2D static problem.

3. *Elemental formulation and Assembly:* In Section 6.3 we saw that the discretized elliptic PDE leads to a linear algebraic system, no matter what shape function or element geometry was employed. This feature carries over to the semidiscretization of linear parabolic PDEs,

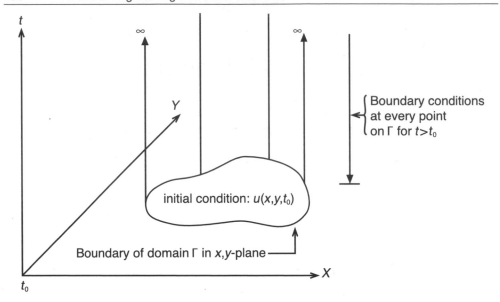

Figure 7.1   Domain of a parabolic PDE: two spatial variables plus time (Burnett, 1987)

as given by Equation (7.1). Just as in the case of one spatial variable, the unconstrained set of differential equations is given by:

$$\mathbf{A\bar{u}} + \mathbf{B\dot{\bar{u}}} = \mathbf{q} \tag{7.6}$$

4. *Boundary conditions:* The constraints arise from the boundary conditions specified in Equation (7.2), i.e. that part of the boundary where a Dirichlet condition is prescribed. The values of the constraints are determined from the u_D function, with the constrained value of u at a node on Γ_D being taken as the value of u_D at that point. Augmenting the unconstrained set of equations (7.6) with these constraints, as in Section 5.1.1, leads to the final global constrained set of linear first order differential equations:

$$\mathbf{M\dot{u}} + \mathbf{Ku} = \mathbf{f} \tag{7.7}$$

5. *Solution:* The system of ordinary initial value problems resulting from semidiscretization of a 2D parabolic problem is precisely of the same form and character as the one resulting from a semidiscretization of a 1D problem. The initial conditions are taken from Equation (7.4):

$$\tilde{u}(t_0) = u(x, y, t_0) \tag{7.8}$$

Just as in the case of 1D parabolic problems, the above system can be solved with the techniques introduced in Section 3.2.9.

6. *Computation of derived variables:* Since the derived variables are time dependent, they need to be computed at each time-step. At each time-step the same comments hold as for the 2D static problem.

Table 7.1   Data for a put on a basket

Parameter	Value
First asset price	18
Weight first asset	1
Second asset price	20
Weight second asset	1
Strike price	40
Interest rate	0.1
Dividend yields	0.0
Correlation	0.5

7.2   CASE STUDIES

7.2.1   Various rainbow options

Put on a basket

For options on baskets, at present there is no known analytical solution (Hunziker and Koch-Medina, 1996, p. 161). Therefore, this option has to be priced with a numerical device or an approximation (Huynh, 1994; Milevsky and Posner, 1998; Zhang, 1998b, Chapter 27). The basic idea of these approximations is to combine the volatilities of the underlying and their correlations to a single volatility of the basket. This basket is then treated as a single underlying. Using this approach, the problem of pricing an option on a basket is reduced to pricing an option on a single equity. Accordingly, the models to price options with exotic features can also be applied to options on baskets. Precise error estimates are generally not provided (Hunziker and Koch-Medina, 1996, p. 163). Here, however, we price options on baskets using a multidimensional PDE. For a plain vanilla put we first derive the boundary conditions. As one or both underlyings become worth much more than the strike, the price of the options goes to zero. As the price of the first underlying is zero, while the second is positive, the value of the option behaves like the value of a plain vanilla put on a single equity. Therefore, the boundary conditions at $S_1 = 0$ and $S_2 = 0$ are the (time-dependent) solution to the basic Black–Scholes problem of pricing a put (Hunziker and Koch-Medina, 1996, p. 162) with strikes at $\frac{E}{w_2}$ and $\frac{E}{w_1}$, respectively. Together with the data shown in Table 7.1, this becomes the following PDE problem.

$$\frac{1}{2}\sigma_1^2 S_1^2 \frac{\partial^2 V}{\partial S_1^2} + \frac{1}{2}\sigma_2^2 S_2^2 \frac{\partial^2 V}{\partial S_2^2} + \rho\sigma_1\sigma_2 S_1 S_2 \frac{\partial^2 V}{\partial S_1 \partial S_2} + \qquad (7.9)$$

$$(r - D_1)S_1 \frac{\partial V}{\partial S_1} + (r - D_2)S_2 \frac{\partial V}{\partial S_2} = rV - \frac{\partial V}{\partial t}$$

$$V(S_1, S_2, T) = \max(0, E - (w_1 S_1 + w_2 S_2)) \qquad (7.10)$$

$$V(S_1, 0, t) = g(S_1, \frac{E}{w_2}, t) \qquad (7.11)$$

$$V(0, S_2, t) = g(S_2, \frac{E}{w_1}, t) \qquad (7.12)$$

$$V(100, S_2, t) = 0 \qquad (7.13)$$

$$V(S_1, 100, t) = 0 \qquad (7.14)$$

Table 7.2 Results for a put option on a basket computed on a square domain

Volatility		Time to maturity			
σ_1	σ_2	0.05	0.5	0.95	Premium
		1.8025	0.9543	0.6043	Tree
	0.1	1.8065	0.9543	0.6035	FEM
		0.2204 %	0.0022 %	0.1352 %	Difference
		1.8333	1.4756	1.2408	Tree
0.1	0.2	1.8341	1.4764	1.2405	FEM
		0.0473 %	0.0512 %	0.0305 %	Difference
		1.9118	2.0186	1.9265	Tree
	0.3	1.9138	2.0187	1.9270	FEM
		0.1034 %	0.0041 %	0.0242 %	Difference
		1.8271	1.4120	1.1607	Tree
	0.1	1.8275	1.4127	1.1601	FEM
		0.0236 %	0.0492 %	0.0489 %	Difference
		1.8859	1.8835	1.7758	Tree
0.2	0.2	1.8856	1.8833	1.7754	FEM
		0.0076 %	0.0125 %	0.0202 %	Difference
		1.9816	2.3941	2.4389	Tree
	0.3	1.9830	2.3942	2.4389	FEM
		0.00602 %	0.0024 %	0.0004 %	Difference
		1.8906	1.8941	1.7649	Tree
	0.1	1.8915	1.8948	1.7647	FEM
		0.0451 %	0.0395 %	0.0108 %	Difference
		1.9683	2.3301	2.3557	Tree
0.3	0.2	1.9687	2.3298	2.3555	FEM
		0.0210 %	0.0138 %	0.0095 %	Difference
		2.0739	2.8112	2.9985	Tree
	0.3	2.0747	2.8119	2.9979	FEM
		0.0360 %	0.0021 %	0.0181 %	Difference

Here, the g functions denote a plain vanilla European put with strikes $\frac{E}{w_2}$ and $\frac{E}{w_1}$ and appropriate volatilities. We compute the cumulative normal distributions in Equations (7.11) and (7.12) with an approximation with four-digit accuracy from Hull (2000). To compare the results, we also price the put on a basket using a two-dimensional binomial tree, as implemented by Haug (1997). This tree can be interpreted as a simple explicit finite difference scheme; see Wilmott *et al.* (1996). The results are shown in Table 7.2.

As an alternative to pricing this option on a square domain, we also price it on a triangle (the results are shown in Table 7.3). Basically, we cut off the section of the domain where the option is totally out of the money and therefore worthless. This, of course, reduces computing time. The boundary conditions (7.11) to (7.14) are replaced by:

$$V(S_1, 0, t) = g(S_1, \frac{E}{w_2}, t) \tag{7.15}$$

$$V(0, S_2, t) = g(S_2, \frac{E}{w_1}, t) \tag{7.16}$$

$$V(S_1, S_2, t) = 0 \text{ on } \overline{S_1^{\max} S_2^{\max}} \tag{7.17}$$

Table 7.3 Results for a put option on a basket computed on a triangular domain

Volatility		Time to maturity			
σ_1	σ_2	0.05	0.5	0.95	Premium
		1.8026	0.9543	0.6043	Tree
	0.1	1.8035	0.9545	0.6035	FEM
		0.0498 %	0.0155 %	0.0938 %	Difference
		1.8333	1.4756	1.2408	Tree
0.1	0.2	1.8334	1.4770	1.2407	FEM
		0.0048 %	0.0899 %	0.0097 %	Difference
		1.9118	2.0186	1.9265	Tree
	0.3	1.9135	2.0187	1.9262	FEM
		0.0919 %	0.0053 %	0.0174 %	Difference
		1.8271	1.4120	1.1607	Tree
	0.1	1.8265	1.4119	1.1604	FEM
		0.0324 %	0.0095 %	0.0247 %	Difference
		1.8859	1.8835	1.7758	Tree
0.2	0.2	1.8850	1.8834	1.7753	FEM
		0.0468 %	0.0033 %	0.0256 %	Difference
		1.9818	2.3941	2.4389	Tree
	0.3	1.9826	2.3940	2.4387	FEM
		0.0426 %	0.0044 %	0.0098 %	Difference
		1.8906	1.8941	1.7649	Tree
	0.1	1.8908	1.8937	1.7644	FEM
		0.0076 %	0.0218 %	0.0283 %	Difference
		1.9683	2.3301	2.3557	Tree
0.3	0.2	1.9679	2.3299	2.3555	FEM
		0.0165 %	0.0076 %	0.0081 %	Difference
		2.0739	2.8120	2.9985	Tree
	0.3	2.0747	2.8120	2.9988	FEM
		0.0368 %	0.0006 %	0.0107 %	Difference

Although it is possible to adjust FD schemes for nonrectangular domains (Ames, 1992, Section 1.9; Bellomo and Preziosi, 1995, p. 258f; Collatz, 1966; Mitchell and Griffiths, 1980, Chapter 2; Morton and Mayer, 1994, Section 3.4), FE is the natural choice. This is even more true for Neumann conditions, which are difficult to integrate into more advanced FD schemes, e.g. in the case of curved boundaries. In an FE setting, however, Neumann conditions are even easier to consider than Dirichlet conditions.

Call on a basket

Rubinstein (1991b) reports results on pricing a call on a basket with the following data, using a two-dimensional binomial tree. These results are reprinted in Table 7.4. With a hundred time-steps he achieves the results given in Table 7.5. In order to achieve higher accuracy, we redo the example with 200 time-steps. These computations were performed with an implementation of the two-dimensional binomial tree by Haug (1997). We compare these results to FE results using the Black–Scholes formula for calls at $S_1 = 0$ and $S_2 = 0$ as boundary conditions. We solve the appropriate PDE on a triangular domain assuming a 90 degree outward pointing gradient on the third side. The results are given in Table 7.6. Redoing the above example

Table 7.4   Data for Rubinstein's example

Parameter	Value
First asset price	100
Weight first asset	1
Second asset price	100
Weight second asset	1
Strike price	200
Interest rate	0.0953102
Dividend yield first asset	0.0487902
Dividend yield second asset	0.0
Correlation	0.5

Table 7.5   Rubinstein's example with 100 time-steps

Volatility		Time to maturity		
σ_1	σ_2	0.05	0.5	0.95
	0.1	1.92	8.97	14.70
0.1	0.2	2.72	11.22	17.45
	0.3	3.58	13.70	20.59
	0.1	2.72	11.15	17.28
0.2	0.2	3.45	13.33	20.13
	0.3	4.24	15.72	23.25
	0.1	3.57	13.56	20.25
0.3	0.2	4.24	15.65	23.08
	0.3	4.99	17.94	26.16

assuming a zero gradient at $S_1 = 0$ and $S_2 = 0$ leads to slightly less accurate results. This Neumann condition is obviously the easiest to apply. These results are given in Table 7.7.

Single barrier knock-out call on a basket: without rebate

Rainbow options with knock-out barriers lead to nonrectangular domains. Examples of this product type are barrier options with various knock-out features. Common knock-out features in the OTC market are up-and-out barriers and down-and-out barriers on the basket with continuous and discrete monitoring. It is possible, however, to construct even more exotic barrier structures with individual barriers on the single underlyings, or combinations which go into the basket (Pooley et al., 2000).

For the knock-out call on a basket, the boundaries for $S_1 = 0$ and $S_2 = 0$ first have to be computed numerically or analytically. The third boundary is the rebate $R = 10$. This leads to the following system of PDEs. Equations (7.18) to (7.21) denote a barrier call on S_1 with $S_2 = 0$. Accordingly, Equations (7.22) to (7.25) denote a barrier call on S_2 with $S_1 = 0$

$$\frac{\partial V_1}{\partial t} + rS\frac{\partial V_1}{\partial S_1} + \frac{1}{2}\sigma_1^2 S_1^2 \frac{\partial^2 V_1}{\partial S_1^2} = rV_1 \tag{7.18}$$

$$V(T, S_1) = \max(S_1 - E, 0) \tag{7.19}$$

$$V(t, 0) = 0 \tag{7.20}$$

Table 7.6 Rubinstein's example with Dirichlet boundary conditions

Volatility		Time to maturity			
σ_1	σ_2	0.05	0.5	0.95	Method
		1.9202	8.9685	14.6980	Tree
	0.1	1.9198	8.9708	14.6979	FEM
		0.02229 %	0.02548 %	0.00055 %	Difference
0.1		2.7244	11.2200	17.4460	Tree
	0.2	2.7316	11.2200	17.4451	FEM
		0.2643 %	0.0007 %	0.0050 %	Difference
		3.5738	13.6926	20.5925	Tree
	0.3	3.5805	13.6928	20.5872	FEM
		0.1857 %	0.0014 %	0.0258 %	Difference
		2.7222	11.1506	17.2815	Tree
	0.1	2.7317	11.1497	17.2804	FEM
		0.3368 %	0.0087 %	0.0061 %	Difference
		3.4482	13.3308	20.1281	Tree
0.2	0.2	3.4563	13.3295	20.1273	FEM
		0.2357 %	0.0093 %	0.0038 %	Difference
		4.2420	15.7150	23.2494	Tree
	0.3	4.2473	15.7138	23.2472	FEM
		0.1252 %	0.0079 %	0.0097 %	Difference
		3.5689	13.5534	20.2429	Tree
	0.1	3.5779	13.5501	20.2406	FEM
		0.2530 %	0.0244 %	0.0110 %	Difference
0.3		4.2398	15.6421	23.0708	Tree
	0.2	4.2453	15.6408	23.0686	FEM
		0.1310 %	0.0090 %	0.0095 %	Difference
		4.9846	17.9382	26.1506	Tree
	0.3	4.9878	17.9362	26.1459	FEM
		0.0640 %	0.0110 %	0.01798 %	Difference

$$V(t, S_1^{\max}) = R \tag{7.21}$$

$$\frac{\partial V_2}{\partial t} + rS\frac{\partial V_2}{\partial S_2} + \frac{1}{2}\sigma_2^2 S_2^2 \frac{\partial^2 V_2}{\partial S_2^2} = rV_2 \tag{7.22}$$

$$V(T, S_2) = \max(S_2 - E, 0) \tag{7.23}$$

$$V(t, 0) = 0 \tag{7.24}$$

$$V(t, S_2^{\max}) = R \tag{7.25}$$

$$\frac{1}{2}\sigma_1^2 S_1^2 \frac{\partial^2 V_3}{\partial S_1^2} + \frac{1}{2}\sigma_2^2 S_2^2 \frac{\partial^2 V_3}{\partial S_2^2} +$$
$$\rho\sigma_1\sigma_2 S_1 S_2 \frac{\partial^2 V_3}{\partial S_1 \partial S_2} + \tag{7.26}$$
$$(r - q_1)S_1 \frac{\partial V_3}{\partial S_1} + (r - q_2)S_2 \frac{\partial V_3}{\partial S_2} = rV_3 - \frac{\partial V_3}{\partial t}$$

$$V_3(S_1, S_2, T) = \max(0, E - (w_1 S_1 + w_2 S_2)) \text{ in } D \tag{7.27}$$

$$V_3(S_1, 0, t) = V_1(S_1, t) \tag{7.28}$$

$$V_3(0, S_2, t) = V_2(S_2, t) \tag{7.29}$$

$$V_3(S_1, S_2, t) = R \text{ on } \overline{S_1^{\max} S_2^{\max}} \tag{7.30}$$

Table 7.7 Rubinstein's example with Neumann boundary conditions

Volatility		Time to maturity			
σ_1	σ_2	0.05	0.5	0.95	Method
		1.9202	8.9685	14.6980	Tree
	0.1	1.9198	8.9712	14.6982	FEM
		0.0223 %	0.0303 %	0.0020 %	Difference
0.1		2.7244	11.2199	17.4460	Tree
	0.2	2.7316	11.2210	17.4461	FEM
		0.2644 %	0.0097 %	0.0004 %	Difference
		3.5738	13.6926	20.5925	Tree
	0.3	3.5805	13.6942	20.5889	FEM
		0.1855 %	0.0112 %	0.0177 %	Difference
		2.7222	11.1507	17.2815	Tree
	0.1	2.7314	11.1506	17.2814	FEM
		0.3368 %	0.0005 %	0.0006 %	Difference
		3.4482	13.3308	20.1281	Tree
0.2	0.2	3.4563	13.3306	20.1286	FEM
		0.2358 %	0.0013 %	0.0028 %	Difference
		4.2420	15.7150	23.2494	Tree
	0.3	4.2473	15.7152	23.2488	FEM
		0.1253 %	0.0016 %	0.0024 %	Difference
		3.5689	13.5534	20.2429	Tree
	0.1	3.5779	13.5517	20.2421	FEM
		0.2530 %	0.0129 %	0.0035 %	Difference
0.3		4.2398	15.6421	23.0708	Tree
	0.2	4.2454	15.6420	23.0704	FEM
		0.1313 %	0.0006 %	0.0017 %	Difference
		4.9846	17.9382	26.1506	Tree
	0.3	4.9878	17.9376	26.1477	FEM
		0.0640 %	0.0031 %	0.0109 %	Difference

The parameters are again taken from Rubinstein's example. We solve this problem as a system, although it could be solved sequentially. Solving this problem as a system avoids having to feed the numerical solutions back into the program. We do not have a direct way of checking the results (shown in Table 7.8), but the premiums should be below the ones from Rubinstein's example due to the knock-out feature (which they are).

Capped call on a basket: analytical boundary conditions

The pricing of this product leads to the PDE (7.10) with the following initial and boundary conditions:

$$V(S_1, S_2, 0) = \min(\text{cap}, \max(0, E - (w_1 S_1 + w_2 S_2))) \tag{7.31}$$

$$V(S_1, 0, t) = g\left(S_1, \frac{E}{w_2}, t\right) - g(S_1, \text{cap}, t) \tag{7.32}$$

$$V(0, S_2, t) = g\left(S_2, \frac{E}{w_1}, t\right) - g(S_2, \text{cap}, t) \tag{7.33}$$

$$V(S_1, S_2, t) = 0 \quad \text{on } \overline{S_1^{\max} S_2^{\max}} \tag{7.34}$$

Table 7.8 Results for a knock-out call on a basket without rebate

Volatility		Time to maturity		
σ_1	σ_2	0.05	0.5	0.95
	0.1	1.7416	0.4645	0.1771
0.1	0.2	1.5198	0.1532	0.0581
	0.3	1.0738	0.0643	0.0246
	0.1	1.5218	0.1568	0.0608
0.2	0.2	1.1074	0.0727	0.0273
	0.3	0.7361	0.0375	0.0140
	0.1	1.0199	0.0665	0.0264
0.3	0.2	0.7368	0.0381	0.0144
	0.3	0.5108	0.0225	0.0083

Table 7.9 Results for a capped call on a basket with analytical boundary conditions

Volatility		Time to maturity		
σ_1	σ_2	0.05	0.5	0.95
	0.1	1.9062	5.3075	6.1865
0.1	0.2	2.5549	4.9940	5.4814
	0.3	3.0139	4.7564	5.0111
	0.1	2.5534	5.0002	5.5096
0.2	0.2	2.9701	4.8713	5.1819
	0.3	3.2831	4.7214	4.8849
	0.1	3.0124	4.7643	5.0459
0.3	0.2	3.2825	4.7260	4.9005
	0.3	3.4980	4.6436	4.7225

The data, again, are taken from Rubinstein's example with an additional cap of 10 on the basket. Again, this PDE can be solved either on a square or triangular domain. For reasons outlined above, we chose the triangle. The boundary conditions at $S_1 = 0$ and $S_2 = 0$ represent the prices of capped European call options with strike prices of $\frac{E}{w_2}$ and $\frac{E}{w_1}$, respectively. This capped call can be priced either by entering a bull spread and pricing its parts individually with the analytical formula for European calls, or numerically.

Again, we do not have a direct way of checking the results (shown in Table 7.9). By arbitrage considerations, however, each premium should be worth more than the example without rebate of Table 7.8.

Capped call on a basket: gradient boundary conditions

An alternative to Equations (7.32) and (7.33) is to assume a zero gradient on $S_1 = 0$ and $S_2 = 0$. The results differ only slightly, see Table 7.10.

7.2.2 Modeling volatility as a risk factor

Contradicting the Black–Scholes model, empirical tests show that σ is not constant. One way to adapt the model is to introduce a local volatility function $\sigma(T, t, S, E)$, as described in

Table 7.10   Results for a capped call on a basket with gradient boundary conditions

Volatility		Time to maturity		
σ_1	σ_2	0.05	0.5	0.95
0.1	0.1	1.9073	5.3112	6.1885
	0.2	2.5604	4.9959	5.4823
	0.3	3.0206	4.7575	5.0116
0.2	0.1	2.5559	5.0002	5.5110
	0.2	2.9735	4.8723	5.1825
	0.3	3.2873	4.7220	4.8853
0.3	0.1	3.0138	4.7654	5.0467
	0.2	3.2844	4.7266	4.9011
	0.3	3.5009	4.6442	4.7223

Table 7.11   The Stein–Stein model: the data

Parameter	Symbol	Value
Asset price	S	100
Strike price	E	90, 95, 100, 105, 110, 115, 120
Interest rate	r	$\ln(1.1) \approx 0.0953$
Maturity	T	1, 3, 6 (months)
Volatility	σ	0.25
Mean reversion level	θ	0.25
Mean reversion speed	δ	16
Volatility of volatility	γ	0.4
Market price of risk	λ	0.0

Section 10.2.4. Another possibility is to add a second risk factor, i.e. to assume that volatility is stochastic as well. Several suggestions have been made for this approach (Lipton, 2001); (Oztukel and Wilmott, 1998):

$$d\sigma = \alpha\sigma dt + \gamma\sigma dE \tag{7.35}$$

$$d\sigma = (\alpha - \beta\sigma)\sigma dt + \gamma\sigma dE \tag{7.36}$$

$$d\sigma = (\beta\alpha - \beta\sigma)dt + \gamma dE \tag{7.37}$$

$$d\sigma = (\frac{\alpha}{\sigma} - \beta\sigma)dt + \gamma dE \tag{7.38}$$

$$d\sigma = \left(\alpha - \beta\sigma^2\right)\sigma dt + \gamma\sigma^2 dE \tag{7.39}$$

$$d\sigma = \alpha^2\sigma^{2\gamma-1}\left(\gamma - \frac{1}{2} - \frac{\ln\frac{\sigma}{\bar{\sigma}}}{2\beta^2}\right)dt + \alpha\sigma^\gamma dE \tag{7.40}$$

The first three models go back to Hull and White (1987), Eisenberg and Jarrow (1991) and Stein and Stein (1991), respectively. It is often more convenient to model the evolution of the variance $v = \sigma^2$. Since there is some empirical evidence that volatility is mean-reverting, the following model with two risk factors has been suggested (Heston, 1993):

$$dS = (r - D)Sdt + \sqrt{v}d\acute{E}_1 \tag{7.41}$$

$$dv = \kappa(\bar{v} - v) + \epsilon\sqrt{v}dE_2 \tag{7.42}$$

with $\bar{v}$ being the mean-reversion level.

Table 7.12 The Stein–Stein model: results

TtM (months)	Strike	Black–Scholes			Stein–Stein				
		Price	Δ	Γ	Analytical	Approximate	FE	Δ_{FE}	Γ_{FE}
1	90	10.8820	0.9459	0.0152	10.92	10.91	10.9181	0.9408	0.0130
1	95	6.5630	0.8042	0.0383	6.61	6.61	6.6077	0.8012	0.0349
1	100	3.2800	0.5581	0.0547	3.31	3.34	3.3187	0.5560	0.0544
1	105	1.3121	0.2981	0.0480	1.35	1.37	1.3522	0.2980	0.0496
1	110	0.4137	0.1201	0.0277	0.46	0.45	0.4586	0.1255	0.0290
1	115	0.1029	0.0367	0.0111	0.14	0.12	0.1334	0.0435	0.0124
1	120	0.0204	0.0087	0.0032	0.04	0.03	0.0360	0.0134	0.0434
3	90	13.0334	0.8634	0.0175	13.13	13.13	13.1318	0.8597	0.0166
3	95	9.2622	0.7465	0.0594	9.37	9.39	9.3718	0.7448	0.0244
3	100	6.1912	0.5999	0.0309	6.30	6.34	6.3056	0.5993	0.0304
3	105	3.8828	0.4454	0.0316	4.00	4.04	3.9978	0.4457	0.0316
3	110	2.2844	0.3052	0.0280	2.40	2.43	2.4019	0.3071	0.0282
3	115	1.2631	0.1935	0.0220	1.38	1.38	1.3736	0.1974	0.0222
3	120	0.6583	0.1140	0.0154	0.76	0.75	0.7581	0.1201	0.0157
6	90	15.9352	0.8300	0.0143	16.09	16.10	16.0939	0.8256	0.0136
6	95	12.4430	0.7415	0.0183	12.62	12.65	12.6261	0.7387	0.0175
6	100	9.4539	0.6398	0.0212	9.65	9.68	9.6540	0.6387	0.0206
6	105	6.9906	0.5327	0.0225	7.20	7.23	7.1981	0.5334	0.0220
6	110	5.0345	0.4281	0.0222	5.24	5.28	5.2435	0.4308	0.0218
6	115	3.5355	0.3326	0.0206	3.74	3.76	3.7357	0.3369	0.0204
6	120	2.4246	0.2503	0.0180	2.61	2.63	2.6143	0.2562	0.0179

It has been shown by Stein and Stein (1991) that, under the assumption of Equation (7.37) for the volatility (while the rest remains the same as in the Black–Scholes model; see Section 2.2), any option V must satisfy

$$\frac{1}{2}\sigma^2 S^2 \frac{\partial^2 V}{\partial S^2} + rS\frac{\partial V}{\partial S} + \frac{1}{2}\gamma^2 \frac{\partial^2 V}{\partial \sigma^2} + (-\beta(\sigma - \alpha) - \lambda\gamma)\frac{\partial V}{\partial \sigma} - rV = \frac{\partial V}{\partial t} \qquad (7.43)$$

subject to appropriate initial and boundary conditions. There is no mixed term $V_{S\alpha}$ in equation (7.43) because the two risk factors driving S and σ are assumed to be not correlated. Because volatility is not a traded asset and therefore cannot be used to hedge, risk preferences of the investor cannot be eliminated. This is why λ as the market price of (volatility) risk enters the PDE. If we set $\lambda = 0$ (as we will do for the rest of the section) this implies that risk neutrality prevails. It does not mean that the PDE holds no matter what the attitude of the subjects towards risk is. For a more detailed discussion of this point see Stein and Stein (1991), Wiggins (1987) and Wilmott (1998). The PDE is defined on the semi-closed domain $(S, \sigma) \in [0, \infty) \times [0, \infty)$. For a plain vanilla European call the authors Stein and Stein derive a messy solution which has to be integrated numerically. The interested reader is referred to the original publication (Stein and Stein, 1991). They also derive an analytical approximation which is somewhat easier. The analytical solution derived by the authors consists only of algebraically simple functions, so that an implementation of that complicated formula seems to be worthwhile when speed is crucial. In order to explore the difference in pricing of their model to the Black–Scholes model, various scenarios with different parameter choices are set up. The data and results are shown in Tables 7.11 and 7.12.

A problem with stochastic volatility models is their long-term behavior (Rebonato, 1999, p. 88). Empirically, there is a big smile for short maturities, which diminishes for longer maturities. Stochastic volatility models, however, tell you the opposite. To overcome this weakness, one would have to make the model nonautonomous, which would not be very plausible. Practitioners prefer the Heston model to the Stein–Stein model because negative volatility can occur with a positive probability. An FE implementation of the Heston model has been produced by Apel *et al.* (2002).

8

Static 3D Problems

8.1 DERIVATION OF ELEMENT EQUATIONS: THE COLLOCATION METHOD

We now present the orthogonal collocation method for solving the PDE:

$$d_{11}u_{xx} + d_{22}u_{yy} + d_{33}u_{zz} + d_{12}u_{xy} + d_{12}u_{xz} + d_{13}u_{yz} + \tag{8.1}$$

$$b_1 u_x + b_2 u_y + b_3 u_z + c\, u(x, y, z) = f(x, y, z), \quad (x, y, z) \in \Omega \tag{8.2}$$

$$u(x, y, z) = u_D(x, y, z), \quad (x, y, z) \in \Gamma_D \tag{8.3}$$

At this point, we only consider Dirichlet conditions. For other types of BC see Sewell (2000).

1. *Discretization:* Since we only consider approximate solutions for 3D boxes ($[x^{\min}, x^{\max}] \times [y^{\min}, y^{\max}] \times [z^{\min}, z^{\max}]$) generated by quadrilateral elements, the discretization does not require any sophisticated techniques. It is sufficient to discretize any spatial direction separately. Each direction can be discretized independently, so that it is possible to have a fine net with respect to one variable and a coarse net with respect to another. There are n_x gridlines associated with x, so that $x^{\min} = x_1$ and $x^{\max} = x_{n_x}$. Similarly, n_y and n_z gridlines are associated with y and z, respectively. Nonuniform grid spacing is allowed. This discretization leads to $(n_x - 1)(n_y - 1)(n_z - 1)$ subrectangles. Associated with each subrectangle are eight collocation points with the following coordinates:

$$[x, y, z] = \left[x_i + \beta_{1,2}(x_{i+1} - x_i),\, y_j + \beta_{1,2}(y_{j+1} - y_j),\, z_k + \beta_{1,2}(z_{k+1} - z_k) \right] \tag{8.4}$$

$$\beta_1 = 0.5 - \frac{0.5}{\sqrt{3}} \tag{8.5}$$

$$\beta_2 = 0.5 + \frac{0.5}{\sqrt{3}} \tag{8.6}$$

Using the Gaussian points as above leads to higher accuracy compared to using grid nodes (Prenter, 1975).

2. *Interpolation:* The approximate solution is constructed as a linear combination of eight $n_x n_y n_z$ basis functions:

$$
\begin{array}{llr}
H_i(x)\, H_j(y)\, H_k(z), & H_i(x)\, H_j(y)\, S_k(z) & (8.7) \\
H_i(x)\, S_j(y)\, H_k(z), & H_i(x)\, S_j(y)\, S_k(z) & (8.8) \\
S_i(x)\, H_j(y)\, H_k(z), & S_i(x)\, H_j(y)\, S_k(z) & (8.9) \\
S_i(x)\, S_j(y)\, H_k(z), & S_i(x)\, S_j(y)\, S_k(z) & (8.10) \\
i = 1, \ldots, n_x, \quad j = 1, \ldots, n_y & k = 1, \ldots, n_z & (8.11)
\end{array}
$$

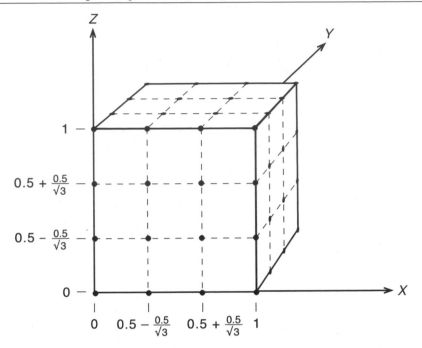

Figure 8.1   Boundary collocation points on a one-element solution

with H_i and S_i as defined in Equations (4.289) and (4.290), respectively. This choice of basis function results in C^∞ continuity with the element and C^1 continuity across element borders. Consequently, the numerical solution is C^1 everywhere in the domain, as required by the collocation method.

3. *Assembly:* Since there are eight $n_x n_y n_z$ basis functions, there are just as many unknowns in the linear system. Eight $(n_x - 1)(n_y - 1)(n_z - 1)$ equations result from the interior collocation points, the rest from boundary collocation points.

4. *Boundary conditions:* The BCs are enforced at the boundary collocation points. Their location follows the distribution of the interior collocation points. This is best explained with the help of Figures 8.1 and 8.2, where the intersections of dotted lines represent boundary collocation points.

5. *Solution:* Linear systems generated by collocation FE methods do not have a symmetric nonzero structure. In general, the matrix of coefficients is not diagonally dominant and many of its diagonal entries are zero. Consequently, it is not possible to employ standard solvers to exploit the sparse structure. Although this might present a tremendous problem for technical applications, many problems from finance and economics can be solved conveniently. With ten elements in each direction, i.e. $n_x = n_y = n_z = 11$, the discretized model consists of $8 \times 11^3 = 10\,648$ equations. Problems of this size can easily be computed on a modern PC, even with inverting the matrix of coefficients.

6. *Computation of derived variables:* As in the 1D collocation approach, analytical derivatives of the approximate solution $\tilde{u}(x, y, z)$ can be computed easily. This also holds for cross-derivatives such as $\tilde{u}_{xy}(x, y, z)$.

Table 8.1   First exit time of a 3D purely Brownian motion

x	y	z	u(x, y, z)
0.0	0.5	0.5	0.00000
0.1	0.5	0.5	0.04713
0.2	0.5	0.5	0.07865
0.3	0.5	0.5	0.09835
0.4	0.5	0.5	0.10904
0.5	0.5	0.5	0.11242
0.6	0.5	0.5	0.10904
0.7	0.5	0.5	0.09835
0.8	0.5	0.5	0.07865
0.9	0.5	0.5	0.04713
1.0	0.5	0.5	0.00000

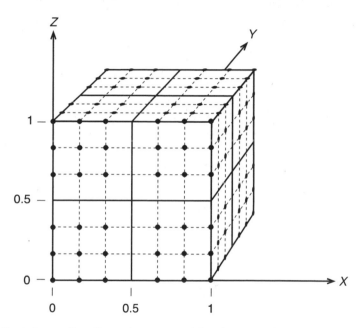

Figure 8.2   Boundary collocation points on an eight-element solution

8.2   CASE STUDIES

8.2.1   First exit time of purely Brownian motion

To demonstrate the collocation method in a 3D setting we consider the first exit time of a 3D Brownian motion on a square given by $[0, 1] \times [0, 1] \times [0, 1]$. The PDE is given by

$$\frac{u_{xx}}{2} + \frac{u_{yy}}{2} + \frac{u_{zz}}{2} = -1 \tag{8.12}$$

Together with $u(x, y, z) = 0$ on the entire boundary, this problem admits a unique solution by Theorem 26 (see Section A.3.6). Data are shown in Table 8.1.

In a one-element solution there are eight interior collocation points and 56 boundary collocation points, as shown in Figure 8.1. So altogether there are 64 equations to determine the 64 unknowns in the linear algebraic system resulting from the cubic Hermite shape functions. A one-element model implies $n_x = n_y = n_z = 2$, so that the approximate solution is given by:

$$\tilde{u}(x, y, z) = a_1 H_1(x)H_1(y)H_1(z) + a_2 H_1(x)S_1(y)H_1(z) + a_3 S_1(x)H_1(y)H_1(z) + \quad (8.13)$$
$$a_4 S_1(x)S_1(y)H_1(z) + a_5 H_1(x)H_1(y)S_1(z) + a_6 H_1(x)S_1(y)S_1(z) +$$
$$a_7 S_1(x)H_1(y)S_1(z) + a_8 S_1(x)S_1(y)S_1(z) + \ldots +$$
$$a_{56} H_2(x)H_2(y)H_2(z) + a_{57} H_2(x)S_2(y)H_2(z) + a_{58} S_2(x)H_2(y)H_2(z) +$$
$$a_{59} S_2(x)S_2(y)H_2(z) + a_{60} H_2(x)H_2(y)S_2(z) + a_{62} H_2(x)S_2(y)S_2(z) +$$
$$a_{63} S_2(x)H_2(y)S_2(z) + a_{64} S_2(x)S_2(y)S_2(z)$$

Doubling the number of gridlines in each direction, we end up with an eight-element model. There are 152 collocation points on the boundary, as shown in Figure 8.2. Besides this, there are 64 interior collocation points, so that altogether there are 216 collocation points to determine the a_i in the approximate solution:

$$\tilde{u}(x, y, z) = a_1 H_1(x)H_1(y)H_1(z) + \ldots + a_{216} S_3(x)S_3(y)S_3(z) \quad (8.14)$$

The solution with an $8 \cdot 8 \cdot 8 = 512$-element model is plotted in Figure 8.3. This plot is a contour 'surface' plot, i.e. each net represents the surface corresponding to a fixed value of u (as indicated in the key). So they are just like the contour plots of a function of two variables, except that for $u(x, y, z)$ the constant values are surfaces, rather than curves, as is the case when $u(x, y)$ is plotted using a contour plot.

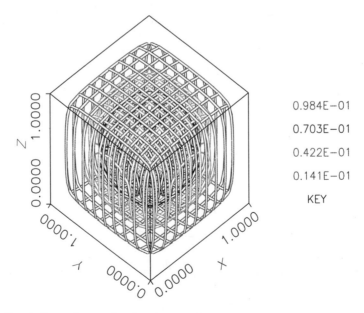

0.984E−01

0.703E−01

0.422E−01

0.141E−01

KEY

Figure 8.3 Purely Brownian motion leaving a cube

Table 8.2 First exit time of geometric Brownian motion

Parameter	Symbol	Value
Lower bound S_1	S_1^{min}	80
Upper bound S_1	S_1^{max}	120
Lower bound S_2	S_2^{min}	80
Upper bound S_2	S_2^{min}	120
Lower bound S_3	S_3^{min}	80
Upper bound S_3	S_3^{min}	120
Volatility of S_1	σ_1	0.1
Volatility of S_2	σ_2	0.3
Volatility of S_3	σ_3	0.5
Interest rate	r	0.1

Table 8.3 First exit time of a 3D geometric Brownian motion

x	y	z	$u(x, y, z)$
80	104	96	0.0000
84	104	96	0.1268
88	104	96	0.1316
92	104	96	0.1298
96	104	96	0.1310
100	104	96	0.1309
104	104	96	0.1309
108	104	96	0.1294
112	104	96	0.1266
116	104	96	0.1071
120	104	96	0.0000

8.2.2 First exit time of geometric Brownian motion

We assume three uncorrelated assets S_1, S_2 and S_3, governed by geometric Brownian motions. Dynkin's equation (B.114) describes the expected first exit time of each one of the three assets leaving a prescribed region. Data are shown in Tables 8.2 and 8.3, and a 3D example, in which the prescribed region is a cube, is shown in Figures 8.4 and 8.5. Questions of this type arise in the context of rainbow options with a separate barrier for each underlying.[1] We also assume that no dividends are paid and that the volatilities are constant. Then Dynkin's equation is given by:

$$\frac{1}{2}\left(\sigma_1^2 S_1^2 \frac{\partial^2 u}{\partial S_1^2} + \sigma_2^2 S_2^2 \frac{\partial^2 u}{\partial S_2^2} + \sigma_3^2 S_3^2 \frac{\partial^2 u}{\partial S_3^2}\right) + r S_1 \frac{\partial u}{\partial S_1} + r S_2 \frac{\partial u}{\partial S_2} + r S_3 \frac{\partial u}{\partial S_3} = -1 \quad (8.15)$$

[1] The reader has to keep in mind that the following computations have to be performed in the real measure, while the computations for pricing an option take place in the risk-neutral measure.

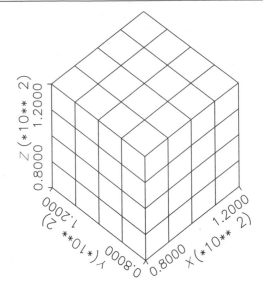

Figure 8.4   Grid for geometric Brownian motion leaving a cube

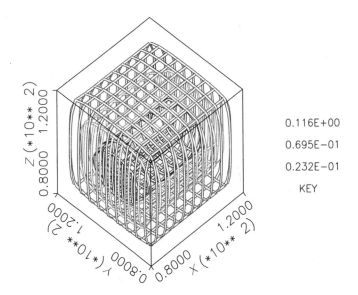

0.116E+00

0.695E−01

0.232E−01

KEY

Figure 8.5   Geometric Brownian motion leaving a cube

The domain of the above PDE is the prescribed region the assets are not to leave:

$$\Omega = \left[S_1^{\min}, S_1^{\max}\right] \times \left[S_2^{\min}, S_2^{\max}\right] \times \left[S_3^{\min}, S_3^{\max}\right] \qquad (8.16)$$

On the entire boundary of the domain Γ_Ω, the BC $u(S_1, S_2, S_3) = 0$ holds.

8.3 NOTES

Despite the overwhelming popularity of the Galerkin approach, the collocation method offers some advantages. It is easier to set up and does not need integration. This comes at the cost of a linear algebraic system which does not possess features that can easily be exploited, as in the Galerkin method. Solving the linear system resulting from cubic Hermite collocation discretization was an active area of research in the 1990s; see Lai *et al.* (1994a, 1994b) and the numerous references given there. PDEs which cannot be put into divergence form (compare Appendix A.3.6) cannot be solved with the Galerkin approach. The collocation method, however, is applicable.

Here, we only discussed domains in the form of boxes. Most FE work today, however, is done with irregular domains. Discretizing an irregular domain like a car door is a task which, in practice, cannot be performed automatically by a computer. A lot of manual meshing and remeshing is still necessary. In option pricing, however, the domains are usually infinite or semi-infinite, so that they can be approximated nicely with the help of boxes. Irregular domains arise in the pricing of basket options with barriers applied to the basket, as in Figure 9.2.

9

Dynamic 3D Problems

9.1 DERIVATION OF ELEMENT EQUATIONS: THE COLLOCATION METHOD

Next, we turn our attention to the orthogonal collocation method for solving the parabolic PDE:

$$d_{11}u_{xx} + d_{22}u_{yy} + d_{33}u_{zz} + d_{12}u_{xy} + d_{12}u_{xz} + d_{13}u_{yz} + \tag{9.1}$$

$$b_1 u_x + b_2 u_y + b_3 u_z + c\, u(x, y, z) - f(x, y, z) = u_t(x, y, z), \quad (x, y, z) \in \Omega \tag{9.2}$$

$$u(x, y, z) = u_D(x, y, z, t)$$

$$(x, y, z) \in \Gamma_D \tag{9.3}$$

$$u(x, y, z, t_0) = u_0(x, y, z) \tag{9.4}$$

$$[x^{\min}, x^{\max}] \times [y^{\min}, y^{\max}] \times [z^{\min}, z^{\max}] = \Omega \tag{9.5}$$

Only Dirichlet conditions will be considered. For other types of BC, see Sewell (2000).

1. *Discretization:* Although there are numerous types of element to discretize arbitrary domains in three spatial directions, we will only consider boxes, as in the 3D static problems. The discretization is exactly the same as in Section 8.1.

2. *Interpolation:* Again, a semidiscretization is employed so that spatial variables and time are separated:

$$\tilde{u}(x, y, z, t) = \sum_{i=1}^{8n_x n_y n_z} a_i(t)\phi_i(x, y, z) \tag{9.6}$$

$$\phi_1(x, y, z) = H_1(x)H_1(y)H_1(z) \tag{9.7}$$

$$\vdots$$

$$\phi_{8n_x n_y n_z}(x, y, z) = S_{n_x}(x)S_{n_y}(y)S_{n_z}(z) \tag{9.8}$$

The basis functions $\phi_i(x, y, z)$ were introduced in Equations (8.7) to (8.11).

3. *Assembly* and *Boundary conditions:* The linear algebraic system associated with the 3D static problem with the given discretization and interpolation, consists of eight $n_x n_y n_z$ equations. In complete analogy to the 1D and 2D problems, the dynamic counterpart to the 3D problem consists of a system of eight $n_x n_y n_z$ ordinary linear initial value problems. The integration of the BCs via the boundary collocation points is performed in a similar fashion as in Section 5.1.2:

$$\mathbf{M}\dot{\bar{\mathbf{u}}} = \mathbf{N}\bar{\mathbf{u}} + \mathbf{f} \tag{9.9}$$

Just as in the 1D and 2D cases, the initial conditions to the above problem are provided by the initial condition (9.4) of the PDE under inspection.

4. *Solution:* The above system (9.9) is stiff and needs to be solved with the appropriate techniques, as outlined in Section 3.2.

5. *Computation of derived variables:* Since the basis function is a cubic polynomial, meaningful expressions for the derivatives and cross-derivatives can be computed analytically up to third order. These derivatives are time-dependent and need to be computed at each time-step. For values between the time-steps, interpolation becomes necessary or the time stepping has to be adjusted.

9.2 CASE STUDIES

9.2.1 Pricing and hedging a basket option

Basket options are quite popular in the OTC market with currencies, equity and indices as underlyings. In spite of their popularity, no closed-form solution has been found so far. There are several approximations (Gentle, 1995; Huynh, 1994; Krekel *et al.*, 2004; Milevsky and Posner, 1998; Turnbull and Wakemann, 1991), all of which suffer from the fact that they do not provide error bounds. The pricing equation for a three-asset basket is given by:

$$
\frac{\partial V}{\partial t} + \frac{1}{2}\sigma_1^2 S_1^2 \frac{\partial^2 V}{\partial S_1^2} + \frac{1}{2}\sigma_2^2 S_2^2 \frac{\partial^2 V}{\partial S_2^2} + \frac{1}{2}\sigma_3^2 S_3^2 \frac{\partial^2 V}{\partial S_3^2}
$$

$$
+ \rho_{12}\sigma_1\sigma_2 S_1 S_2 \frac{\partial^2 V}{\partial S_1 S_2} + \rho_{13}\sigma_1\sigma_3 S_1 S_3 \frac{\partial^2 V}{\partial S_1 S_3} + \rho_{23}\sigma_2\sigma_3 S_2 S_3 \frac{\partial^2 V}{\partial S_2 S_3} \qquad (9.10)
$$

$$
+ (r - D_1)S_1 \frac{\partial V}{\partial S_1} + (r - D_2)S_2 \frac{\partial V}{\partial S_2} + (r - D_3)S_3 \frac{\partial V}{\partial S_3} = 0
$$

Here, this PDE is solved on a cube with the following final and boundary conditions:

$$ V(0, S_2, S_3, t) = 0 \qquad (9.11) $$

$$ V(S_1, 0, S_3, t) = 0 \qquad (9.12) $$

$$ V(S_1, S_2, 0, t) = 0 \qquad (9.13) $$

$$ V(300, S_2, S_3, t) = V - (n_1 S_1 + n_2 S_2 + n_3 S_3) - E \exp(-rt) \qquad (9.14) $$

$$ V(S_1, 300, S_3, t) = V - (n_1 S_1 + n_2 S_2 + n_3 S_3) - E \exp(-rt) \qquad (9.15) $$

$$ V(S_1, S_2, 300, t) = V - (n_1 S_1 + n_2 S_2 + n_3 S_3) - E \exp(-rt) \qquad (9.16) $$

$$ V(S_1, S_2, S_3, T) = \max(n_1 S_1 + n_2 S_2 + n_3 S_3 - E, 0) \qquad (9.17) $$

Other boundary conditions are possible with the final condition determining the type of option. Data for our example are given in Table 9.1.

Table 9.1 Data for a call on a basket of three equities (Randall and Tavella, 2000, p. 143)

Parameter	Symbol	Value		
Second asset	S_2		100	
Third asset	S_3		100	
Strike	E		100	
Time to maturity	T		1 year	
Interest rate (cont. comp.)	r		0.045	
Return/Weight/Volatility of first asset	$D_1/n_1/\sigma_1$	0.05	0.38	0.25
Return/Weight/Volatility of second asset	$D_2/n_2/\sigma_2$	0.07	0.22	0.35
Return/Weight/Volatility of third asset	$D_3/n_3/\sigma_3$	0.04	0.40	0.20
Correlations	$[\rho]_{ij}$	$\begin{pmatrix} 1.00 & -0.65 & 0.25 \\ -0.65 & 1.00 & 0.50 \\ 0.25 & 0.50 & 1.00 \end{pmatrix}$		

Table 9.2 Results for a call on a basket of three equities

Method	S_1						
	90	94	96	100	104	106	110
Gamma-approx. (Milevsky and Posner, 1998)	3.8660	4.4556	4.7731	5.4540	6.1963	6.5902	7.4230
Lognormal-approx (Huynh, 1994)	3.8631	4.4667	4.7914	5.4868	6.2432	6.6440	7.4898
FD (Randall and Tavella, 2000)	n.a.	n.a.	n.a.	5.4850	n.a.	n.a.	n.a.
FE	3.8781	4.4669	4.7861	5.4750	6.2288	6.6299	7.4792
MC	3.8609	4.4624	4.7837	5.4754	6.2288	6.6297	7.4718
Variance of MC	0.0016	0.0017	0.0017	0.0019	0.0020	0.0020	0.0022
Difference	0.4459%	0.1017%	0.0506%	−0.0082%	−0.0013%	0.0032%	0.1002%

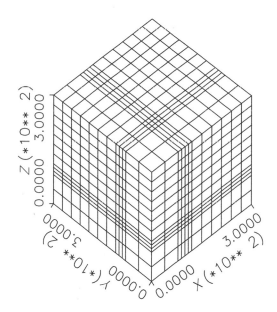

Figure 9.1 FE grid of a basket option

The results are shown in Table 9.2. The MC results were obtained without sophistication, except for the antithetic variable technique (Hull, 2000) using one million paths. For the FE approximation, a tighter FE net close to the strike, as in Figure 9.1, leads to exact values close to the strike and to less exact values away from the strike. For the FD computations, a regular net with 50 nodes in each direction and 25 equidistant time-steps were employed (Randall and Tavella, 2000 p. 143). For the FE computations, seven elements in each spatial

Table 9.3 $\Delta_1 = \frac{\partial V}{\partial S_1}$ of a call on a basket of three equities

		S_1						
	Method	90	94	96	100	104	108	110
$\Delta_1 = \frac{\partial V}{\partial S_1}$	FE	0.1392	0.1554	0.1639	0.1803	0.1966	0.2124	0.2200
$\Delta_1 = \frac{\partial V}{\partial S_1}$	MC	0.1423	0.1577	0.1654	0.1807	0.1960	0.2108	0.2180
$\Delta_2 = \frac{\partial V}{\partial S_2}$	FE	0.0874	0.0951	0.0990	0.1071	0.1151	0.1230	0.1268
$\Delta_2 = \frac{\partial V}{\partial S_2}$	MC	0.0850	0.0927	0.0965	0.1042	0.1119	0.1194	0.1232
$\Delta_3 = \frac{\partial V}{\partial S_3}$	FE	0.1656	0.1808	0.1884	0.2038	0.2188	0.2331	0.2400
$\Delta_3 = \frac{\partial V}{\partial S_3}$	MC	0.1644	0.1800	0.1877	0.2032	0.2184	0.2332	0.2404

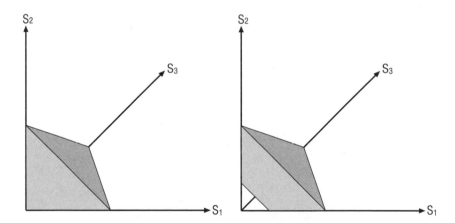

Figure 9.2   Domains of three-asset single and double barrier knock-out options

direction and cubic Hermite basis functions were used. Time integration is performed with a backward Euler method using 100 time-steps. Values of Δ for this example are shown in Table 9.3.

9.2.2 Basket options with barriers

Knock-out barriers on the basket result in nonrectangular domains, as in Figure 9.2. Knock-out barriers on the individual assets in the basket lead to rectangular domains as discussed above. Here, we will present numerical results for the latter type of basket option and compare our results to MC results from the literature (Shevchenko, 2003) with a down-and-out barrier on each asset.

Table 9.4 Data for a call on a basket with barriers

Parameter		Value	
Second asset		100	
Third asset		100	
Strike		100	
Time to maturity		1 year	
Interest rate (cont. comp.)		0.05	
Return/Weight/Volatility of first asset	0.0	0.333	0.4
Return/Weight/Volatility of second asset	0.0	0.333	0.4
Return/Weight/Volatility of third asset	0.0	0.333	0.4
Correlations		$\begin{pmatrix} 1.0 & 0.5 & 0.5 \\ 0.5 & 1.0 & 0.5 \\ 0.5 & 0.5 & 1.0 \end{pmatrix}$	

Table 9.5 Results for a call on a basket with barriers

Method	S_1					
	94	96	100	104	108	110
V_{MC}	n.a.	n.a.	7.600	n.a.	n.a.	n.a.
V_{FE}	6.124	6.671	7.592	8.339	8.996	9.300
$\Delta_1 = \frac{\partial V}{\partial S_3}$	0.290	0.258	0.205	0.173	0.156	0.148
$\Delta_2 = \frac{\partial V}{\partial S_1}$	0.148	0.169	0.205	0.235	0.260	0.271
$\Delta_3 = \frac{\partial V}{\partial S_1}$	0.148	0.169	0.205	0.235	0.260	0.271

The PDE (9.10) and the associated IC remain unchanged, while the BCs change their location. For the lower boundaries $S_1^{\min} = S_2^{\min} = S_3^{\min} = 80$ for the down-and-out barrier. The upper boundaries are moved somewhat to account for higher volatility: $S_1^{\max} = S_2^{\max} = S_3^{\max} = 700$. The discretization uses $13 \times 13 \times 13$ elements for the spatial variables and 500 equidistant time-steps (backward Euler). Data for our example are shown in Table 9.4, while the results are reported in Table 9.5.

10

Nonlinear Problems

10.1 INTRODUCTION

Nonlinear models arise naturally since many phenomena cannot be modeled adequately by linear differential equations. Often, originally linear models are refined by weakening assumptions, thus resulting in nonlinear models. A typical example is the Evans model from Section 2.1 that turns into a nonlinear ODE by introducing nonlinear cost and demand functions. Some refinements of the Black–Scholes model introduced in Section 2.2, such as transaction costs (Wilmott, 2000, Chapter 21) or early exercise costs (Section 2.3) can lead to nonlinear PDEs. Other models are nonlinear from the very beginning, such as optimal consumption models or pricing models for passport options (Section 10.2.3). The latter group comprises nested models which are linear PDEs. From a numerical point of view, the most striking difference is the FE discretization leading to a *nonlinear* system of equations. This fact was introduced with Equations (4.138ff). It will become clearer with the help of an example. Consider the following nonlinear two-point boundary value problem:

$$u_{xx} = -(u_x)^2 \tag{10.1}$$

$$u(1) = 0 \tag{10.2}$$

$$u(2) = \ln(2) \approx 0.693147181 \tag{10.3}$$

This problem has the true solution $u(x) = \ln(x)$. A one-element solution with a cubic shape function is to be derived, both with the Galerkin and with the collocation method:

$$\tilde{u}(x) = a_1 + a_2 x + a_3 x^2 + a_4 x^3 \tag{10.4}$$

Integrating the BC as in Section 4.2:

$$\tilde{u}(x) = 2(a_3 + 3a_4) - \ln(2) + [\ln(2) - 3a_3 - 7a_4] x + a_3 x^2 + a_4 x^3 \tag{10.5}$$

$$= (x - 1) \ln(2) + a_3 \underbrace{(x^2 - 3x + 2)}_{\phi_1} + a_4 \underbrace{(x^3 - 7x + 6)}_{\phi_2} \tag{10.6}$$

$$= (x - 1) \ln(2) + a_3 \phi_1 + a_4 \phi_2 \tag{10.7}$$

$$\tilde{u}'(x) = \ln(2) + 3a_4 x^2 + 2a_3 x - 3a_3 - 7a_4 \tag{10.8}$$

$$\tilde{u}''(x) = 6a_4 x + 2a_3 \tag{10.9}$$

The residual is given by:

$$R(x, a_3, a_4) = \tilde{u}'' + (\tilde{u}')^2 \tag{10.10}$$

Applying the Galerkin criterion:

$$\int_1^2 R \, \phi_1 \overset{!}{=} 0 \tag{10.11}$$

$$\int_1^2 R \, \phi_2 \overset{!}{=} 0 \tag{10.12}$$

This is a nonlinear system in the two variables a_3 and a_4:

$$\frac{[\ln(2)]^2}{6} - \frac{a_4 \ln(2)}{30} + \frac{7a_3^2 + 7a_3(9a_4 - 10) + a_4(143a_4 - 315)}{210} \overset{!}{=} 0 \quad (10.13)$$

$$\frac{3[\ln(2)]^2}{4} + \frac{a_3 \ln(2)}{30} + \frac{126a_3^2 + 4a_3(284a_4 - 315) + 21a_4(123a_4 - 272)}{840} \overset{!}{=} 0 \quad (10.14)$$

Applying the collocation method (with equidistant collocation points) leads to a different nonlinear system:

$$R\left(\frac{4}{3}, a_3, a_4\right) \overset{!}{=} 0 \quad (10.15)$$

$$R\left(\frac{5}{3}, a_3, a_4\right) \overset{!}{=} 0 \quad (10.16)$$

$$\frac{2(a_3 + 5a_4)\ln(2)}{3} - \frac{a_3^2 + 2a_3(5a_4 - 9) + a_4(25a_4 - 72)}{9} \overset{!}{=} [\ln(2)]^2 \quad (10.17)$$

$$\frac{2(a_3 + 4a_4)\ln(2)}{3} - \frac{a_3^2 + 2a_3(4a_4 - 9) + 2a_4(8a_4 - 45)}{9} \overset{!}{=} [\ln(2)]^2 \quad (10.18)$$

These systems are usually solved with some variant of the Newton–Raphson method.

A further numerical difficulty arises from some of the integrals becoming more complicated in the Galerkin method. A third difference is more on an abstract level. While for self-adjoint linear differential equations it is always possible to find an equivalent problem from the calculus of variations, for nonlinear differential equations a variational formulation may or may not exist. The Ritz method applied to a variational problem results in the same approximation as the Galerkin method applied to the differential equation, provided the same shape functions are employed. For economic and financial models, this is of little relevance since usually these models are not self-adjoint.

So far only nonlinear boundary value problems have been considered. Nonlinear time-dependent problems to be discussed here can always be formulated in the following way:

$$u_t = L(u_{xx}, u_x, u) \quad (10.19)$$

with $L(\cdot)$ being some nonlinear operator. The coefficients of $L(\cdot)$ are still allowed to be functions of time; $L(\cdot)$ itself, however, is not allowed to depend on time directly. The extension to problems with more than one spatial variable is straightforward. In complete analogy to linear problems, a semidiscretization of a nonlinear differential equation, as given by Equation (10.19), and the counterparts in higher dimensions results in a system of nonlinear ordinary first order initial value problems:

$$\dot{\mathbf{u}} = \begin{pmatrix} g_1(\mathbf{u}, a_1, \ldots, a_N) \\ \vdots \\ g_N(\mathbf{u}, a_1, \ldots, a_N) \end{pmatrix} \quad (10.20)$$

Problems of the above form can be solved with the techniques outlined in Section 3.2.9. Note that these are the same techniques as applied to systems of ordinary linear initial value problems. This comes from the fact that it is common practice to apply linear recurrence relations, as in Equation (3.20), to linear and also nonlinear initial value problems. For some comments on time stepping methods for nonlinear problems, see Section 10.1.3 of Sluzalec (1992). Pure FD schemes for nonlinear PDEs like Equation (10.19) are discussed in Ames (1965) and Hoggard *et al.* (1994).

10.2 CASE STUDIES

10.2.1 Penalty methods

In this section a technique is investigated which can be used to price American options, as introduced in Section 2.3. A penalty term is added to the pricing equation for the European option, as discussed in Section 2.3. Before we present this technique for PDEs, a simple example with an ODE will be discussed. The problem to be solved is a two-point boundary value problem:

$$u''(x) = 1 \quad \text{with} \quad x_{\text{free}} \leq x \leq 2 \tag{10.21}$$

The right-hand BC is $u(2) = 0$, while the left-hand BC at x_{free} has to be determined as part of the solution:

$$u(x) = 1 - x, \quad 0 \leq x \leq x_{\text{free}} \tag{10.22}$$
$$u(x_{\text{free}}) = 1 - x_{\text{free}} \tag{10.23}$$
$$u'(x_{\text{free}}) = -1 \tag{10.24}$$

The penalty formulation of this problem is:

$$v''(x) = \frac{[v(x) - (1 - x)] + \epsilon}{[v(x) - (1 - x)]^+ + \epsilon} \tag{10.25}$$
$$v(0) = 1 \tag{10.26}$$
$$v(2) = 0 \tag{10.27}$$

with ϵ some small parameter. The analytical solution is given by:

$$x_{\text{free}} = 2 - \sqrt{2} \tag{10.28}$$
$$u(x) = \begin{cases} 1 - x, & 0 \leq x \leq x_{\text{free}} \\ \left(\sqrt{2} - 1\right)(x - 2) + \frac{1}{2}(x - 2)^2, & x_{\text{free}} < x \leq 2 \end{cases} \tag{10.29}$$

Data for the numerical and analytical solutions are shown in Table 10.1.

We solve Equation (10.25) subject to the BCs with a collocation FE approach with two elements of equal length and cubic Hermite basis functions. The system of six nonlinear equations resulting from this discretization was solved with five iterations employing Newton's method.

10.2.2 American options

In Section 2.3 it was shown that the problem of pricing an American option can be formulated as a nonlinear PDE with appropriate BCs and FCs. It can be shown that even the multidimensional option pricing problem with early exercise is well-posed (Villeneuve, 1999b, Section 1.3.1). As a first example, the fair values of an American put with market data, as given in Table 10.2, will be computed and compared to the results of the binomial method. The results are given in Table 10.3 and Figures 10.1 and 10.2. The figures show the differences to a European put

Table 10.1 Penalty function for a static problem

x	Analytical solution	Numerical solution	Percentage difference
0.1	0.9000000	0.8999999	0.000011%
0.2	0.8000000	0.7999999	0.000013%
0.3	0.7000000	0.6999999	0.000014%
0.4	0.6000000	0.5999999	0.000017%
0.5	0.5000000	0.4999999	0.000020%
0.6	0.4001010	0.4001011	−0.000027%
0.7	0.3065224	0.3065225	−0.000033%
0.8	0.2229437	0.2229438	−0.000043%
0.9	0.1493651	0.1493652	−0.000059%
1	0.0857864	0.0857865	−0.000089%
1.1	0.0322078	0.0322079	−0.000215%
1.2	−0.0113708	−0.0113708	0.000544%
1.3	−0.0449495	−0.0449494	0.000119%
1.4	−0.0685281	−0.0685281	0.000068%
1.5	−0.0821068	−0.0821067	0.000047%
1.6	−0.0856854	−0.0856854	0.000036%
1.7	−0.0792641	−0.0792640	0.000029%
1.8	−0.0628427	−0.0628427	0.000025%
1.9	−0.0364214	−0.0364213	0.000020%

Table 10.2 Data for an American put

Parameter	Value
Asset price	9
Strike price	10
Interest rate	0.12
Volatility	0.5
Dividend yield	0.0

Table 10.3 Results for the American put

Time to maturity (months)	Binomial method Time-steps		FEM Time-steps		
	128	256	128	256	1000
1	1.1317	1.1316	1.1260	1.1262	1.1263
3	1.3814	1.3805	1.3737	1.3743	1.3747
6	1.6185	1.6178	1.6104	1.6112	1.6110
9	1.7847	1.7817	1.7745	1.7756	1.7764
12	1.9106	1.9094	1.9013	1.9025	1.9034

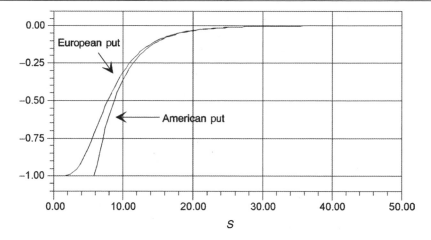

Figure 10.1 Delta of an American and a European put

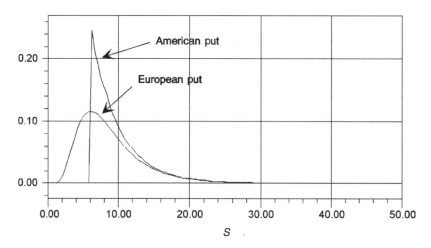

Figure 10.2 Gamma of an American and a European put

with the same market and contract data (maturity one year). Early exercise is enforced with the penalty function Equation (2.42) with $c_{\text{penalty}} = 10^5$. For all maturities the same number of equidistant time-steps (using the backward Euler method) and a regular mesh of 100 elements was used. The BCs were specified to be:

$$\frac{\partial V(0)}{\partial S} = 1 \tag{10.30}$$

$$V(100) = 0 \tag{10.31}$$

Since cubic Hermite shape functions were used, the final condition was replaced by a least-squares approximation to avoid a spike at the strike.

As a second example, a barrier option is chosen that was computed with FE in an early publication on the use of FE in finance (Tomas, 1996).[1] Tomas used a Galerkin FE method with

---

[1] The author is unaware of any earlier work.

Table 10.4   Data for an American up-and-out put

Parameter	Value
Strike price	100
Up-and-out barrier	108
Interest rate	0.1
Volatility	0.2
Dividend yield	0.0
Maturity	0.25 year

Table 10.5   Premiums for the American up-and-out put

S	90	95	100	105
Binomial tree (Rubinstein, 1991a)	10.0000	5.7751	2.8836	0.9547
Galerkin FE (Quadratic Lagrange) (Tomas, 1996)	10.0023	5.7516	2.8496	0.9264
Galerkin FE (Cubic Lagrange) (Tomas, 1996)	9.9743	5.7513	2.8484	0.9242
Galerkin FE (Quintic Hermite) (Tomas, 1996)	9.9973	5.7720	2.8855	0.9878
Collocation FE (Cubic Hermite)	9.9765	5.7538	2.8401	0.8865
Collocation FE (Cubic Hermite)	9.9748	5.7432	2.8265	0.8801

Table 10.6   Deltas for the American up-and-out put

S	90	95	100	105
Binomial tree (Rubinstein, 1991a)	−0.9840	−0.7042	−0.4683	−0.3165
Galerkin FE (Quadratic Lagrange) (Tomas, 1996)	−0.9997	−0.7039	−0.4312	−0.3175
Galerkin FE (Cubic Lagrange) (Tomas, 1996)	−0.9749	−0.7064	−0.4677	−0.3157
Galerkin FE (Quintic Hermite) (Tomas, 1996)	−1.0283	−0.7026	−0.4657	−0.3104
Collocation FE (Cubic Hermite)	−0.9832	−0.7045	−0.4729	−0.3230
Collocation FE (Cubic Hermite)	−0.9828	−0.7061	−0.4724	−0.3209

various shape functions. However, this work was done on the Black–Scholes PDE transformed to the heat equation. Early exercise is integrated by checking the nodal solution values at each time-step and replacing this value if it violates the early exercise constraint. Our approach is the same as in the previous example, i.e. with a penalty function. The same routine as for the previous example was employed, with the details of the relevant data, premiums and deltas given in Tables 10.4–10.6 respectively, and details on the steps in space and time given in Table 10.7. The BCs were specified to be:

$$\frac{\partial V(0)}{\partial S} = 1 \qquad (10.32)$$

$$V(108) = 0 \qquad (10.33)$$

Note the difference to the previous example. The BC on the RHS in the second example states the up-and-out knock-out condition. In the first example, the spatial domain is cut off at $S = 100$ and at that point it is assumed that an option being so far out of the money becomes worthless.

Table 10.7 Numerical details for the American up-and-out put

S	Time-steps	Elements
Binomial tree (Rubinstein, 1991a)	500	—
Galerkin FE (Quadratic Lagrange) (Tomas, 1996)	50	50
Galerkin FE (Cubic Lagrange) (Tomas, 1996)	50	50
Galerkin FE (Quintic Hermite) (Tomas, 1996)	50	50
Collocation FE (Cubic Hermite)	1000	108
Collocation FE (Cubic Hermite)	50	50

10.2.3 Passport options

Passport options are a new kind of financial instrument introduced by Bankers Trust in 1997. They are used to protect trading accounts. The basic passport option allows the holder to take the profit from a trading account while any losses are covered by the writer of the option. The maximal amount a trader can go, either long or short, is limited to some prespecified amount. To make passport options cheaper, or to reduce the risk to the writer, certain exotic features such as caps, floors and barriers have been employed.

The concept of passport options has been extended to general options on trading accounts where the limits for going short or long do not necessarily have to be equal anymore (Shreve and Večeř, 2000b). This concept of an option on a trading account contains many special cases, such as plain vanilla European and American options, passport options and Asian options. The latter is of great practical importance since it offers an efficient way to compute the fair value and the Greeks of an Asian option with *discrete* sampling.

Passport options can be used to:

- protect the trading account for a variety of reasons, such as inexperienced traders, employment of new strategies, etc.;
- price life insurance claims contingent on the performance of a reference fund (Berti, 2000);
- develop new commodity hedging strategies (Henderson *et al.*, 2001).

Passport options are usually not applied directly to the trading account of a trader. The account that is protected by the passport option is usually run by the writer of the option. This account is only virtual. The buyer of the option informs the writer about his moves via phone[2] or internet. At the end of a prespecified period of time, the writer of the option has to pay the payoff to the holder of the passport option (Figure 10.3). Since usually not all holders of passport options follow the same strategy, the writer of several passport options can aggregate all positions on the same underlying and take advantage of netting long and short positions. Some of these positions should cancel each other out, so that only a smaller residual risk needs to be hedged.

The pricing model for passport options

The starting point is the Black–Scholes framework, in which the underlying S follows the following stochastic differential equation; see Anderson and Brotherton-Ratcliffe (1998):

$$\frac{\mathrm{d}S(t)}{S(t)} = (r - \gamma)\, \mathrm{d}t + \sigma\, \mathrm{d}W(t) \tag{10.34}$$

---

[2] The number of trades that the writer of the option is allowed is often limited. In this case, the price of the passport options, which allow continuous rebalancing, denotes the upper limit of the fair value. How to integrate limitations on the number of trades is discussed in Ahn *et al.* (1999b) and Anderson and Brotherton-Ratcliffe (1998).

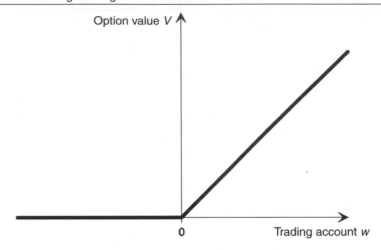

Figure 10.3 Payoff of a passport option

Consider an investor at t_i holding $u(t_i) \in [-1, 1]$ in this underlying. From t_i to t_{i+1}, the investor gains $u(t_i)\,[S(t_{i+1}) - S(t_i)]$. Summing up over all periods, the investor's total gain w is

$$w = \sum_{i=0}^{H-1} u(t_i)\,[S(t_{i+1}) - S(t_i)] \tag{10.35}$$

No interest is paid on assets in the trading account. Assuming continuous trading, i.e. $\lim_{i \to 0}(t_i - t_{i+1}) = 0$, the gain can be expressed as:

$$w(t) = \int_0^t u(s)\mathrm{d}S(s) \tag{10.36}$$

$$\Longleftrightarrow \; \mathrm{d}w(t) = u(t)\mathrm{d}S(t) \;\text{ with }\; w(0) = 0 \tag{10.37}$$

The European passport option gives the holder the right, but not the obligation, to receive w in T. In the case where $w < 0$, the rational investor is not interested in delivery, so the payoff equals

$$[w(T)]^+ \equiv \max[0, w(T)] \tag{10.38}$$

For deriving the pricing equation we will use a similar argument as Black and Scholes: an instantaneously riskless portfolio Π consists of one passport option and $-k$ units of the underlying:

$$\Pi = V - kS \tag{10.39}$$

Within the time interval $(t, t + \mathrm{d}t)$ the value of this portfolio changes by

$$\mathrm{d}\Pi = \mathrm{d}V - k(\mathrm{d}S + \gamma S \mathrm{d}t) \tag{10.40}$$

We assume the existence of an optimal strategy q^* and the derivatives V_{SS}, V_{ww} and V_{Sw}. We also presuppose that the holder of the option maximizes his revenues without being hindered from taking q^* by hedging necessities or other superimposed circumstances. Then the following holds:

$$dV = \frac{\partial V}{\partial t}dt + \frac{\partial V}{\partial S}dS + \frac{\partial V}{\partial w}dw + \frac{1}{2}\frac{\partial^2 V}{\partial S^2}(dS)^2 + \frac{\partial^2 V}{\partial S \partial w}(dS\ dw) + \frac{1}{2}\frac{\partial^2 V}{\partial w^2}(dw)^2$$

$$(10.41)$$

To simplify this equation, two more results are needed. Squaring Equation (10.34) gives:

$$(dS)^2 = \sigma^2 S^2 dt \tag{10.42}$$

The profit-maximizing behavior of the holder turns Equation (10.37) into:

$$dw = q^* dS \tag{10.43}$$

Substituting these results into Equation (10.41) leads to:

$$dV = \frac{\partial V}{\partial t}dt + \left(\frac{\partial V}{\partial S} + q^*\frac{\partial V}{\partial w}\right)dS + \frac{1}{2}\left(\frac{\partial^2 V}{\partial S^2} + 2q^*\frac{\partial^2 V}{\partial S \partial w} + (q^*)^2\frac{\partial^2 V}{\partial w^2}\right)\sigma^2 S^2 dt$$

$$(10.44)$$

The parameter k has to be chosen for the portfolio Π to become instantaneously riskless:

$$k = \frac{\partial V}{\partial S} + q^*\frac{\partial V}{\partial w} \tag{10.45}$$

Because of the absence of arbitrage, the riskless portfolio Π has to grow by the same rate as a money market account r.

$$d\Pi = r\Pi dt \tag{10.46}$$

Combining the above results gives:

$$\frac{\partial V}{\partial t} + \frac{\sigma^2 S^2}{2}\left(\frac{\partial^2 V}{\partial S^2} + 2q^*\frac{\partial^2 V}{\partial S \partial w} + (q^*)^2\frac{\partial^2 V}{\partial w^2}\right) + (r - \gamma)S\left(\frac{\partial V}{\partial S} + q^*\frac{\partial V}{\partial w}\right) = rV$$

$$(10.47)$$

with the following final condition:

$$V(T, S, w) = w^+ \tag{10.48}$$

The change of variables, $x \equiv w/S$, reduces the dimensions of the problem by one. Using this substitution generates the following PDE:

$$\frac{\partial v}{\partial t} + (q^* - x)(r - \gamma)\frac{\partial v}{\partial x} + \frac{1}{2}(q^* - x)^2\sigma^2\frac{\partial^2 v}{\partial x^2} = \gamma v \tag{10.49}$$

with

$$q^* = \text{sgn}\left((r - \gamma)\frac{\partial v}{\partial x} - x\sigma^2\frac{\partial^2 v}{\partial x^2}\right) \tag{10.50}$$

where the payoff function $v(x, T)$ has to be monotonically increasing and convex in x. The above expression can be derived by differentiating Equation (10.49) with respect to q^* and setting equal to zero. Equivalent formulations of Equation (10.50) are:

$$\frac{\partial v}{\partial t} - x(r - \gamma)\frac{\partial v}{\partial x} + \frac{1}{2}(1 + x^2)\sigma^2\frac{\partial^2 v}{\partial x^2} + q^*\left((r - \gamma)\frac{\partial v}{\partial x} - x\sigma^2\frac{\partial^2 v}{\partial x^2}\right) = \gamma v \tag{10.51}$$

and:

$$\frac{\partial v}{\partial t} - x(r - \gamma)\frac{\partial v}{\partial x} + \frac{1}{2}(1 + x^2)\sigma^2\frac{\partial^2 v}{\partial x^2} + \left|(r - \gamma)\frac{\partial v}{\partial x} - x\sigma^2\frac{\partial^2 v}{\partial x^2}\right| = \gamma v \qquad (10.52)$$

with:

$$v(T, x) = v_T(x) \qquad (10.53)$$

as a final condition. This is the setting applicable to most real-world passport options. For the sake of completeness, we will also state the pricing PDE for a more general case. Instead of Equation (10.43):

$$dw = q^* dS \qquad (10.54)$$
$$= q^*(r - \gamma)S dt + q^* \sigma S dW(t) \qquad (10.55)$$

we allow for different interest rates and dividend yields within and outside the trading account. Outside the trading account, the interest rate r and the dividend yield γ are the same as above; inside the trading account, r_w is paid on money. The cost of carry within the trading account is v, so that the dynamics of the trading account are given by:

$$dw = \left[r_w w + q^*(r - \gamma - v)S\right] dt + q^* \sigma S dW(t) \qquad (10.56)$$

Obviously, the above equation reduces to Equation (10.55) with $r_w = v = 0$. The resulting pricing PDE is:

$$-\frac{\partial v}{\partial t} = \left[(r - \gamma - v)q^* - (r - \gamma - r_w)x\right]\frac{\partial v}{\partial x} + \frac{1}{2}(x - q^*)^2\sigma^2\frac{\partial^2 v}{\partial x^2} + \gamma v \qquad (10.57)$$

with

$$q^* = \text{sgn}\left((r - \gamma - v)\frac{\partial v}{\partial x} - x\sigma^2\frac{\partial^2 v}{\partial x^2}\right) \qquad (10.58)$$

which can be rewritten as:

$$-\frac{\partial v}{\partial t} = \gamma v + \left[-(r - \gamma - r_w)x\right]\frac{\partial v}{\partial x} + \frac{1}{2}(1 + x^2)\sigma^2\frac{\partial^2 v}{\partial x^2}$$
$$+ \left|(r - \gamma - v)\frac{\partial v}{\partial x} - x\sigma^2\frac{\partial^2 v}{\partial x^2}\right| \qquad (10.59)$$

The literature discusses various nested models:

- The symmetric case has been defined by Hyer *et al.* (1997) to be $v = r - \gamma$ and $r_w = r$. However, Anderson and Brotherton-Ratcliffe (1998) call the case of $r_w = v = 0$ 'symmetric' (which is asymmetric in terms of the above definition). This, i.e. $r_w = v = 0$, is the setting on which this paper has its focus, since most real-world passport options can be found in this framework.

- In Hyer *et al.* (1997), an analytical solution for the symmetric case and $\gamma = 0$ is derived, i.e. the PDE

$$\frac{\partial v}{\partial t} + \frac{1}{2}(1 + |x|)^2\sigma^2\frac{\partial^2 v}{\partial x^2} = 0 \qquad (10.60)$$

  is solved with the appropriate boundary data for a passport option.

- In Anderson and Brotherton-Ratcliffe (1998), a special asymmetric case with $r_w = v = 0$ is discussed, and within this framework, an analytical solution for $r = \gamma$ is derived. Inserting $r = \gamma$ into Equation (10.49) leads to:

$$\gamma v = \frac{\partial v}{\partial t} + \underbrace{\frac{\sigma^2}{2}(q^* - x)^2}_{=A} \frac{\partial^2 v}{\partial x^2} \qquad (10.61)$$

$$q^* = \mathrm{sgn}\left(-x\underbrace{\sigma^2\frac{\partial^2 v}{\partial x^2}}_{>0}\right) = -\mathrm{sgn}(x) \qquad (10.62)$$

Now consider two distinct cases:
1. $x > 0$: $A = (-1 - x)^2 = [-(1 + x)]^2 = (1 + |x|)^2$
2. $x \leq 0$: $A = (1 - x)^2 = (1 + |x|)^2$

so that $A = (1 + |x|)^2 \ \forall \ x$. Consequently, PDE (10.47) can be rewritten as:

$$\frac{\partial v}{\partial t} + \frac{1}{2}(1 + |x|)^2\sigma^2\frac{\partial^2 v}{\partial x^2} = \gamma v \qquad (10.63)$$

This PDE is solved analytically with the appropriate boundary data for a passport option in Anderson and Brotherton-Ratcliffe (1998). Note that a special case of PDE (10.63) with $\gamma = 0$ is the same as PDE (10.60) but with different interpretations: PDE (10.63) with $\gamma = 0$ implies $r_w = r = v = \gamma = 0$, while PDE (10.60) results from $r_w = r = v$ and $\gamma = 0$.

- Ahn, Penaud and Wilmott (Ahn et al., 1999b, 1999c; Wilmott, 1998) discuss the symmetric case with $\gamma = 0$, implying $v = r = r_w$.

Equation (10.52) is the PDE for the passport option value, given a strategy q^*. It is also possible to view this problem from a different perspective by looking for an equation that defines the optimal strategy μ^*. Via the principles of dynamic programming, a PDE, called the *Hamilton–Jacobi–Bellmann equation* (HJB), which defines this optimal strategy, can be derived (Hyer et al., 1997):

$$-V_t = -rV + \underbrace{(r - \gamma)SV_S}_{A} \qquad (10.64)$$

$$+ \max_{|\mu| \leq 1}\left\{\underbrace{[r_w w + (r - \gamma - v)\mu S]V_w}_{B} + \underbrace{\frac{\sigma^2 S^2}{2}\left[V_{SS} + 2\mu V_{Sw} + \mu^2 V_{ww}\right]}_{C}\right\}$$

For contracts involving no barrier on either S or w or both, the transformation $x \equiv w/S$ is possible and results in a simpler PDE. Using the substitution:

$$V(t, S, w) = Sv\left(t, \frac{w}{S}\right) = Sv(t, x) \qquad (10.65)$$

we get the following derivatives:

$$\frac{\partial V}{\partial t} = S\frac{\partial v}{\partial t} \qquad (10.66)$$

$$\frac{\partial V}{\partial S} = v + \frac{\partial}{\partial x}v(t, \frac{w}{S})S\frac{(-w)}{S^2} = v - x\frac{\partial v}{\partial x} \qquad (10.67)$$

$$\frac{\partial V}{\partial w} = \frac{\partial v}{\partial x} \tag{10.68}$$

$$\frac{\partial^2 V}{\partial S \partial w} = \frac{\partial}{\partial S}\left(\frac{\partial v}{\partial x}\right) = -\frac{w}{S^2}\frac{\partial^2 v}{\partial x^2} = -\frac{x}{S}\frac{\partial^2 v}{\partial x^2} \tag{10.69}$$

$$\frac{\partial^2 V}{\partial w^2} = \frac{\partial}{\partial w}\left(\frac{\partial v}{\partial x}\right) = \frac{1}{S}\frac{\partial^2 v}{\partial x^2} \tag{10.70}$$

$$\frac{\partial^2 V}{\partial S^2} = \frac{\partial}{\partial S}\left[v - x\frac{\partial v}{\partial x}\right] = -\frac{w}{S^2}\frac{\partial v}{\partial x} + x\frac{w}{S^2}\frac{\partial^2 v}{\partial x^2} + \frac{w}{S^2}\frac{\partial v}{\partial x} = \frac{x^2}{S}\frac{\partial^2 v}{\partial x^2} \tag{10.71}$$

Inserting these expressions into Equation (10.65):

$$A = (r - \gamma)S\left(v - x\frac{\partial v}{\partial x}\right) \tag{10.72}$$

$$= (r - \gamma)Sv - (r - \gamma)Sxv_x \tag{10.73}$$

$$= rSv - \gamma Sv + S(\gamma x v_x - r x v_x) \tag{10.74}$$

$$B = [r_w w v_x + (r - \gamma - v)\mu S]v_x \tag{10.75}$$

$$= r_w x S v_x + (r - \gamma - v)\mu S v_x \tag{10.76}$$

$$C = \frac{\sigma^2 S^2}{2}\left[\frac{x^2}{S}v_{xx} + 2\mu\left(-\frac{x}{S}v_{xx}\right) + \frac{\mu^2}{S}v_{xx}\right] \tag{10.77}$$

$$= \frac{\sigma^2 S}{2}\left[x^2 v_{xx} - 2\mu x v_{xx} + \mu^2 v_{xx}\right] \tag{10.78}$$

$$= \frac{\sigma^2 S}{2}\left[(x - \mu)^2 v_{xx}\right] = \frac{\sigma^2 S}{2}\left[(\mu - x)^2 v_{xx}\right] \tag{10.79}$$

$$\Rightarrow -Sv_t = -rSv + A + \max_{|\mu|\le 1}\{B + C\} \tag{10.80}$$

$$= -rSv + [rSv - \gamma Sv + S(\gamma x v_x - r x v_x)] + r_w x S v_x$$
$$+ \max_{|\mu|\le 1}\left\{(r - \gamma - v)\mu S v_x + \frac{\sigma^2 S}{2}\left[(x - \mu)^2 v_{xx}\right]\right\} \tag{10.81}$$

$$\Leftrightarrow -v_t = -\gamma v - (r - \gamma - r_w)x v_x$$
$$+ \max_{|\mu|\le 1}\left\{(r - \gamma - v)\mu v_x + \frac{\sigma^2}{2}\left[(x - \mu)^2 v_{xx}\right]\right\} \tag{10.82}$$

The symmetric case (i.e. $v = r - \gamma$ and $r_w = r$) leads to further simplifications:

$$-v_t = -\gamma v + (r - \gamma - r)x v_x + \max_{|\mu|\le 1}\left\{\frac{\sigma^2}{2}[(x - \mu)^2 v_{xx}]\right\} \tag{10.83}$$

$$= -\gamma v - \gamma x v_x + \frac{\sigma^2}{2}\max_{|\mu|\le 1}\left\{(x - \mu)^2 v_{xx}\right\} \tag{10.84}$$

The term in brackets is maximized by:

$$\mu^* = \begin{cases} -1 \; \forall \, x > 0 \\ 1 \; \forall \, x \le 0 \end{cases} \tag{10.85}$$

so that Equation (10.84) reduces to:

$$-v_t = -\gamma v + \gamma x v_x + \frac{\sigma^2}{2}(|x| + 1)^2 v_{xx} \qquad (10.86)$$

With $\gamma = 0$, PDE (10.86) further reduces to PDE (10.63). The two Hamilton–Jacobi–Bellmann equations (10.64) and (10.84) can be used as a basis for numerical computations when q^* and μ^* are not known. Such problems arise when caps or barriers on either w and/or S are introduced.

The question of existence and uniqueness of a solution to Equations (10.64) and (10.63) can only be answered with sophisticated methods which will not be employed here. It can be shown that the problem may not possess a classical solution, but only a viscosity solution (Chan, 1999), i.e. the solution is uniformly continuous and bounded but may not be differentiable, see Section A.1.1. The latter property is somewhat surprising for the solution of a differential equation.

The *hedge ratio k* is slightly different to the Black–Scholes framework. The basic idea is that a portfolio Π consisting of a long call C and a short position in k shares S

$$\Pi = C - kS \qquad (10.87)$$

is riskfree for an infinitesimal amount of time.[3] The hedge parameter in the Black–Scholes model is

$$k = \frac{\partial C}{\partial S} = \Delta \qquad (10.88)$$

Portfolios containing passport options can be immunized against infinitesimal changes of the share price via Equation (10.87). For k we have (compare Anderson and Brotherton-Ratcliffe, 1998, pp. 33f):

$$k = \frac{\partial V}{\partial S} + q^* \frac{\partial V}{\partial w} = v + (q^* - x)\frac{\partial v}{\partial x} \qquad (10.89)$$

This implies that numerical difficulties arising in computing Δ are also present in computing k. Finite elements provide approximate solutions to the entire domain, consisting of simple algebraic functions. Whenever $q^*(S)$ changes its sign, the hedge ratio $k(x)$ shows a jump. This also implies that the writer of the options needs to know the strategy of the buyer in order to be able to hedge the option. Based on this fact it becomes clear why passport options protect *virtual* trading accounts (on the computer of the option's writer) instead of real trading accounts (which are, by their very nature, on the computer of the option's buyer).

Nonconvex payoffs: a correction

According to Proposition 5 in Anderson and Brotherton-Ratcliffe (1998), the result of the previous section can be generalized to nonconvex payoffs by changing the control to $q^* \in [-1, 1]$:

$$q^* = \begin{cases} \psi & \text{if } \psi(x, t) \in [-1, 1] \text{ and } \frac{\partial^2 v}{\partial x^2} < 0 \\ \text{sgn}(\psi(x, t)) & \text{otherwise} \end{cases} \qquad (10.90)$$

---

[3] To be exact: Black and Scholes (1973) consider a portfolio *short* in the call and *long* in the share, as in Section 2.2. In order to keep conformity to Anderson and Brotherton-Ratcliffe (1998), we have switched the positions.

with

$$\psi(x, t) = x - \frac{r - \gamma}{\sigma^2} \frac{\frac{\partial v}{\partial x}}{\frac{\partial^2 v}{\partial x^2}} \tag{10.91}$$

With the help of a simple counterexample it can be shown that this proposition is wrong. Convex payoffs are supposed to be a special case of Equation (10.90). This is not the case. We consider the special case of $r = \gamma$, which can easily be extended to $r \neq \gamma$. The unique optimal control q^* for convex payoffs is, according to Equation (10.50), (see also Andersen and Brotherton-Ratcliffe, 1998, Proposition 2):

$$q^*(x, t) = \text{sgn} \left((r - \gamma) \frac{\partial v}{\partial x} - x \sigma^2 \frac{\partial^2 v}{\partial x^2} \right) \tag{10.92}$$

Inserting $r = \gamma$ simplifies the expression:

$$q^*(x, t) = \text{sgn} \left(-x \sigma^2 \frac{\partial^2 v}{\partial x^2} \right) \tag{10.93}$$

Convexity of v in x implies $\frac{\partial^2 v}{\partial x^2} \geq 0$. Together with $\sigma > 0$ it has to hold for $t \in [0, T]$

$$\text{sgn} \left(-x \sigma^2 \frac{\partial^2 v}{\partial x^2} \right) = \text{sgn}\,(-x) \tag{10.94}$$

Inserting $r = \gamma$ into the general payoff Equation (10.90) gives:

$$q^* = \text{sgn}(\psi) = \text{sgn}(x) \tag{10.95}$$

Therefore, the convex payoff is not a special case of the general payoff. This shows that Proposition 5 in Anderson and Brotherton-Ratcliffe (1998) is wrong.

General payoffs

We will present the general control q^* for arbitrary payoffs first in this section. Then we will deduce the special controls for convex and concave payoffs.

The general control is:

$$q^* = \begin{cases} \psi & \text{if } \psi(x, t) \in [-1, 1] \text{ and } \frac{\partial^2 v}{\partial x^2} < 0 \\ \text{sgn}(-\psi \frac{\partial^2 v}{\partial x^2}) & \text{otherwise} \end{cases} \tag{10.96}$$

For strictly concave payoff functions ($\frac{\partial^2 v}{\partial x^2} < 0$), Equation (10.96) simplifies to:

$$q^* = \begin{cases} \psi & \text{if } \psi(x, t) \in [-1, 1] \\ \text{sgn}(\psi) & \text{otherwise} \end{cases} \tag{10.97}$$

For (strictly and simply) convex payoff functions, the following control function holds:

$$q^* = \text{sgn} \left(-\psi \frac{\partial^2 v}{\partial x^2} \right) \tag{10.98}$$

$$= \text{sgn} \left(\left[x - \frac{r - \gamma}{\sigma^2} \frac{\frac{\partial v}{\partial x}}{\frac{\partial^2 v}{\partial x^2}} \right] \frac{\partial^2 v}{\partial x^2} \right) \tag{10.99}$$

$$= \mathrm{sgn}\left(\frac{1}{\sigma^2}\left[(r-\gamma)\frac{\partial v}{\partial x} + \sigma^2 x \frac{\partial^2 v}{\partial x^2}\right]\right) \tag{10.100}$$

$$= \mathrm{sgn}\left((r-\gamma)\frac{\partial v}{\partial x} + \sigma^2 x \frac{\partial^2 v}{\partial x^2}\right) \tag{10.101}$$

since $\sigma^2 \geq 0$. This is Equation (18) from Anderson and Brotherton-Ratcliffe (1998).

Integrating early exercise

Early exercise of the option can be integrated with a penalty function p. This function p ensures that in areas of the (t, S)-space, where early exercise is optimal, the pricing equation (10.52) is forced to take on the intrinsic value of the option, while it vanishes on the rest of the domain. For details of this technique and various specifications of p see Nielsen et al. (2002) and Zvan et al. (1998b).

$$\gamma v = \frac{\partial v}{\partial t} - x(r-\gamma)\frac{\partial v}{\partial x} + \frac{1}{2}(1+x^2)\sigma^2\frac{\partial^2 v}{\partial x^2} + \left|(r-\gamma)\frac{\partial v}{\partial x} - x\sigma^2\frac{\partial^2 v}{\partial x^2}\right| + p \tag{10.102}$$

$$p = c_{\mathrm{penalty}}\left\{\min\left[v - \max\left(S,0\right),0\right]^2\right\} \tag{10.103}$$

According to Zvan et al. (1998b), c_{penalty} depends on the type of element; according to our experience, it suffices to choose c_{penalty} sufficiently large, such as $c_{\mathrm{penalty}} = 10^8$. Computing the value of American passport options can be greatly simplified when the interest rate for money within and outside the trading account is the same, i.e. $r_w = r$. In this case, one can always emulate early exercise with a European passport option by entering a zero position in the underlying, so the American option is worth the same. A Finite Differences approach to the American passport options is studied in Chan (1999).

Extending passport options to general options on trading accounts

The defining property of an option on a trading account is that the control is not restricted to the closed interval $[-1, 1]$ anymore (Shreve and Večeř, 2000a, 2000b). Working in the symmetric setting, as given by Equation (10.56), the pricing PDE becomes:

$$-V_t = -rV + (r-\gamma)SV_S \tag{10.104}$$

$$+ \max_{\mu \in [a,b]}\left\{[r_w w + (r-\gamma-v)\mu S]V_w + \frac{\sigma^2 S^2}{2}\left[V_{SS} + 2\mu V_{Sw} + \mu^2 V_{ww}\right]\right\}$$

The plain vanilla contract is defined by the payoff function

$$V(T, S, w) = w^+ \tag{10.105}$$

The optimal control for the plain vanilla contract is known to be:

$$q^* = \alpha - \mathrm{sgn}\left(\frac{w}{S} - \alpha\right)\beta \tag{10.106}$$

$$\alpha = \frac{b-a}{2} \tag{10.107}$$

$$\beta = \frac{b+a}{2} \tag{10.108}$$

Table 10.8 Data for a European passport option (1)

Parameter	Value
Spot S	100.0
Dividend yield γ	0.0
Interest rate r	0.0
Volatility σ	0.3
Time to maturity	1 year

Table 10.9 Results for a European passport option (1)

		Time steps							
		25	50	100	200	400	800	1600	3200
	25	13.0824	13.1120	13.1268	13.1342	13.1379	13.1398	13.1407	13.1412
	50	13.1860	13.2150	13.2290	13.2369	13.2405	13.2423	13.2433	13.2437
Spatial	100	13.1047	13.1341	13.1488	13.1561	13.1598	13.1617	13.1626	13.1631
steps	200	13.0853	13.1148	13.1295	13.1369	13.1406	13.1425	13.1433	13.1439
	400	13.0806	13.1101	13.1249	13.1322	13.1359	13.1378	13.1387	13.1392
	800	13.0795	13.1089	13.1236	13.1311	13.1348	13.1366	13.1376	13.1380
	1600	13.0791	13.1086	13.1234	13.1308	13.1345	13.1363	13.1373	13.1377

It can be shown that European-style and American-style plain vanilla options, vacation options (i.e. either $a = 0$ or $b = 0$) and also Asian options are special cases of the general option on a trading account. Two approaches for Asian options leading to simple PDEs have been explored by Večeř (2001, 2002); see also Section 5.2.3.

Computations

Most of the problems in these sections, such as Equation (10.104), cannot be put into the divergence form, so the Galerkin method requiring integration by parts drops out as a candidate for a numerical method. All problems in these sections are solved with the collocation method and cubic Hermite basis functions. For the numerical solution of the PDE (10.60), besides the initial condition given by Equation (10.48), two boundary conditions are necessary, which turn the unbounded domain of the PDE $-\infty < x < \infty$ into a compact one. Anderson and Brotherton-Ratcliffe (1998) suggest

$$v(t, -e^{h\sigma\sqrt{T}}) = 0 \tag{10.109}$$

$$\frac{\partial v(t, e^{h\sigma\sqrt{T}})}{\partial x} = e^{-r(T-t)} \tag{10.110}$$

$$h = 4 \tag{10.111}$$

We solve this problem repeatedly using different numbers of elements and time-steps to observe the convergence behavior. Data are shown in Table 10.8.

For the computations shown in Tables 10.9 and 10.10, we employ 400 spatial steps (i.e. 399 elements) and 800 time-steps. For the time integration, a first order backward difference method is used.

Table 10.10 Results for a European passport option (2)

w	Analytical solution	Numerical solution	Error
100	100.15660	100.1575	−0.0009 %
50	51.58181	51.5833	−0.0029 %
20	25.88757	25.8878	−0.0009 %
10	18.88084	18.8881	−0.0385 %
0	13.13810	13.1378	0.0023 %
−10	8.88084	8.8806	0.0027 %
−20	5.88757	5.8878	−0.0039 %
−50	1.58181	1.5833	−0.0942 %
−100	0.15660	0.1575	−0.5747 %

Table 10.11 Data for a European passport option (2)

Parameter	Value
Spot S	100.0
Dividend yield γ	5.0
Interest rate r	4.5
Volatility σ	0.3
Time to maturity	2 years

Table 10.12 Results for the European passport option (3)

w	FD	FE (Galerkin)	FE	Difference	Hedge ratio k FE (Galerkin)	k_{FE}	Difference
20	28.2277	28.2249	28.2295	−0.0064 %	−0.4674	−0.4679	−0.1070 %
10	22.3741	22.3734	22.3760	−0.0085 %	−0.3724	−0.3729	−0.1343 %
0	17.4323	17.4423	17.4438	−0.0660 %	—	—	—
−10	13.5100	13.5113	13.5135	−0.0259 %	0.5180	0.5176	0.0772 %
−20	10.4261	10.4293	10.4320	−0.0566 %	0.4302	0.4300	0.0465 %

A nonlinear model

Here we solve Equation (10.51) with Equation (10.53). The data of Table 10.11 are taken from Anderson and Brotherton-Ratcliffe (1998). This problem has been solved already with Finite Differences by Anderson and Brotherton-Ratcliffe (1998) and a Galerkin Finite Element method by Topper (2001a). For $w = 0$ no result can be given because of the jump in this case.

In the column 'Difference' of Table 10.12, the two finite element approaches are compared: a Galerkin method with quadratic shape functions vs. a collocation FEM with cubic Hermite basis functions. Next, we introduce early exercise to the nonlinear example from above (see Table 10.13). Again, this problem has already been solved with Finite Differences by Anderson and Brotherton-Ratcliffe (1998) and a Galerkin Finite Element method by Topper (2001a).

Relative exotics

So-called *relative exotics* in the world of passport options have the exotic feature on x; i.e. the cap, floor and/or barrier(s) is/are applied to the ratio of wealth and stock $x = \frac{w}{S}$. *Absolute*

Table 10.13 Results for the American passport option

w	FD	FE (Galerkin)	FE	Difference	Hedge ratio k (Galerkin)	k_{FE}	Difference
20	29.1764	29.2110	29.2135	−0.0086 %	−0.5042	−0.5041	0.0198 %
10	23.0050	23.0272	23.0298	−0.1078 %	−0.3974	−0.3871	2.5918 %
0	17.8418	17.8648	17.8668	−0.0112 %	—	—	—
−10	13.7776	13.7873	13.7902	−0.0210 %	0.5330	0.5331	−0.0188 %
−20	10.6031	10.6124	10.6150	−0.0245 %	0.4406	0.4406	0.0000 %

Table 10.14 Results for the European passport option with knock-out barrier (1)

w	FE
20	20
10	16.5737
0	13.4722
−10	10.8090
−20	8.5884

exotics bear the exotic feature(s) on w and/or S individually. These contracts are discussed further below.

As a numerical example, we add a knock-out barrier at $x = 20$ to the nonsymmetric contract from above (see Table 10.14). Note that by introducing a rebate of $R = 20$, the payoff function does not lose its convexity.

Absolute exotics

Absolute exotics refer to passport options with a cap, floor or barrier put on either the current value of the trading account $w(t)$ or the underlying S. Putting a barrier or cap on $w(t)$ seems natural, because beyond some limit the interest in additional hedging should be small. Besides, it makes the option cheaper (Ahn *et al.*, 1999b). For this kind of option, the HJB equation has to be solved directly, since no control is known *a priori*. In this section, we outline the idea of the algorithm, which involves some basic ideas from (static) optimization. Then we apply this algorithm to a problem which has already been solved above.

The starting point is the HJB equation, as in Equation (10.64):

$$(r - \gamma)SV_S + rwV_w - rV - \frac{\sigma^2 S^2}{2} \max_{-1 \leq \mu \leq 1} \left[V_{SS} + 2\mu V_{Sw} + \mu^2 V_{ww} \right] = V_t \qquad (10.112)$$

We generalize this problem slightly by allowing the control to vary within $[a, b]$. Then, one has to solve the following optimization problem for each node in each time-step:

$$\max_{a \leq \mu \leq b} (V_{SS} + 2\mu V_{Sw} + \mu^2 V_{ww}) \qquad (10.113)$$

This is just a quadratic polynomial in μ, so the existence of a maximum is guaranteed. The maximum has to occur at either $\mu = a$, $\mu = b$ or $\mu = -\frac{V_{Sw}}{V_{ww}}$ (if $a < \mu < b$), the latter point being the vertex of the quadratic polynomial. All three values are calculated and the maximum

Table 10.15   Results for the European passport option (4)

		Numerical solution	
w	Analytical solution	Equation (10.47)	Equation (10.64)
100	100.1566	100.1566	100.1581
50	51.5818	51.5804	51.5835
20	25.8876	25.8878	25.8904
10	18.8808	18.8820	18.8840
0	13.1381	13.1397	13.1414
−10	8.8808	8.8820	8.8840
−20	5.8876	5.8878	5.8904
−50	1.5818	1.5804	1.5836
−100	0.1566	0.1566	0.1581
Time Steps	—	31	99

is taken. Obviously, in case we do not have a constant solution or have a bang-bang solution, i.e. a control switching back and forth between its possible extremes, accurate approximations of the second derivatives are needed. As a first test, we recompute the problem from Table 10.8. This problem cannot test all features of the proposed algorithm since it is known that its optimal control is of the bang-bang type. The routine should, therefore, never be forced to compute second derivatives. Since this problem has two spatial dimensions, we also provide a numerical solution based on Equation (10.47), so that we have two numerical solutions based on two spatial variables. For both problems, the following boundary conditions have been applied:

$$V(0, w, t) = \max(0, w) \tag{10.114}$$

$$\frac{\partial^2 V(0, w, t)}{\partial S^2} = 0 \tag{10.115}$$

$$\frac{\partial^2 V(200, w, t)}{\partial S^2} = 0 \tag{10.116}$$

$$V(S, -200, t) = 0 \tag{10.117}$$

$$V(S, 200, t) = w \tag{10.118}$$

With a rectangular mesh of 19 elements in S and 42 in w, and an adaptive second order Adams–Moulton method for the integration of time, the results of Table 10.15 are produced. The algorithm that computes the optimal control implicitly gives slightly less accurate results for the same mesh, while using about three times more time-steps.

Knock-out barriers can be applied to both stock and/or the account. In the second case, the rationale is that when the loss becomes too large, the coverage is turned off; in the first case, the usual rationale for making the option cheaper is applied. As an example, we take the problem from above and introduce a knock-out barrier on w:

$$V(S, w_{\text{barrier}}, t) = 0 \tag{10.119}$$

With a rectangular mesh of 19 elements in S and 42 in w, and an adaptive second order Adams–Moulton method for the time integration, the results of Table 10.16 were produced. Figure 10.4 illustrates the case with the knock-out barrier at $w = -20$.

Table 10.16   Results for the European passport option with knock-out barrier (2)

	Location of barrier		
w	w= −20	w= −30	w= −40
100	100.0000	100.1408	100.1602
50	51.4286	51.5643	51.5798
20	25.3353	25.7934	25.8769
10	18.0489	18.7373	18.8666
0	11.7753	12.9275	13.1203
−10	6.0061	8.3566	8.8363
−20	0.0000	4.3371	5.6642
−50	0.0000	0.0000	0.0000
−100	0.0000	0.0000	0.0000
Time steps	64	60	104

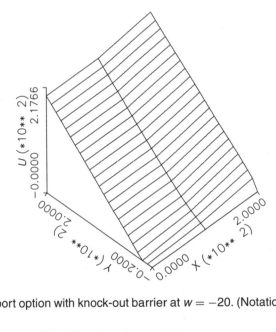

Figure 10.4   Passport option with knock-out barrier at $w = -20$. (Notation: $Y = w$ and $X = S$)

10.2.4   Uncertain volatility: Best and worst cases

The model

In the Black–Scholes model, all input parameters are observable, except for volatility.[4] When both underlying and option are traded, these two prices can be used to compute volatility implicitly. Often, there are several options with different strikes on the same underlying, so that it is possible to compute several values for the implied volatility. One regularly observes that the values for implied volatility are the lowest at the money of the option and increase for both in the money and out of the money. This phenomenon is usually called the *volatility smile*. Besides this, implied volatility also depends on the time to maturity of the option, with

---

[4] Discrete dividends are often not known either, especially in the setting of long-dated options in the equity market. For some details and references, see Section 5.2.5.

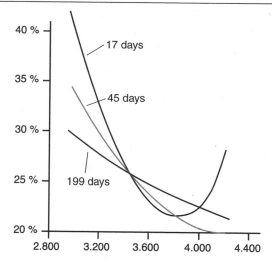

Figure 10.5 Implied volatilities of the DAX on 9/2/2003 for 17, 45 and 199 days until maturity (Leschhorn and Eibl, 2004)

longer maturities usually having higher implied volatilities. This phenomenon is usually called the *term structure of volatility*. There are several ways to adjust the Black–Scholes model to cope with both phenomena:

- *Implied volatility*. Quoted option prices with various strikes and maturities on the same underlying are used to construct a volatility surface. In order to price exotic options (which are not traded) the appropriate volatility is taken from the volatility surface and used as input in the Black–Scholes model. Although this approach is widely used, it has been criticized heavily:

  > A smiley implied volatility is the wrong number to put in the wrong formula to obtain the right price (Rebonato, 1999).

  A typical volatility smile is shown in Figure 10.5.

- *Local volatility*. This approach assumes that the underlying follows

$$dS = \mu S dt + \sigma(S, \tau) S dX \tag{10.120}$$

The function $\sigma(S, \tau)$ has to be constructed from quoted option prices. Various approaches have been suggested (Andersen and Brotherton-Ratcliffe, 1998; Dumas *et al.*, 1998). It has to be emphasized that $\sigma(S, \tau)$ does not represent implied volatility, although there is a relationship between local and implied volatility (Wilmott *et al.*, 1996).

- *Stochastic volatility*. Another stylized fact of volatility is that it is not stable and seems to be mean-reverting (Stein and Stein, 1991). This feature is especially important for long-term options. A common approach to capturing the random evolution of volatility is to add a second risk factor to the Black–Scholes model:

$$dS = \mu S dt + \sigma S dX \tag{10.121}$$

$$d\sigma = A(\sigma) dt + B(\sigma, S) dX \tag{10.122}$$

Several stochastic processes for the evolution of volatility have been suggested; see Section 7.2.2.

The Uncertain Volatility Model (UVM) takes a completely different approach. Let us start with a quote that nicely describes the essence of this approach (Buff, 2002a, 2000b, p. 2):

> The original uncertain volatility model in Avellaneda *et al.* (1994) is a worst-case scenario for the sell-side. By maximizing the portfolio value and charging accordingly, sellers are guaranteed coverage against adverse market behavior if the realized volatility belongs to the candidate set. Buyers, on the other hand, would use the uncertain volatility scenario to compute a minimal portfolio value and be covered against adverse market behavior if they pay no more than the obtained value for the portfolio.

UVM is based on the work of Avellaneda and Parás (1994), Avellaneda *et al.* (1994), Lyons (1995) and Oztukel and Wilmott (1998). Our derivation of the pricing equation follows Wilmott *et al.* (1996). Based on an educated guess, lower and upper bounds for volatility are fixed:

$$\sigma^- < \sigma < \sigma^+ \tag{10.123}$$

Both σ^- and σ^+ might be functions of t and/or S: $\sigma^+ = \sigma^+(t, S)$ and $\sigma^- = \sigma^-(t, S)$. Following the usual Black–Scholes argument, we construct a *Delta*-hedge:

$$\Pi = V - \Delta S \tag{10.124}$$

The asset is assumed to follow the usual geometric Brownian motion:

$$dS = \mu S dt + \sigma S dX \tag{10.125}$$

with σ being unknown. An infinitesimal change in the portfolio Π can be expressed with the help of Itô's Lemma:

$$d\Pi = \left(\frac{\partial V}{\partial t} + \frac{1}{2}\sigma^2 S^2 \frac{\partial^2 V}{\partial S^2} \right) dt + \left(\frac{\partial V}{\partial S} - \Delta \right) dS \tag{10.126}$$

Choosing $\Delta = \partial V / \partial S$, Equation (10.126) simplifies to

$$d\Pi = \left(\frac{\partial V}{\partial t} + \frac{1}{2}\sigma^2 S^2 \frac{\partial^2 V}{\partial S^2} \right) dt \tag{10.127}$$

Now we have to deviate from the Black–Scholes argument. Based on the fact that we do not know σ, the value of $d\Pi$ is unknown as well. We assume that σ takes on a value that minimizes the RHS of Equation (10.127). The financial interpretation of this assumption is presupposing the worst possible infinitesimal increase in the portfolio. The worst-case increase is set equal to the riskfree rate (as in the Black–Scholes argument):

$$\min_{\sigma^- < \sigma < \sigma^+} (d\Pi) = r\Pi dt \tag{10.128}$$

$$\Leftrightarrow \min_{\sigma^- < \sigma < \sigma^+} \left(\frac{\partial V}{\partial t} + \frac{1}{2}\sigma^2 S^2 \frac{\partial^2 V}{\partial S^2} \right) dt = r \left(V - S \frac{\partial V}{\partial S} \right) dt \tag{10.129}$$

After rearranging, we find that the UVM pricing PDE for the worst case looks like:

$$\frac{\partial V^-}{\partial t} + \frac{1}{2}\sigma(\Gamma)^2 S^2 \frac{\partial^2 V^-}{\partial S^2} + rS \frac{\partial V^-}{\partial S} - rV^- = 0 \tag{10.130}$$

$$\Gamma = \frac{\partial^2 V^-}{\partial S^2} \tag{10.131}$$

$$\sigma(\Gamma) = \begin{cases} \sigma^+ & \text{if } \Gamma < 0 \\ \sigma^- & \text{if } \Gamma > 0 \end{cases} \tag{10.132}$$

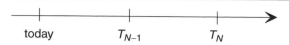

Figure 10.6 Direction of (natural) time t: $\longrightarrow$; direction of time to expiry $\tau = T_i - t$: $\longleftarrow$

Accordingly, the UVM pricing PDE for the best case looks like:

$$\frac{\partial V^+}{\partial t} + \frac{1}{2}\sigma(\Gamma)^2 S^2 \frac{\partial^2 V^+}{\partial S^2} + rS\frac{\partial V^+}{\partial S} - rV^+ = 0 \qquad (10.133)$$

$$\Gamma = \frac{\partial^2 V^+}{\partial S^2} \qquad (10.134)$$

$$\sigma(\Gamma) = \begin{cases} \sigma^+ & \text{if } \Gamma > 0 \\ \sigma^- & \text{if } \Gamma < 0 \end{cases} \qquad (10.135)$$

Both PDEs are nonlinear, parabolic PDEs of second order and backward in time. Well-posedness of this problem can be established; see Lyons (1995) and references therein. While the worst-case pricing PDE can be interpreted as the lowest possible price for the sell-side, the best-case pricing PDE determines the lowest possible value for the buy-side. Or, equivalently, the worst-case PDE represents the worst-case value of a long position, while the best-case PDE represents the negative of the worst-case price of a short position.

Pricing portfolios: horizontal integration

For this paragraph we only consider options which are *not* path dependent. Path-dependent options are discussed below. Whenever the payoff function $V(S, T)$ of the option under consideration is convex in S, i.e. $\frac{\partial^2 V}{\partial S^2} > 0$,[5] Γ does not change its sign and so the worst-case price is equal to the Black–Scholes price with σ^-. When there are several options, all payoff functions have to be summed up. This sum constitutes the final condition for the pricing PDE. In the case where this final condition happens to be convex, again the worst-case price is equal to the Black–Scholes price with σ^-. This way, a portfolio of options with the same maturity can be treated as one option.

Different times to expiry: vertical integration

As observed by Avellaneda and Parás (1994), options expiring at different points in time can be combined into one portfolio in the following way. There are N different expiry dates with $T_1 < t < T_{N-1} < T_N$, as in Figure 10.6. The options maturing at T_1 are the next to mature; the options maturing at T_N have the longest times to expiry. Accordingly, $V(S, T_I)$ represents the sum of all non-path-dependent options maturing at T_I. To integrate different maturities, the PDE has to be manipulated at each time to maturity. $V_{\text{Portfolio}}(S, T_N)$ is the value of the portfolio containing all options. It can be constructed in the following way:

- at T_N: $V_{\text{Portfolio}}(S, T_N) \leftarrow V(S, T_N)$;
- at T_{N-1}: $V_{\text{Portfolio}}(S, T_{N-1}) \leftarrow V_{\text{Portfolio}}(S, T_{N-1}) + V(S, T_{N-1})$;
- at T_{N-2}: $V_{\text{Portfolio}}(S, T_{N-2}) \leftarrow V_{\text{Portfolio}}(S, T_{N-2}) + V(S, T_{N-2})$;
- etc. up to $I = 1$.

with '$\leftarrow$' denoting that the LHS has to be replaced by the RHS.

---

[5] For plain vanilla options this (necessary) criterion of convexity does not hold, since $V(S, T)$ is not differentiable in S at the strike. However, in that case convexity can be checked easily via its definition.

Table 10.17   Data for a butterfly spread

Parameter	Symbol	Value
Asset price	S	100
1. Strike	K_1	90
2. Strike	K_2	110
Interest rate	r	0.1
Maximum volatility	σ_{max}	0.25
Minimum volatility	σ_{max}	0.15
Maturity	T	0.25 year

Table 10.18   Results for the butterfly spread: FE vs. FD

Nodes	Time-step	FD (fully implicit)	FD (4-step Rannacher)	Crank–Nicolson	FE (Crank–N.)	FE (Euler)
61	25	2.3501	2.3041	1.7246	3.0248	2.4393
121	50	2.3250	2.2996	1.5713	2.7843	2.3339
241	100	2.3116	2.2982	1.4622	n.a.	n.a.
481	200	2.3047	2.2978	1.3806	n.a.	n.a.
961	400	2.3012	2.2977	1.3264	n.a.	n.a.

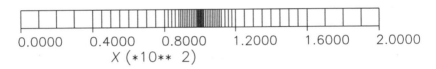

0.0000	0.4000	0.8000	1.2000	1.6000	2.0000

$X\ (\ast 10\ast\ast\ 2)$

Figure 10.7   Discretization in S with 61 nodes

As an example we recompute a butterfly spread from the literature (Pooley *et al.*, 2003a). The data are given in Table 10.17. The payoff of a butterfly leads to the following final condition:

$$V(S, T) = (S - K_1)^+ - 2\left(S - \frac{(K_1 + K_2)}{2}\right)^+ + (S - K_2)^+ \tag{10.136}$$

Since the payoff of this instrument is not convex, it needs to be computed numerically. We compare three FD implementations from the literature (Pooley *et al.*, 2003a) with two collocation FE solutions. The third column of Table 10.18 reports the results achieved by a fully implicit FD scheme initially unequal in space but uniformly refined (as in Figures 10.7 and 10.8) with equidistant intervals in time. The fourth column reports results based on so-called Rannacher time stepping, which switches from a fully implicit scheme to a Crank–Nicolson scheme after four time-steps (Rannacher, 1984). The fifth column reports the results from a pure Crank–Nicolson scheme. Obviously, this scheme does not lead to the right solution (which is due to the lack of smoothness in the final condition, as explained by Pooley *et al.* (2003a)). The spaces between nodes in the direction of the spatial variable are chosen as indicated by Figures 10.7 and 10.8.

Table 10.19 Results for the butterfly spread: FE with various equidistant time-steps

Nodes	Steps	FE (Crank–N.)	FE (Euler)
61	100	2.8994	2.3306
61	200	2.8649	2.3130
61	400	2.6920	2.3045
61	800	2.6256	2.3006
61	1600	2.2981	2.2989
61	3200	2.2973	2.2980
121	100	2.7843	2.3340
121	200	2.8021	2.3151
121	400	2.9375	2.3059
121	800	2.8636	2.3015
121	1600	2.6995	2.2994
121	3200	2.6101	2.2984

$$0.0000 \qquad 0.4000 \qquad 0.8000 \qquad 1.2000 \qquad 1.6000 \qquad 2.0000$$
$$X\ (*10** \ 2)$$

Figure 10.8 Discretization in S with 121 nodes

Since FE solutions need less nodes for the spatial variable than FD schemes, we only increase the number of time-steps. The results are given in Table 10.19. Here, again the Crank–Nicolson time-discretization fails, so that we will not use it for further computations. Within the UVM framework, portfolios of options are known to be subadditive for the sell-side (and superadditive for the buy-side) (Avellaneda and Parás, 1994). We will use this relationship as a (necessary but by far not sufficient) test for the correctness of our approach. The butterfly spread can be decomposed into

- a long call with strike K_1: $C(K_1, \ldots)$;
- a long call with strike K_2: $C(K_2, \ldots)$;
- two short calls with strike $\frac{K_1+K_2}{2}$: $-2C\left(\frac{K_1+K_2}{2}, \ldots\right)$.

Since all payoffs of vanilla options are convex, the worst-case price of each option can be computed analytically:

$$C(K_1, \sigma^-, \ldots) + C(K_2, \sigma^-, \ldots) - 2C\left(\frac{K_1+K_2}{2}, \sigma^+ \ldots\right) \tag{10.137}$$

$$\approx 12.3391 + 0.7277 - 2*4.3515 \tag{10.138}$$

$$= 0.5578 < 2.29 \tag{10.139}$$

Barrier options: a knock-out call

As an example we will take an up-and-out call with a barrier at $S = S_u$. The data are given in Table 10.20. It is obvious that the Γ of this reverse barrier option changes its sign, so the PDE

Table 10.20 Data for an up-and-out call

Parameter	Value
Strike price	100
Interest rate	0.05
Lower bound of volatility	0.17
Upper bound of volatility	0.23
Maturity	1 year

Source: Wilmott *et al.,* 1996

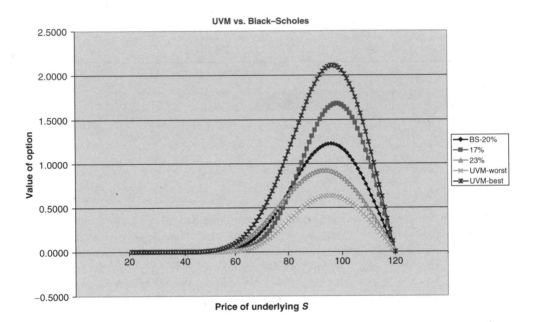

Figure 10.9 Up-and-out call

(10.133) has to be solved numerically. The boundary conditions are:

$$V(S, T) = \max(S - E, 0) \tag{10.140}$$

$$V(0, t) = 0 \tag{10.141}$$

$$V(S_u, t) = 0 \tag{10.142}$$

The results are plotted in Figure 10.9.

A knock-in call

As an example we will take a down-and-in call with a barrier at $S = S_d$. The data are given in Table 10.21. The boundary conditions for the Black–Scholes model are:

$$V(S, T) = 0 \tag{10.143}$$

$$V(0, t) = 0 \tag{10.144}$$

$$V(S_d, t) = C(S_d, t, \sigma) \tag{10.145}$$

Table 10.21   Data for a down-and-in call

Parameter	Value
Strike price	40
Interest rate	0.1
Lower bound of volatility	0.17
Upper bound of volatility	0.23
Volatility	0.20
Maturity	0.5 year
Barrier	50

Table 10.22   Results for the down-and-in call

	Black–Scholes		UVM
Underlying	Analytical solution		FE (Euler)
25	0.0000	0.0000	0.0000
30	0.0079	0.0080	0.0008
35	0.2299	0.2299	0.0808
40	1.7963	1.7962	1.1931
45	6.0415	6.0419	5.4320
49	10.7893	10.7896	10.0638

with $C(\cdot)$ denoting the value of a plain vanilla European call. Equations (10.143) and (10.144) remain unchanged for the UVM model. Equation (10.145) has to be replaced by

$$V(S_d, t) = C(S_d, t, \sigma^-) \tag{10.146}$$

because the Γ of $C(\cdot)$ is single-signed so that the Black–Scholes value corresponding to σ^- can be used as the boundary condition. There are several ways to solve this problem:

- *Hierarchically.* First one computes $C(S_d, t)$ either numerically or analytically and feeds them into the FE program when required.

- *As a system of PDEs.* The PDEs for the knock-in option and for the plain vanilla call are computed simultaneously. Since the location of the BC is different for both options, a simple device has to be constructed, since the PDE solver used for this book (PDE2D) can only deal with a common domain for all PDEs within a system.

Since an analytical solution for single barrier options has been derived by Reiner and Rubinstein (1991) and implemented by Haug (1997), we can easily check the numerical results. The numerical solution was obtained with 50 equidistant time-steps and 100 equidistant nodes. The results in Table 10.22 show that the numerical values for the Black–Scholes value deviate only marginally from the analytical solution. The UVM values are below the Black–Scholes values, as expected.

10.2.5   Worst-case pricing of rainbow options

Deriving the model

Assuming that a lower and an upper bound for correlation is

$$-1 \le \rho^- \le \rho \le \rho^+ \le 1 \tag{10.147}$$

we shall use a similar argument as Black and Scholes (1973) to derive a model for worst-case pricing. Both assets follow the usual SDE:

$$dS_1 = \mu_1 \, S_1 \, dt + \sigma_1 \, S_1 \, dX \tag{10.148}$$
$$dS_2 = \mu_2 \, S_2 \, dt + \sigma_2 \, S_2 \, dX \tag{10.149}$$

Both assets are correlated

$$E\left(dX_i dX_j\right) = \rho \, dt \tag{10.150}$$

Setting up a portfolio consisting of a long position in one option and short positions in both underlyings leads to:

$$\Pi = V(S_1, S_2, t) - \Delta_1 S_1 - \Delta_2 S_2 \tag{10.151}$$

With Itô's Lemma, an infinitesimal change in this portfolio can be expressed as:

$$d\Pi = \left(\frac{\partial V}{\partial t} + \frac{1}{2}\sum_{i=1}^{2}\sum_{j=1}^{2}\sigma_i\sigma_j\rho_{ij}S_iS_j\frac{\partial^2 V}{\partial S_i\partial S_j}\right)dt + \sum_{i=1}^{2}\left(\frac{\partial V}{\partial S_i} - \Delta_i\right)dS_i \tag{10.152}$$

This expression can be simplified because $i, j = 2$. We also know that $\rho_{ii} = 1$ and $\rho_{ij} = \rho_{ji}$. To simplify the notation we set $\rho_{12} = \rho_{21} = \rho$.

$$d\Pi = \left(\frac{\partial V}{\partial t} + \sigma_1\sigma_2\rho S_1 S_2\frac{\partial^2 V}{\partial S_1\partial S_2} + \frac{1}{2}\sigma_1^2 S_1^2\frac{\partial^2 V}{\partial S_1^2} + \frac{1}{2}\sigma_2^2 S_2^2\frac{\partial^2 V}{\partial S_2^2}\right)dt$$
$$+ \left(\frac{\partial V}{\partial S_1} - \Delta_1\right)dS_1 + \left(\frac{\partial V}{\partial S_2} - \Delta_2\right)dS_2 \tag{10.153}$$

Choosing $\Delta_i = \frac{\partial V}{\partial S_i}$ for $i = 1, 2$ eliminates the risk, just as in the classical argument when deriving the multidimensional Black–Scholes equation.

$$d\Pi = \left(\frac{\partial V}{\partial t} + \sigma_1\sigma_2\rho S_1 S_2\frac{\partial^2 V}{\partial S_1\partial S_2} + \frac{1}{2}\sigma_1^2 S_1^2\frac{\partial^2 V}{\partial S_1^2} + \frac{1}{2}\sigma_2^2 S_2^2\frac{\partial^2 V}{\partial S_2^2}\right)dt$$

In the Black–Scholes model it can be argued that from knowing V we also know $d\Pi$. Here, however, this is not true, since ρ is unknown. Since we want to derive a worst-case scenario model, we want to be extremely pessimistic. In every infinitesimal time-step, we assume a correlation that leads to the smallest growth in the portfolio. This implies that the return on this worst-case portfolio is equal to the riskfree rate.

$$\min d\Pi \stackrel{!}{=} r\, \Pi \, dt \tag{10.154}$$

The minimization is to be performed with respect to ρ. Using Equation (10.151) the right-hand side of Equation (10.154) can be rewritten as

$$r\, \Pi \, dt = r\left(V(S_1, S_2, t) - \frac{\partial V}{\partial S_1}S_1 - \frac{\partial V}{\partial S_2}S_2\right)dt \tag{10.155}$$

The left-hand side can be expressed as

$$\min_{\rho} \left[\left(\frac{\partial V}{\partial t} + \underbrace{\sigma_1 \sigma_2 \rho S_1 S_2 \frac{\partial^2 V}{\partial S_1 \partial S_2}}_{A} + \frac{1}{2} \sigma_1^2 S_1^2 \frac{\partial^2 V}{\partial S_1^2} + \frac{1}{2} \sigma_2^2 S_2^2 \frac{\partial^2 V}{\partial S_2^2} \right) dt \right] \tag{10.156}$$

Because of the minimization with respect to $\rho \in [\rho^-, \rho^+]$, we can distinguish two cases:

1. $\frac{\partial V}{\partial S_1 \partial S_2} = \Gamma_{\text{cross}} > 0 \Rightarrow A = \rho^- \sigma_1 \sigma_2 S_1 S_2 \Gamma_{\text{cross}}$
2. $\frac{\partial V}{\partial S_1 \partial S_2} = \Gamma_{\text{cross}} < 0 \Rightarrow A = \rho^+ \sigma_1 \sigma_2 S_1 S_2 \Gamma_{\text{cross}}$

Combining these results and taking dividends into account, we arrive at the worst-case pricing equation:

$$\frac{1}{2} \sigma_1^2 S_1^2 \frac{\partial^2 V}{\partial S_1^2} + \frac{1}{2} \sigma_2^2 S_2^2 \frac{\partial^2 V}{\partial S_2^2} + \rho(\Gamma_{\text{cross}}) \sigma_1 \sigma_2 S_1 S_2 \frac{\partial^2 V}{\partial S_1 \partial S_2} \tag{10.157}$$

$$+ (r - D_1) S_1 \frac{\partial V}{\partial S_1} + (r - D_2) S_2 \frac{\partial V}{\partial S_2} = rV - \frac{\partial V}{\partial t}$$

$$\rho(\Gamma_{\text{cross}}) = \begin{cases} \rho^- & \text{if } \Gamma_{\text{cross}} > 0 \\ \rho^+ & \text{if } \Gamma_{\text{cross}} < 0 \end{cases} \tag{10.158}$$

$$\Gamma_{\text{cross}} = \frac{\partial^2 V}{\partial S_1 \partial S_2} \tag{10.159}$$

These results can be extended to ρ^- and ρ^+ as they are deterministic functions of t. Obviously, it has to hold that $\rho^-(t) < \rho^+(t)$. It is also possible to combine this model with other models to adjust for the uncertainty of the other unknown parameters.

Example 23 (capped call on a basket)
As a first example of a rainbow option, we consider a capped call on a basket of two equities. The data are given in Table 10.23. This is a common OTC product. Sometimes it is traded at exchanges as well. For pricing this call we have to solve the PDEs (10.157)–(10.159) with the following final and boundary conditions:

$$V(T) = \min [cap, \max (0, w_1 S_1 + w_2 S_2 - E)] \tag{10.160}$$

$$\Omega = [S_1^{\min}; S_1^{\max}] \times [S_2^{\min}; S_2^{\max}] \tag{10.161}$$

$$V(S_1, S_2^{\min}) = BS \left(S_1, \frac{E}{w_1}, t \right) - BS (S_1, cap, t) \tag{10.162}$$

$$\frac{\partial V(S_1, S_2^{\max})}{\partial S_2} = 0 \tag{10.163}$$

$$V(S_1^{\min}, S_2) = BS \left(S_2, \frac{E}{w_1}, t \right) - BS (S_2, cap, t) \tag{10.164}$$

$$\frac{\partial V(S_1^{\max}, S_2)}{\partial S_1} = 0 \tag{10.165}$$

The function $BS(x, y, t)$ denotes the price of a European call option with an underlying price of x and a strike price of y, as given by the Black–Scholes model (Black and Scholes, 1973).[6] The results for this example are shown in Table 10.24.

---

[6] For a discussion of the boundary conditions and alternative specifications see Topper (1999).

Table 10.23   Data for a capped call on a basket

Parameter	Symbol	Value
First asset price	S_1	100
Weight first asset	w_1	1
Second asset price	S_2	100
Weight second asset	w_2	1
Strike price	E	200
Interest rate	r	0.0953102
Dividend yield first asset	D_1	0.0487902
Dividend yield second asset	D_2	0.0
Time to maturity	T	0.5
	S_1^{min}	0
Domain	S_1^{max}	200
	S_2^{min}	0
	S_2^{max}	200
Cap	cap	10

Table 10.24   Results for the capped call on a basket

Volatility		Scenario 1	Scenario 2	Scenario 3	
σ_1	σ_2	$\rho = 0.5$	$\rho^- = 0.4 \leq \rho \leq 0.6 = \rho^+$	$\rho^- = 0.3 \leq \rho \leq 0.7 = \rho^+$	δ_{ρ^-}
	0.1	5.3676	5.1953	5.0833	−0.050
0.1	0.2	4.9923	4.8832	4.7742	−0.050
	0.3	4.7555	4.6612	4.5677	−0.045
	0.1	4.9625	4.8878	4.7767	−0.050
0.2	0.2	4.9989	4.7348	4.5998	−0.065
	0.3	4.7120	4.5884	4.4569	−0.065
	0.1	4.7632	4.6669	4.5706	−0.050
0.3	0.2	4.7246	4.5686	4.4584	−0.070
	0.3	4.6329	4.5006	4.3591	−0.070

☐

Example 24 (two-asset barrier options)
In October 1993, Bankers Trust structured a call option on a basket of Belgian stocks that would knock out if the Belgian franc – not the Belgian stock – appreciated by more than 3.5 %. So we have two spatial variables entering the pricing equation: one that determines the payoff, treating the basket as a single risk factor, and one that determines whether the option is still alive. This kind of product went through a boom in the OTC market in 1997, with a strike in the exchange rate of USD vs. DEM and a barrier in the exchange rate in USD vs. FRF. Back then, this product was employed to profit from the high difference in historical and implied correlation. Normally, this product comes with an FX rate on the knock-out barrier, so that an adjustment for discrete monitoring of the barrier is not necessary. This product can be generalized for knock-in and part-time barriers, so that the barrier is only valid for parts of the life of the option (Haug, 1997).

Table 10.25   Data for a two-asset barrier option product

Parameter	Symbol	Value
First asset price	S_1	100
Weight first asset	w_1	1
Weight second asset	w_2	1
Strike price	E	100
Interest rate	r	0.1
Dividend yield first asset	D_1	0.0
Dividend yield second asset	D_2	0.0
Volatility first asset	σ_1	0.2
Volatility second asset	σ_2	0.3
Time to maturity	T	0.5
Correlation	ρ	varies
	$S_1^{\min}$	varies
Domain	$S_1^{\max}$	300
	$S_2^{\min}$	0
	$S_2^{\max}$	300

We again solve the PDEs (10.157)–(10.159) with the following final and boundary conditions and the data from Table 10.25:

$$V(T) = \max(0, w_2 S_2 - E) \tag{10.166}$$

$$\Omega = \left[S_1^{\min}; S_1^{\max}\right] \times \left[S_2^{\min}; S_2^{\max}\right] \tag{10.167}$$

$$\frac{\partial V\left(S_1, S_2^{\min}\right)}{\partial S^2} = 0 \tag{10.168}$$

$$\frac{\partial V\left(S_1, S_2^{\max}\right)}{\partial S_2} = 1 \tag{10.169}$$

$$V\left(S_1^{\min}, S_2\right) = 0 \tag{10.170}$$

$$\frac{\partial V\left(S_1^{\max}, S_2\right)}{\partial S_1} = 0 \tag{10.171}$$

The results for this example are shown in Table 10.26.   □

Example 25 (basket options)
It holds for both calls and puts that the higher the correlation, the higher the price of the option. Consequently,

$$\frac{\partial V}{\partial \rho} = \kappa \geq 0 \tag{10.172}$$

From the following relationship (Reiss and Wystup, 2001, Equation (100))

$$\kappa = \underbrace{\sigma_1 \sigma_2 \tau S_1 S_2}_{\geq 0} \Gamma_{12} \tag{10.173}$$

we can conclude that Γ_{12} is always positive. Therefore, the worst-case price of a basket option only depends on ρ^-. Since no closed-form solution is known for basket options, the price of this type of option has to be computed numerically or by using an analytical approximation. □

Table 10.26 Results for the European two-asset barrier option

	Location of barrier					
	95			99		
	$\rho = 0.5$		$0.4 \leq \rho \leq 0.6$	$\rho = 0.5$		$0.4 \leq \rho \leq 0.6$
S_2	Analytical	FE	FE	Analytical	FE	FE
80	1.3373	1.3353	1.2257	0.3570	0.3568	0.3143
85	2.1369	2.1320	1.9643	0.5572	0.5560	0.4945
90	3.1633	3.1566	2.9205	0.8080	0.8064	0.7233
95	4.4008	4.3939	4.0836	1.1039	1.1026	0.9969
100	5.8210	5.8155	5.4230	1.4376	1.4368	1.3090
105	7.3901	7.3874	6.9287	1.8007	1.8008	1.6524
110	9.0741	9.0753	8.5487	2.1859	2.1870	2.0199
115	10.8430	10.8485	10.2603	2.5868	2.5890	2.4051
120	12.6721	12.6821	12.9316	2.9984	3.0017	2.8030

10.3 NOTES

Stochastic optimal control for Markov diffusion processes leads to second order parabolic PDEs. Examples for PDEs resulting from dynamic programming can be found in Fleming and Rishel (1975). The links between dynamic programming and contingent claim valuation are discussed by Knudsen et al. (1999). Models applying (singular) stochastic control to problems of portfolio selection with transaction costs and optimal investment can be found in Chapter 8 of Fleming and Soner (1993). Numerous nonlinear PDEs arising from dynamic optimization techniques applied to financial problems can be found in Merton (1998) and Ferguson and Lim (1998) and the references therein.

The UVM can be generalized in various directions. The extension of uncertain correlation to more than two assets is not straightforward. A simple intuitive explanation is that the correlation matrix has a special structure. It is always symmetric and positive-definite. Integrating path dependency is discussed in several papers:

- For barrier options, see Buff (2002b) and Lyons (1995).
- American and Bermudan options are not discussed here but in Avellaneda and Buff (1999), Buff (2000) and Chapter 8 of Buff (2002b).
- Various other exotics, such as Asian options, are discussed in Lyons (1995) and the references given there.

The UVM *term structure* model by Wilmott (Wilmott, 1998, Chapter 40; Penaud, 2002, Chapter 9) is a nonlinear *hyperbolic* PDE; see Section A.3.4 (while all other PDEs in this book are parabolic). The UVM can be used to construct an optimal hedge (Avellaneda et al., 1994; Avellaneda and Parás, 1996). This approach requires the combination of optimization tools and PDE solvers.

Part III
Outlook

11

Future Directions of Research

In the previous chapters it was outlined how to compute approximate solutions with FE for a wide range of differential equations arising in modern finance. More general stochastic processes allow a more realistic modeling of the risk factors. Jump diffusion processes lead to parabolic integro-differential equations which can be solved with FE as well. In some cases, more advanced models lead to systems of PDEs. Examples for linear systems are the models on speculating with options (Korn and Wilmott, 1996; Korn and Wilmott, 1998; Wilmott, 1998, Section 28.3.2) and jump volatility (Wilmott, 1998, Section 28.3). Examples for nonlinear models are the credit risk models (Huge and Lando, 1999; Duffie and Huang, 1996). The techniques introduced in the previous chapters can easily be extended to systems of differential equations.

In engineering, it is common practice to combine FE solvers with optimization tools. In finance, there are basically two settings where an optimizer and a PDE solver can be coupled. Either the optimizer is called after one or several numerical solutions have been computed and these numerical results are used for the next iteration in the optimization, or the optimizer is called from within the PDE solver. An example of the latter has been given in Section 10.2.3 with the solution of Equation (10.64). At each node, the following problem needs to be solved:

$$\max_{a<\mu<b} (V_{SS} + 2\mu V_{Sw} + \mu^2 V_{ww}) \tag{11.1}$$

In Section 10.2.3 this problem has been solved analytically. In the case of several underlyings $S_1, \ldots, S_n$ and/or additional trading constraints on μ, S and w, it might not be possible anymore to solve the restricted nonlinear optimization problem analytically, so that a numerical method needs to be invoked. Since one optimization problem is associated with each node at each time-step, it is highly recommendable to choose fast optimizers. Other problems do not require solution of an optimization problem within the solution of the PDE, but need the solution of a PDE as input for each iteration in the solution of an optimization problem. A typical application is the calibration of a local volatility surface. The local volatility is given by $\sigma(\tau, S, a_1, \ldots, a_n)$. The parameters a_i are to be determined by minimizing the difference between model option prices and observed prices in some norm. This leads to an unrestricted nonlinear optimization problem. In each iteration of the optimization routine, the PDE has to be solved with a new set of parameters a_i. Similar problems arise in the context of dynamic asset allocation and optimal consumption of the Merton type.

A further interesting topic to investigate is adaptivity in the spatial variables. This could produce large savings in computing time, since in most pricing problems a large part of the solution is very smooth, so that it could be approximated with some large elements. There are various techniques for spatial adaptivity. Briefly stated, h-type adaption models add new elements into a mesh to improve the discretization; p-type adaption models increase the degree of approximation used for the elements in a mesh to improve the discretization; hp-type adaption models apply a combination of both of these procedures; and r-type adaption models reposition element vertices within the mesh to improve solution accuracy.

Part IV
Appendices

Appendix A

Some Useful Results from Analysis

A.1 IMPORTANT THEOREMS FROM CALCULUS

A.1.1 Various concepts of continuity

Before uniform continuity is defined, we give a definition of continuity for the sake of completeness:

Definition 2 (Continuity)
Given a function $f : U \to \mathbb{R}^n$, $x \to f(x)$, $U \subset \mathbb{R}^m$. Provided the following holds, this function is called continuous:

$$\forall x \in U \quad \forall \varepsilon > 0 \quad \exists \delta > 0 \quad \forall y \in U \quad such\ that\ ||x - y||_2 = \sqrt{\sum_{i=1}^{m}(x_i - y_i)^2} < \delta :$$
$$||f(x) - f(y)||_2 < \varepsilon \qquad \square$$

Definition 3 (Uniform Continuity)
Given a function $f : U \to \mathbb{R}^n$, $x \to f(x)$, $U \subset \mathbb{R}^m$. Provided the following holds, this function is called uniformly continuous:

$$\forall \varepsilon > 0 \quad \exists \delta > 0 \quad \forall x, y \in U \quad such\ that\ ||x - y||_2 < \delta :$$
$$||f(x) - f(y)||_2 < \varepsilon \qquad \square$$

Theorem 2
Any uniformly continuous function is continuous. $\qquad \square$

The converse is not necessarily true, see Example 26. However, there is an important class of domains for which continuity and uniform continuity are equivalent.

Theorem 3
Assume that U is compact. Then the following holds:

$$f \text{ is uniformly continuous} \Leftrightarrow f \text{ is continuous} \qquad (A.1)$$
$$\square$$

Example 26
Consider the function $f(x) = sin(1/x)$, $U = (0, \infty)$. This function is continuous but not uniformly continuous. Given an arbitrarily small δ, we find $x, y \in U$ such that $||x - y||_2 < \delta$ but $||f(x) - f(y)||_2 > 1$. However, on any compact subset $[a, b] \subset U$ the function $f(x)$ is uniformly continuous. $\qquad \square$

Definition 4 (Lipschitz Continuity)
Let $U \subset \mathbb{R}^n$ be open, let $T = (a, b) \subset \mathbb{R}$, and let $f : U \times T \to \mathbb{R}^m$.

1. f is locally Lipschitz continuous with respect to x:
$\forall (x, t) \in U \times T \ \exists \varepsilon_1, \varepsilon_2, L > 0 \ \forall (y, s) \in U \times T$ such that $||x - y||_2 < \varepsilon_1$ and $|t - s| < \varepsilon_2$:

$$||f(x, s) - f(y, s)||_2 \leq L||x - y||_2 \qquad (A.2)$$

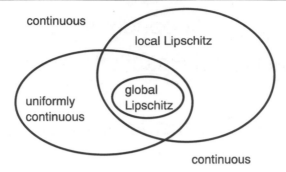

continuous

local Lipschitz

global Lipschitz

uniformly continuous

continuous

Figure A.1 Different concepts of continuity. All Lipschitz continuous functions are also continuous.

2. *f is globally Lipschitz continuous with respect to x:*
$\exists L > 0 \ \forall (x, t), (y, t) \in U \times T :$

$$\|f(x, t) - f(y, t)\|_2 \leq L \|x - y\|_2 \tag{A.3}$$

□

Local Lipschitz continuity is a weaker criterion than global Lipschitz continuity, which practically means global boundedness of the first order derivatives of the function under inspection. Local and global Lipschitz continuity can usually be established easily with the help of the following theorem.

Theorem 4 (sufficient condition for Lipschitz continuity)
Let $U \subset \mathbb{R}^n$ be open, let $T = (a, b) \subset \mathbb{R}$, and let $f : U \times T \to \mathbb{R}^m$ be continuous.

1. *Let $\frac{\partial f}{\partial x_i}$ exist, and be continuous for all $1 \leq i \leq n$. Then, f is locally Lipschitz continuous with respect to x.*
2. *Let $\frac{\partial f}{\partial x_i}$ exist, and be continuous and bounded for all $1 \leq i \leq n$. Then, f is globally Lipschitz continuous with respect to x.* □

Example 27
The function $f(x) = x^2$ is locally but not globally Lipschitz continuous in x. □

A final remark: in Figure A.1, only continuous functions are considered. However, continuity does not imply Lipschitz continuity. There are functions which are continuous but not Lipschitz continuous, for instance $f(x) = \sqrt[3]{x}$, which is continuous for all $x \in \mathbb{R}$ but not Lipschitz continuous at $x = 0$.

A.1.2 Taylor's theorem

Taylor's theorem for univariate functions

Taylor's famous existence theorem is used widely for approximations in all fields of science.

Theorem 5 (Taylor)
Suppose $f \in C^n$ and $f^{(n+1)}$ exists in a closed interval $[x_{\min}, x_{\max}]$. Let $x_0 \in [x_{\min}, x_{\max}]$. For every $x \in [x_{\min}, x_{\max}]$, there exists a value $\xi(x)$ between x_0 and x such that

$$f(x) = P_n(x) + R_n(x) \tag{A.4}$$

with $R_n(x) = f(x) - P_n(x)$ being the truncation error and $P_n(x)$ being the nth Taylor polynomial:

$$P_n(x) = f(x_0) + \frac{f'(x_0)}{1!}(x - x_0) + \frac{f''(x_0)}{2!}(x - x_0)^2 + \cdots + \frac{f^{(n)}(x_0)}{n!}(x - x_0)^n$$

$$= \sum_{i=0}^{n} \frac{f^{(i)}(x_0)}{i!}(x - x_0)^i \tag{A.5}$$

$\square$

There are numerous expressions for the remainder $R_n(x)$. One of them was derived by Lagrange:

$$R_n(x) = \frac{f^{(n+1)}(\xi(x))}{(n + 1)!}(x - x_0)^{n+1} \tag{A.6}$$

It follows directly that, when $R_n(x)$ is small,

$$f(x) \approx P_n(x) \tag{A.7}$$

This approximation is widely used, often with $n = 2$. An example will show how to use Taylor's theorem.

Example 28
The value of $\ln(x)$ *for* $x = 1.1$ *is to be approximated and its accuracy is to be determined with* $x_0 = 1$ *and* $n = 3$. *First, we need the first four derivatives of* $f(x)$:

$$\begin{array}{rcl rcl}
f(x) = & \ln x & \Longrightarrow & f(1) = 0 \\
f'(x) = & x^{-1} & \Longrightarrow & f'(1) = 1 \\
f''(x) = & x^{-2} & \Longrightarrow & f''(1) = 1 \\
f'''(x) = & 2x^{-3} & \Longrightarrow & f'''(1) = 2 \\
f^{(4)}(x) = & -6x^{-4} & \Longrightarrow & f^{(4)}(z) = -6z^{-4}
\end{array}$$

By Taylor's theorem, Equation (A.4),

$$\ln(x) = 0 + \frac{1}{1!}(x - 1) - \frac{1}{2!}(x - 1)^2 + \frac{2}{3!}(x - 1)^3 - \frac{6z^{-4}}{4!}(x - 1)^4$$

$$= (x - 1) - \frac{1}{2}(x - 1)^2 + \frac{1}{3}(x - 1)^3 - \frac{1}{4z^4}(x - 1)^4$$

$$\ln(1.1) = \underbrace{0.1 - \frac{0.1^2}{2} + \frac{0.1^3}{3}}_{\approx 0.0953} - \frac{0.1^4}{4z^4}$$

with $1 < z < 1.1$. *To estimate the accuracy of this approximation, the remainder needs to be computed:*

$$R_3(1.1) = \frac{(0.1)^4}{4z^4} = \frac{1}{z^4}\frac{0.0001}{4} = \frac{1}{z^4}0.000025$$

Since

$$z > 1$$
$$\Longrightarrow \frac{1}{z} < 1$$
$$\Longrightarrow \frac{1}{z^4} < 1$$

it follows that

$$R_3(1.1) < 0.000025 \qquad \text{(A.8)}$$

Consequently, the approximation $\ln(1.1) \approx 0.0953$ *is accurate to four decimal digits. A better approximation (with all given decimal places being correct) of* $\ln(1.1)$ *is 0.095310179804 324860043. This confirms our results.*   □

Taylor's theorem for bivariate functions

For bivariate and multivariate functions, Taylor's theorem has to take the cross-derivatives into account as well.

Theorem 6 (Taylor)
If f and all its partial derivatives of order n and less are continuous in $\{(x, y)|x^{\min} < x < x^{\max}, y^{\min} < y < y^{\max}\}$, and $x^{\min} < x + h < x^{\max}$ and $y^{\min} < y + k < y^{\max}$, then

$$f(x + h, y + k) = f(x, y) + \left[h \frac{\partial f(x, y)}{\partial x} + k \frac{\partial f(x, y)}{\partial y} \right]$$

$$+ \left[\frac{h^2}{2} \frac{\partial^2 f(x, y)}{\partial x^2} + hk \frac{\partial^2 f(x, y)}{\partial x \partial y} + \frac{k^2}{2} \frac{\partial^2 f(x, y)}{\partial y^2} \right] + \cdots$$

$$+ \frac{1}{n!} \sum_{j=0}^{n} \binom{n}{j} h^{n-j} k^j \frac{\partial^n f(x, y)}{\partial x^{n-j} \partial y^j} + R_n(x, y) \qquad \text{(A.9)}$$

□

Again, various expressions for the remainder $R_n(x, y)$ exist, these will not be given here.

Example 29
The first order Taylor approximation of $f(x, y) = yx^2 + \ln(y)$ around $(2, 1)$ can be computed as follows:

$$f(x + h, y + k)\Big|_{x=2, y=1} \approx P_1(x + h, y + k)|_{x=2, y=1} \qquad \text{(A.10)}$$

$$= f(x, y) + \left[h \frac{\partial f(x, y)}{\partial x} + k \frac{\partial f(x, y)}{\partial y} \right] \Big|_{x=2, y=1} \qquad \text{(A.11)}$$

$$= \left\{ (yx^2 + \ln(y)) + \left[h \cdot 2xy + k \cdot \left(x^2 + \frac{1}{y} \right) \right] \right\}_{x=2, y=1} \qquad \text{(A.12)}$$

$$= \left\{ yx^2 + \ln(y) + 2hxy + kx^2 + \frac{k}{y} \right\}_{x=2, y=1} \qquad \text{(A.13)}$$

$$= 4 + 4h + 5k \qquad \text{(A.14)}$$

We want to approximate $f(2.1, 1.1)$, i.e. $h = k = 0.1$. The function value is $f(2.1, 1.1) \approx 4.9463$, while the above Taylor approximation yields 4.9.   □

A.1.3   Mean value theorems

Theorem 7 (mean value theorem (Swokowski, 1979, p. 145))
If a real-valued function $f : [a, b] \to \mathbb{R}$ is continuous on a closed interval $[a, b]$ and is differentiable on the open interval (a, b), then there exists a number c in (a, b) such that

$$f(b) - f(a) = f'(c)(b - a) \qquad \text{(A.15)}$$

□

It is advisable to rewrite Equation (A.15):

$$f'(c) = \frac{f(b) - f(a)}{b - a} \tag{A.16}$$

Now the meaning of the mean value theorem is obvious: the derivative of f is equal to the difference quotient of f at least at one point c in (a, b).

Example 30

Assume $f(x) = x^2 + 2$ on $[1.0, 3.0]$. Then:

$$f'(x) = 2x \tag{A.17}$$

$$\frac{f(b) - f(a)}{b - a} = \frac{11 - 3}{3 - 1} = 4 = f'(2) \tag{A.18}$$

$\square$

Theorem 8 (the mean value theorem for definite integrals (Swokowski, 1979, p. 242))

If a function f is continuous on a closed interval $[a, b]$ then there exists a number z in the open interval (a, b) such that

$$\int_a^b f(x)\, \mathrm{d}x = f(z)(b - a) \tag{A.19}$$

$\square$

A.1.4 Various theorems

Theorem 9 (Leibnitz's rule for definite integrals)

Assume that f has a continuous derivative f_x for $a \le t \le b$. Then the following holds:

$$\frac{\mathrm{d}}{\mathrm{d}x} \int_a^b f(t, x)\, \mathrm{d}t = \int_a^b f_x(t, x)\, \mathrm{d}t \quad a > 0 \tag{A.20}$$

$\square$

This means that a definite integral can be differentiated with respect to a variable which is neither the variable of integration (here: t) nor a limit of integration (here: a or b) by differentiating through the integral sign.

Example 31

$$\frac{\mathrm{d}}{\mathrm{d}x} \int_a^b x^t\, \mathrm{d}t = \int_a^b \frac{\mathrm{d}}{\mathrm{d}x} x^t\, \mathrm{d}t = \int_a^b t x^{t-1}\, \mathrm{d}t \tag{A.21}$$

$\square$

Theorem 10 (product rule)

Assume that the two functions $g(x)$ and $f(x)$ are differentiable. Then:

$$\frac{\mathrm{d}}{\mathrm{d}x}(fg) = \frac{\mathrm{d}f}{\mathrm{d}x}g + f\frac{\mathrm{d}g}{\mathrm{d}x} \tag{A.22}$$

$\square$

Example 32

The derivative of $k(x) = (2x^3 - 5x)(3x + 1)$ is to be found. Since $k(x) = f(x) \cdot g(x)$, this can be done with the product rule (and, of course, also by first multiplying the two factors and then differentiating the resulting polynomial):

$$f'(x) = 6x^2 - 5 \tag{A.23}$$

$$g'(x) = 3 \tag{A.24}$$

$$k'(x) = 24x^3 + 6x^2 - 30x - 5 \tag{A.25}$$

$\square$

Theorem 11 (integration by parts)

Integrating Equation (A.22) over a closed interval leads to:

$$\int_a^b \frac{\mathrm{d}}{\mathrm{d}x}(fg)\,\mathrm{d}x = \int_a^b \frac{\mathrm{d}f}{\mathrm{d}x}g\,\mathrm{d}x + \int_a^b f\frac{\mathrm{d}g}{\mathrm{d}x}\,\mathrm{d}x \qquad (A.26)$$

For the LHS, the following holds:

$$\int_a^b \frac{\mathrm{d}}{\mathrm{d}x}(fg)\,\mathrm{d}x = \int_a^b \mathrm{d}(fg) = [fg]_a^b = fg|_{x=b} - fg|_{x=a} \qquad (A.27)$$

Thus, Equation (A.26) can be rewritten as:

$$\int_a^b \frac{\mathrm{d}f}{\mathrm{d}x}g\,\mathrm{d}x = [fg]_a^b - \int_a^b f\frac{\mathrm{d}g}{\mathrm{d}x}\,\mathrm{d}x \qquad (A.28)$$

□

A.2   BASIC NUMERICAL TOOLS

A.2.1   Quadrature

Basic quadrature rules

Galerkin FE requires the solution of integrals. In commercial FE packages, this task is usually solved by numerical integration, also called *quadrature*. The numerical approximation of integrals can often be performed with few resources and high accuracy.

The basic idea of the most simple quadrature rule, the *Midpoint rule*, is to approximate the (definite) integral by a rectangle:

$$\int_{x^{\min}}^{x^{\max}} f(x)\,\mathrm{d}x \approx (x^{\max} - x^{\min})\,f\left(\frac{x^{\min} + x^{\max}}{2}\right) \qquad (A.29)$$

The geometric interpretation is shown in Figure A.2.

Another simple quadrature rule is the *Trapezoidal rule*:

$$\int_{x^{\min}}^{x^{\max}} f(x)\,\mathrm{d}x \approx (x^{\max} - x^{\min})\frac{f(x^{\min}) + f(x^{\max})}{2} \qquad (A.30)$$

The geometric interpretation is shown in Figure A.3. It can be shown (Burden and Faires, 1998, Section 4.2) that the following holds:

$$\int_{x^{\min}}^{x^{\max}} f(x)\,\mathrm{d}x$$

$$= (x^{\max} - ax^{\min})f\left(\frac{x^{\min} + x^{\max}}{2}\right) + \frac{f''(\xi_1)}{24}(x^{\max} - x^{\min})^3 \quad \text{(Midpoint rule)} \qquad (A.31)$$

$$\int_{x^{\min}}^{x^{\max}} f(x)\,\mathrm{d}x$$

$$= (x^{\max} - x^{\min})\frac{f(x^{\min}) + f(x^{\max})}{2} - \frac{f''(\xi_2)}{12}(x^{\max} - x^{\min})^3 \quad \text{(Trapezoidal rule)} \qquad (A.32)$$

for some ξ_1, ξ_2 in $[x^{\min}, x^{\max}]$. Both methods are accurate up to third order. Another quadrature formula, *Simpson's rule*, is of order 5. Its idea is to approximate the function f by a quadratic

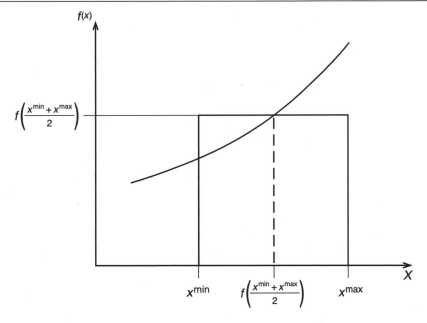

Figure A.2   Midpoint rule

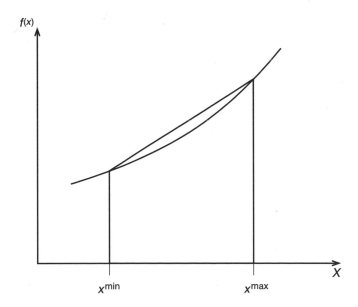

Figure A.3   Trapezoidal rule

polynomial that agrees with f at a, $(a + b)/2$, b:

$$\int_{x^{\min}}^{x^{\max}} f(x) = \frac{x^{\max} - x^{\min}}{6} \left[f(x^{\min}) + 4f \left(\frac{x^{\min} + x^{\max}}{2} \right) + f(x^{\max}) \right]$$
$$- \frac{f^{(4)}(\xi)}{2880} (x^{\max} - x^{\min})^5 \tag{A.33}$$

for some ξ in $[x^{\min}, x^{\max}]$. A straightforward improvement of the above techniques is to divide the interval $[x^{\min}, x^{\max}]$ into equidistant subintervals and apply the quadrature rule to each subinterval. For Simpson's rule, the interval $[x^{\min}, x^{\max}]$ is subdivided into n subintervals:

$$x^{\min} = x_0 < x_1 < x_2 < \cdots < x_{2j-2} < x_{2j-1} < x_{2j} < \cdots < x^{\max} = x_n,$$
$$h = x_j - x_{j-1} \tag{A.34}$$

Then the *Composite Simpson's rule* is given by:

$$\int_{x^{\min}}^{x^{\max}} f(x) = \frac{h}{3} \left[f(x^{\min}) + 2 \sum_{j=1}^{\frac{n}{2}-1} f(x_{2j}) + 4 \sum_{j=1}^{\frac{n}{2}} f(x_{2j-1}) + f(x^{\max}) \right]$$
$$- \frac{(x^{\max} - x^{\min})h^4}{180} f^{(4)}(\xi) \tag{A.35}$$

for some ξ in $[x^{\min}, x^{\max}]$. The Composite Simpson's rule is the most frequently used general purpose quadrature formula. It is a nested form of the Newton–Cotes method.

Gaussian quadrature

All of the above formulae use equally spaced values of the function whose integral is to be approximated. The accuracy of the approximation can be increased significantly if the nodes $x_1, x_2, \ldots, x_n$ in the interval $[x^{\min}, x^{\max}]$ and the coefficients $c_1, c_2, \ldots, c_n$ are specifically chosen to minimize the expected error of the approximation

$$\int_{x^{\min}}^{x^{\max}} f(x)dx \approx \sum_{i=1}^{n} c_i f(x_i) \tag{A.36}$$

As a measure of accuracy, we require that the chosen coefficients and nodes produce exact results for the largest class of polynomials (as test functions to integrate) possible. This gives us $2n$ parameters to choose. If we interpret these as parameters of a polynomial, they specify a polynomial of order $2n - 1$.

But how should we choose the parameters? For this, we consider the (orthogonal) Legendre polynomials. This is a class of polynomials frequently occurring in mathematics and physics. They have a number of representations, including the so-called *Rodrigues* representation

$$P_n(x) = \frac{1}{2^n n!} \frac{d^n}{dx^n} (x^2 - 1)^n$$

where $P_n(x)$ refers to the Legendre polynomial of degree n. The first few in this set are

$$P_0(x) = 1, \quad P_1(x) = x, \quad P_2(x) = x^2 - \frac{1}{3}$$

Table A.1   Roots and coefficients for Gaussian quadrature

n	Roots $x_{n,i}$	Coefficients $c_{n,i}$
2	± 0.5773502692	1.0
3	0.0	0.8888888889
	± 0.7745966692	0.5555555556
4	± 0.8611363116	0.3478548451
	± 0.3399810436	0.6521451549

$$P_3(x) = x^3 - \frac{3}{5}x \quad \text{and} \quad P_4(x) = x^4 - \frac{6}{7}x^2 + \frac{3}{35} \tag{A.37}$$

The roots of the polynomials are distinct, lie in the interval $(-1, 1)$ and are symmetrical with respect to the origin. Furthermore, it can be shown that they provide the correct choice for the interpolation parameters above. Let $x_1, x_2, \ldots, x_n$ be the roots of the nth degree Legendre polynomial, and let

$$c_i = \int_{-1}^{1} \prod_{\substack{j=1 \\ i \neq i}}^{n} \frac{x - x_j}{x_i - x_j} \, dx \tag{A.38}$$

Then, if $P(x)$ is a polynomial of degree less than $2n$,

$$\int_{-1}^{1} P(x)dx = \sum_{i-1}^{n} c_i P(x_i) \tag{A.39}$$

This tells us that if we calculate the coefficients as above and use the roots of the nth degree Legendre polynomial as the evaluation nodes of our function, the approximation to the integral will be accurate up to order $2n - 1$. Both the weights c_i and the roots x_i of the Legendre polynomials are widely tabulated in the literature; for the first few, see Table A.1.

An integral over an arbitrary interval $[x^{min}, x^{max}]$ can be performed via Gaussian quadrature by using a change of variables

$$t = \frac{2x - x^{min} - x^{max}}{x^{max} - x^{min}} \quad \Leftrightarrow \quad x = \frac{1}{2}[(x^{max} - x^{min})t + x^{min} + x^{max}] \tag{A.40}$$

Therefore

$$\int_{x^{max}}^{x^{min}} f(x)dx = \int_{-1}^{1} f(\frac{1}{2}[(x^{max} - x^{min})t + (x^{max} - x^{min})]) \frac{x^{max} - x^{min}}{2} \, dt \tag{A.41}$$

2D problems

Extending the above 1D results to multiple integrals is straightforward. Consider the double integral

$$\int \int_R f(x, y) \, dA \tag{A.42}$$

for $R = \{(x, y)|a \le x \le b, c \le y \le d\}$ being rectangular in the plane. The integral is evaluated by simply writing

$$\int \int_R f(x, y) dA = \int_a^b \left(\int_c^d f(x, y) dy \right) dx \qquad (A.43)$$

and taking x to be constant for consecutive evaluations of the dy integral.

Using Gaussian quadrature with $n = 3$ in every dimension would result in a total of $n^2 = 9$ evaluations of $f(x, y)$. Note that when using Gaussian quadrature, one needs to transform both dimensions onto the interval $[-1, 1]$ individually:

$$R = \{(x, y)|a \le x \le b, c \le y \le d\} \Leftrightarrow \hat{R} = \{(u, v)| -1 \le u \le 1, -1 \le v \le 1\} \qquad (A.44)$$

via

$$u = \frac{2x - a - b}{b - a} \quad \text{and} \quad v = \frac{2y - c - d}{c - d} \qquad (A.45)$$

Using Simpson's rule for the above integral, the error term from the one-dimensional formula (A.33) turns into

$$E = -\frac{(d - c)(b - a)}{180} \left[h^4 \frac{\partial^4 f}{\partial x^4}(\bar{\eta}, \bar{\mu}) + k^4 \frac{\partial^4 f}{\partial y^4}(\hat{\eta}, \hat{\mu}) \right] \qquad (A.46)$$

where h (k) is the step size in the x (y) direction and $(\bar{\eta}, \bar{\mu})$ and $(\hat{\eta}, \hat{\mu})$ are some points in R.

Finally, an integral of the form

$$\int_a^b \left(\int_{c(x)}^{d(x)} f(x, y) dy \right) dx \qquad (A.47)$$

is evaluated in exactly the same way as described above, except that the limits of individual dy integrations vary as a function of the then-fixed x. Note, however, that when employing Gaussian quadrature, one needs to perform the transformation of

$$v = \frac{2y - c(x_i) - d(x_i)}{c(x_i) - d(x_i)} \qquad (A.48)$$

at each value of x_i separately.

A.2.2 Solving nonlinear equations

In the FE framework, we frequently need to solve systems of equations. If they are linear, we can use methods such as Gaussian elimination or more sophisticated methods (see Section C.4). If it is possible to transform the nonlinear system into a linear one, these methods should always be preferred, since one has better control over such things as convergence or finding the complete set of solutions. However, if this is not possible, a standard method we can use is the Newton–Raphson method. Consider the following general problem. Given

$$g : D \subset \mathbb{R}^n \to \mathbb{R}^n \qquad (A.49)$$

we want to find the points $x \in D$ where $g(x) = \mathbf{0}$ for $n \in \mathbb{N}$, i.e. the roots of $g(x)$. This is achieved through linearization. Assuming g to be continuously differentiable, it can be written as a Taylor series

$$g(x + h) = g(x) + g'(x)h + \mathbf{O}(||h||) \quad \text{for} \quad h \to \mathbf{0} \qquad (A.50)$$

where any norm for $O(||h||)$ can be used. Linearization in this context means neglecting the $O(||h||)$ term in (A.50), such that we are left with the linear problem

$$0 = g(x) + g'(x)h \tag{A.51}$$

Furthermore, we assume that x is already a known approximation for the root, and so we calculate a (small) correction h to x. Numerically, the following happens. With known x_k, we solve

$$0 = g(x_k) + g'(x_k)h \tag{A.52}$$

for the unknown h. Let the solution be h_k. We then set

$$x_{k+1} = x_k + h_k \tag{A.53}$$

and repeat the procedure for x_{k+1}. This is the *Newton–Raphson* method for finding the roots of g. In one dimension, we have the formula

$$x_{k+1} = x_k - g(x_k)/g'(x_k) \tag{A.54}$$

For $n > 1$ dimensions, $g'(x)$ is, of course, the Jacobian matrix $J_{ij} = \partial g_i / \partial x_j$. So, for more than one dimension, Equations (A.52) and (A.53) yield

$$x_{k+1} = x_k - J^{-1}g(x_k) \tag{A.55}$$

As an example, let us consider $n = 2$ explicitly. For the above equations, we get

$$z_k = \begin{pmatrix} x_k \\ y_k \end{pmatrix}, \quad g(z_k) = \begin{pmatrix} u(z_k) \\ v(z_k) \end{pmatrix}, \quad g'(z_k) = \begin{pmatrix} u_x(z_k) & u_y(z_k) \\ v_x(z_k) & v_y(z_k) \end{pmatrix} \tag{A.56}$$

As the inverse of the Jacobi matrix we have

$$[g'(z_k)]^{-1} = \frac{1}{u_x(z_k)v_y(z_k) - u_y(z_k)v_x(z_k)} \begin{pmatrix} -v_y(z_k) & u_y(z_k) \\ v_x(z_k) & -u_x(z_k) \end{pmatrix} \tag{A.57}$$

$$= \frac{1}{\det J} (\ldots) \tag{A.58}$$

Therefore, the explicit solution for Newton's method in two dimensions is

$$x_{k+1} = x_k + (-u(z_k)v_y(z_k) + u_y(z_k)v(z_k))/(\det J(z_k)) \tag{A.59}$$

$$y_{k+1} = y_k + (-u_x(z_k)v(z_k) + u(z_k)v_x(z_k))/(\det J(z_k)) \tag{A.60}$$

The power of Newton's method lies in its fast rate of convergence. Let $\{x_k\} \subset \mathbb{R}^n$ be a convergent sequence with limit $\hat{x}$. If there exists a $0 < c < 1$ with

$$||x_{k+1} - \hat{x}|| \le c||x_k - \hat{x}||, \quad k = 0, 1, \ldots$$

we call $\{x_k\}$ linearly convergent. If there exists a $c > 0$ and a real number $q > 1$ with

$$||x_{k+1} - \hat{x}|| \le c||x_k - \hat{x}||^q, \quad k = 0, 1, \ldots$$

we call $\{x_k\}$ convergent of order q. It can be shown that Newton's method is indeed quadratically ($q = 2$) convergent. Near a root, the number of significant digits therefore roughly doubles at each iteration, which makes it the method of choice for problems where the derivative can be evaluated effectively and is continuous and nonzero across the root. This, consequently, also

Ordinary differential equation	order $n=1$	Initial value problem (IVP)	linear	homogeneous
			linear	nonhomogeneous
			nonlinear	
	order $n=2$	IVP: can be transformed to a first order system		
		Boundary value problem	linear	homogeneous
			linear	nonhomogeneous
			nonlinear	
	order $n>2$	IVP: can be transformed to a first order system		
		Boundary value problem: not to be discussed here		
Partial differential equation	order $n=1$	not to be discussed here		
	order $n=2$	linear		parabolic
				elliptic
				hyperbolic
		nonlinear (includes quasilinear)		parabolic
				elliptic
				hyperbolic
	order $n>2$	not to be discussed here		

Figure A.4   Classification of differential equations

displays the weaknesses of the method. The most important ones are:

- If higher order terms in the Taylor series are non-negligible, since Newton's method only takes the first derivative into account, it can produce the wrong results in this case.
- If the first derivative is close to, or even exactly, zero, the method breaks down or oscillates wildly, as can be seen from Equation (A.54) (in more than one dimension this translates into an ill-conditioned or singular Jacobian, which is hard or impossible to invert).
- Obviously, we need the derivative to begin with. If not analytically available, approximating it with the secant is an alternative, but includes the danger of instability.
- In some cases, the function's derivatives at the points in the sequence are unfortunately such that the algorithm enters a nonconvergent cycle.
- Finally, we need an initial guess. Since Newton's method is efficient close to root, we should employ a more robust method, such as bisection, to find the initial guess and then 'polish up' the result with Newton's method.

Besides Newton's method there are many other methods to solve systems of nonlinear equations. The interested reader is referred to Burden and Faires (1998) or Press *et al.* (1992) as starting points to the (vast) literature. In FE analysis, however, usually some variant of the Newton–Raphson method is chosen.

A.3   DIFFERENTIAL EQUATIONS

In this section we will discuss the definitions and classification of the different types of differential equation. A brief summary is shown in Figure A.4.

A.3.1   Definition and classification

Definition 5
A differential equation is an equation involving an unknown function and at least one of its derivatives.   □

Examples of differential equations include:

$$4x + 4 = \frac{df}{dx} \qquad (A.61)$$

$$\sin(f)\frac{d^2 f}{dx^2} + 7\frac{df}{dx} = 2 \qquad (A.62)$$

$$5\frac{d^3 f}{dx^3} + \exp(x)\frac{d^2 f}{dx^2} + 5xf = 0 \qquad (A.63)$$

$$\left(\frac{d^2 f}{dx^2}\right)^3 + 3f\left(\frac{df}{dx}\right)^4 + f^4\left(\frac{df}{dx}\right)^2 = 5x \qquad (A.64)$$

$$\frac{\partial^2 f}{\partial t^2} = \frac{\partial^2 f}{\partial x^2} \qquad (A.65)$$

$$u_{xxxx} = 0 \qquad (A.66)$$

An *ordinary* differential equation (ODE) is an equation involving one or more derivatives of an unknown *univariate* function. Accordingly, a *partial* differential equation (PDE) is an equation involving one or more partial derivatives of an unknown function of *several* variables. Following this definition, all of the above examples are ODEs, except for Equation (A.65) which is a PDE. ODEs are much simpler than PDEs from both an analytical and numerical point of view. Therefore, we start our discussion with ODEs.

ODEs can be distinguished by various criteria:

- *Order.* The *order* of a differential equation is the order of the highest derivative that occurs in the equation. The orders of our example equations are given in Table A.2:
- *Linearity.* A differential equation $F(x, y, z, t, f, f_x, f_y, f_z, f_t, f_{xx}, f_{xy}, \ldots) := L[F(\cdot)] = g(x, y, z, t)$ is said to be *linear* if
  1. All terms are linear in $f, f_x, f_y, \ldots$ algebraically; and
  2. All coefficients of f and its derivatives $f, f_x, f_y, \ldots$ are functions of the independent variables x, y, z, t only.
  A differential equation which is not linear is called *nonlinear*. Consequently, all examples except for Equations (A.62) and (A.64) are linear. Equations (A.62) and (A.64) are *nonlinear*.
- *Homogeneity.* A linear differential equation is *homogeneous* if there are no terms involving either the unknown function f or any of its derivatives. A linear differential equation involving terms which do not depend on f or any of its derivatives is called *nonhomogeneous*. According to this definition, Equations (A.63) and (A.66) are homogeneous, and Equation (A.61) is nonhomogeneous. A linear nonhomogeneous differential equation can be rewritten as

$$L[F(\cdot)] = g(x, y, z, t) \qquad (A.67)$$

$L(F) = 0$ denoting a linear ODE. The function $g(\cdot)$ is called the *perturbation function*. The perturbation function for Equation (A.61) is $g(x) = 4x + 4$ and for Equation (A.64) $g(x) = 5x$. The simplest perturbation function is of the form $g = $ constant.

Table A.2 Orders of various differential equations

Equation number	A.61	A.62	A.63	A.64	A.66
Order	1	2	3	2	4

The *solution* to a differential equation is any sufficiently regular function f that satisfies the differential equation identically.

Example 33
The differential equation

$$f'(x) + 2f(x) = 0 \tag{A.68}$$

is satisfied by $f(x) = f_0 \exp(-2x)$ for every $f_0 \in \mathbb{R}$. No other function satisfies Equation (A.68), as we shall see later.   □

A.3.2 Ordinary initial value problems

The exposition of various types of differential equations starts with *ordinary initial value problems*. For reasons to be understood later, much of the discussion is limited to *ordinary initial value problems of first order*. These are ODEs that describe the evolution of an object under inspection as time goes by, given an *initial value*. A simple example is the growth of capital. Assume you have an amount of capital M. As time goes by, M grows because of interest payments r (which are supposedly continuous) as described by:

$$\frac{dM}{dt} = rM \tag{A.69}$$

To pin down a solution, it needs to be known what the initial amount of money was at the beginning of compounding, M_0. Then the solution of Equation (A.69) is:

$$M = M_0 \exp(rt) \tag{A.70}$$

Especially in derivative pricing, the direction of time is often backwards. When time goes forward, we will use the common symbol t; however, when time goes backward, this will be denoted by τ. Consider a zero bond, which pays a prespecified amount of money $M(T)$ at some prespecified point of time in the future T. Assuming constant interest rates, the value of this zero bond satisfies

$$\frac{dM}{d\tau} = -rM \;\Rightarrow\; M(\tau) = M(\text{T}) \exp(-r\tau) \tag{A.71}$$

Both ODEs are linear. While for these ODEs it is possible to simply guess the solution, this is not possible for more general ODEs.

Systems of ordinary initial value problems

Without loss of generality, ODEs of order n can be formulated as a system of n first order ODEs. Systems consisting of ODEs of higher order can be treated in a similiar fashion. First, each ODE of the system has to be converted into a subsystem. In a second step, all subsystems have to be assembled into one large system. The transformation works both for linear and nonlinear problems. Think of the second order problem

$$\ddot{u} = f(t, u, \dot{u}) \tag{A.72}$$
$$u(0) = a \tag{A.73}$$
$$\dot{u}(0) = b \tag{A.74}$$

By introducing the auxiliary variable $v \equiv \dot{u}$ we get:

$$\dot{u} = v \qquad , \qquad u(x_0) = a \tag{A.75}$$

$$\dot{v} = f(t, u, v) \quad , \qquad v(x_0) = b \tag{A.76}$$

This technique is easily generalized to problems of higher order.

Example 34 (Bronson, 1993, p. 185)

Consider the following second order initial value problem:

$$\ddot{u} + u = 3 \tag{A.77}$$

$$u(\pi) = 1 \tag{A.78}$$

$$\dot{u}(\pi) = 2 \tag{A.79}$$

Define $v \equiv \dot{u}$ so that $\dot{v} = \ddot{u} = 3 - u$:

$$\dot{u} = v \tag{A.80}$$

$$\dot{v} = 3 - u \tag{A.81}$$

$$u(\pi) = 1 \tag{A.82}$$

$$v(\pi) = 2 \tag{A.83}$$

which is a system of first order.   □

Example 35 (Bronson, 1993, p. 187)

Consider the system:

$$\dddot{x} = t\ddot{x} + x - \dot{y} + t + 1 \tag{A.84}$$

$$\ddot{y} = \sin(t)\dot{x} + x - y + t^2 \tag{A.85}$$

$$x(1) = 2 \tag{A.86}$$

$$\dot{x}(1) = 3 \tag{A.87}$$

$$\ddot{x}(1) = 4 \tag{A.88}$$

$$y(1) = 5 \tag{A.89}$$

$$\dot{y}(1) = 6 \tag{A.90}$$

Since this system contains a third order differential equation in x and a second order differential equation in y, we will need five new variables. Defining:

$$x_1(t) = x \tag{A.91}$$

$$x_2(t) = \dot{x} \tag{A.92}$$

$$x_3(t) = \ddot{x} \tag{A.93}$$

$$y_1(t) = y \tag{A.94}$$

$$y_2(t) = \dot{y} \tag{A.95}$$

Thus:

$$\dot{x}_1 = x_2, \quad x_1(1) = 2 \tag{A.96}$$

$$\dot{x}_2 = x_3, \quad x_2(1) = 3 \tag{A.97}$$

$$\dot{x}_3 = tx_3 + x_1 - y_2 + t + 1, \quad x_3(1) = 4 \tag{A.98}$$

$$\dot{y}_1 = y_2, \quad y_1(1) = 5 \tag{A.99}$$

$$\dot{y}_2 = \sin(t)x_2 + x_1 - y_1 + t^2, \quad y_2(1) = 6 \tag{A.100}$$

$\square$

Riccati equations

These are simple examples of nonlinear ordinary initial value problems. The ordinary differential equation

$$f'(t) \equiv \dot{f}(t) = a(t)[f(t)]^2 + b(t)f(t) + c(t) \tag{A.101}$$

is called a (general) Riccati equation. With $a(t) \equiv 0$, it reduces to a simple linear ODE, and with $c(t) \equiv 0$, it becomes the Bernoulli equation. Systems of Riccati equations are called matrix Riccati equations. Matrix Riccati equations arise in various areas of optimal control (Reid, 1972) and in affine-linear term structure models (Duffie and Kan, 1996). In general, Riccati equations do not have closed-form solutions.

Example 36 (J. Bernoulli)
The first Riccati equation appears in a work by J. Bernoulli in 1694:

$$\dot{f} = t^2 + [f(t)]^2 \tag{A.102}$$

$\square$

Example 37 (Heuser, 1995, p. 69)
The Riccati equation

$$\dot{f} = (1 - t)[f(t)]^2 + (2t - 1)f(t) - t \tag{A.103}$$

is solved by

$$f(t) = 1 + \frac{1}{ce^{-t} + t - 2} \tag{A.104}$$

This can be verified easily by inserting Equation (A.104) and its derivative

$$\dot{f} = \frac{ce^{-t} - 1}{(ce^{-t} + t - 2)^2} \tag{A.105}$$

into Equation (A.103). $\square$

Example 38 (Rudolf, 2000, p. 40)
Certain one-factor term structure models can be reduced to solving the following backward system of Riccati equations:

$$\dot{A}(t) - h(t)B(t) + \frac{[B(t)]^2}{2}d(t) = 0 \tag{A.106}$$

$$1 + \dot{B}(t) + g(t)B(t) - \frac{[B(t)]^2}{2}c(t) = 0 \tag{A.107}$$

$$A(T) = B(T) = 0 \tag{A.108}$$

$\square$

Forward versus backward

Differential equations involving time as an independent variable are usually defined for a time interval between t_0 and T, with $t_0 < T$. When going from t_0 to T, this problem is called a *forward problem*. Accordingly, when going from T to t_0, this problem is called a *backward*

Figure A.5 Forward vs. backward problem

problem (see Figure A.5). Problems in economics are usually forward problems, i.e. a current state of the model is known and a future state is to be computed. In finance, however, most problems are backwards in time: the future value of a security is known at its day of maturity; its current value has to be traced backwards from its future value. Time going backwards is called *time to maturity*, $\tau = T - t$. Obviously, problems that are backwards in time are forward in time to maturity. This is an important fact since software can often only deal with forward problems. In order to be able to compute a backward problem formulated in terms of t (instead of τ), t has to be exchanged by $T - \tau$ and the sign in front of $\dot{f}$ reversed:

$$\frac{df}{d\tau} = \frac{df}{dt}\frac{dt}{d\tau} \quad \text{(Chain rule)} \tag{A.109}$$

$$= \frac{df}{dt}\frac{d(T-\tau)}{d\tau} \tag{A.110}$$

$$= -\frac{df}{dt} \tag{A.111}$$

The sign in front of $\ddot{f}$ remains unchanged:

$$\frac{d^2 f}{d\tau^2} = \frac{d}{d\tau}\left(-\frac{df}{dt}\right) \tag{A.112}$$

$$= -\frac{d}{d\tau}\left(\frac{df}{dt}\right) \tag{A.113}$$

$$= (-1)^2 \frac{d^2 f}{dt^2} \tag{A.114}$$

$$= \frac{d^2 f}{dt^2} \tag{A.115}$$

Example 39
Consider the following backward problem:

$$-u_t(x,t) = u_{xx}(x,t) \tag{A.116}$$

First we define:

$$\hat{u}(x,\tau) := u(x, T - \tau) \tag{A.117}$$

Then

$$\hat{u}_\tau(x,\tau) = u_t(x, T - \tau) \cdot (-1) \tag{A.118}$$

$$= u_{xx}(x, T - \tau) \tag{A.119}$$

$$= \hat{u}_{xx}(x,\tau) \tag{A.120}$$

$\square$

Example 40

Consider the following backward problem (Black–Scholes):

$$V_t + \frac{[\sigma(S,t)]^2}{2} S^2 V_{SS} - q(t)SV_S = r(t)[V - SV_S] \qquad (A.121)$$

The associated forward problem is given by:

$$-V_\tau + \frac{[\sigma(S,T-\tau)]^2}{2} S^2 V_{SS} - q(T-\tau)SV_S = r(T-\tau)[V - SV_S] \qquad (A.122)$$

This means that the functional forms of σ, q and r do not change. $\square$

Existence and uniqueness of solutions

From the point of view of epistemology, existence and uniqueness of a solution to a model affirm the *principle of determinism*, 'without which experiments could not be repeated with expectation of consistent data' (Ames, 1992). Although we are not dealing with mechanical or chemical experiments, we expect the same price of a derivative security given the same input parameters. Also, from a more practical point of view, the concepts of existence and uniqueness are of utmost importance. Before we start to find a solution, either numerically or analytically, we want to know that such a solution does actually exist. Whenever possible, it should be assured that a solution exists *before* the problem is tackled numerically, because the numerical algorithm still might compute something, even though it is employed in an incorrect manner. However, for many real-life problems, the existence of a meaningful solution cannot be guaranteed before numerical computations are performed. Here, we will take an engineering approach:

> Numerical method development for partial differential equations has also been based, to a con-
> siderable degree, on intuition and empiricism. Existence and uniqueness is often ignored, as are
> convergence proofs and error estimates. We should not hastily condemn these attempts, for if the
> practitioner waited for existence and uniqueness proofs for his nonlinear boundary value problem
> and convergence proofs and error estimates for any new numerical method, much of engineering
> and most of the computers in use would come to a halt! Further, he cannot be sure if he waited
> that mathematicians *would* or *could* attempt his problem (Ames, 1965, p. 475).

Ordinary initial value problems have been studied for a long time and many results are available. The starting point is a powerful result on the existence of a solution:

Theorem 12 (Peano)

The existence of a local in time solution to an initial value problem of an ODE

$$f'(t) = g(f,t) \qquad (A.123)$$

$$f(t_0) = f_0 \qquad (A.124)$$

$$t_1 < t_0 < t_2 \qquad (A.125)$$

can be concluded from the simultaneous continuity of g in both variables. $\square$

Although g is defined on (t_1, t_2) the solution f, in general, is defined on a smaller interval around t_0. ($t_0 \in (t_0 - \epsilon, t_0 + \epsilon) \subset (t_1, t_2)$, $\epsilon > 0$). The interval $(t_0 - \epsilon, t_0 + \epsilon)$ might be small in some instances, but for linear ODEs with constant coefficients it comprises $(-\infty, \infty)$.

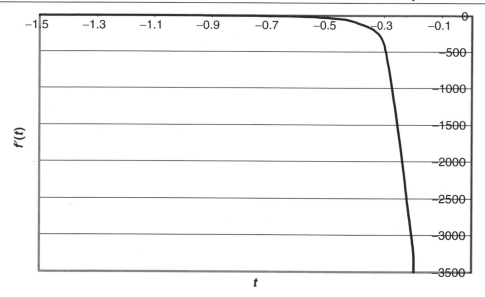

Figure A.6 Example 42 ($\alpha = -0.5$)

Example 41 (Strogatz, 1994, p. 28f)
Consider the following example:

$$\dot{f} = f'(t) = 1 + f^2 \tag{A.126}$$

$$f(0) = 0 \tag{A.127}$$

The solution can be shown to be $f(t) = \tan t$. But this solution only exists for $-\pi/2 < t < \pi/2$ since $f(t) \to \infty$ as $t \to \pi/2$ and $f(t) \to -\infty$ as $t \to -\pi/2$. □

Example 42 (Bellomo and Preziosi, 1995, p. 58ff)
Consider the following example:

$$f'(t) = t^\alpha \tag{A.128}$$

$$t(0) = t_0 \tag{A.129}$$

For $\alpha < 0$, the expression t^α blows up to infinity when $t \to 0$; g is not defined at $t = 0$. Consequently, it is not continuous around $t = 0$, so that existence cannot be established by Theorem 12 (see Figure A.6). □

The conditions establishing existence *and* uniqueness are only slightly less general. For the sake of brevity, we state the theorem for systems of IVPs.

Theorem 13 (Lindelöf)
Let $U \subset \mathbb{R}^n$ be open, let $T = (a, b) \subset \mathbb{R}$, let $(x_0, t_0) \in U \times T$, and let $f : U \times T \to \mathbb{R}^m$ be continuous.

- *Let f be locally Lipschitz continuous w.r.t. x. Then the initial value problem*

$$\dot{u}(t) = f(u(t), t); \quad u(t_0) = x_0 \tag{A.130}$$

has a unique local in time solution u.

- *Let $U = \mathbb{R}^n$, and let f be globally Lipschitz continuous w.r.t. x. Then the initial value problem*

$$\dot{u}(t) = f(u(t), t); \quad u(t_0) = x_0 \tag{A.131}$$

has a unique global in time solution u.

□

Example 43 (Example 41 continued)
Is the solution of Example 41 unique? This is the case, since $g(f, t)$ satisfies the local Lipschitz condition with respect to f, because $g(f, t) = 1 + f^2$ is both continuous in t and f and $g_f(f, t) = 2f$ is continuous w.r.t. f.

□

Example 44 (Example 42 continued)
For $0 < \alpha < 1$, the function $g(f, t) = t^\alpha$ is continuous in both f and t. Since $g_f(f, t) = 0$, local Lipschitz continuity can be established as well.

□

The *local Lipschitz condition* can be replaced by even weaker assumptions (Hsieh and Sibuya, 1999), but in most economic and financial applications it is met. Both theorems state sufficiency conditions, i.e. there are initial value problems not satisfying these conditions although yielding a unique solution. Besides existence and uniqueness, the stability of a solution with respect to perturbations of the initial conditions needs to be assured to establish *well-posedness*.

Definition 7 (Hadamard)
A problem is called well-posed *if the following requirements are satisfied:*

- *A solution exists, at least for the period of time desired.*
- *This solution is unique.*
- *The solution depends continuously on the (initial) data.*

□

The concept of well-posedness has been introduced by Hadamard in various publications (Maz'ya and Shaposhnikova, 1998). The continuous dependence of the solution on the initial data is assured by the following theorem.

Theorem 14 (Gronwall)
Let $\hat{f}$ and $\tilde{f}$ be the two solutions of an ordinary initial value problem $f'(t) = g(f, t)$ with initial conditions $\hat{f}(t_0) = \hat{f}_0$ and $\tilde{f}(t_0) = \tilde{f}_0$, respectively. L denotes the Lipschitz constant as defined in Definition 4. If g is continuous and satisfies the (local) Lipschitz condition, then

$$\|\hat{f}(t) - \tilde{f}(t)\| \leq \|\hat{f}_0 - \tilde{f}_0\| \exp(L|t - t_0|) \tag{A.132}$$

□

This theorem says that for t close to t_0, the difference between $\hat{f}$ and $\tilde{f}$ is close to the difference of the initial conditions $\hat{f}(t_0)$, and $\tilde{f}(t_0)$, because $\exp(L|t - t_0|) \approx 1$. For $t \to \infty$, this difference cannot grow without bounds.

Next, we turn our attention to a special feature of some ordinary first order initial value problems. Numerical analysis usually distinguishes two classes of problem within this family: *stiff* and *nonstiff* problems. The exact solution of an ordinary first order initial value problem usually consists of a *transient* and a *steady-state* solution. In a stiff problem, the transient solution is of the form $e^{-\lambda t}$ ($\lambda \gg 0$), so that it rapidly decays to zero. However, the derivative of the transient solution is $\lambda e^{-\lambda t}$, which does not converge to zero as quickly as the transient solution. Since many numerical schemes do not evaluate the derivative at t, but somewhere in the interval $(0, t)$, this can cause problems. Stiffness is a relative concept, depending on the stability of the numerical method employed, the ODE at hand, the required degree of accuracy and, last but not least, the length of the domain for which an approximate solution is sought. Stiffness is usually associated with systems of initial value problems resulting from various physical models and, more important in this context, the spatial discretization of parabolic PDEs with FE. A (stable) system of ODEs is stiff if the eigenvalues of the Jacobian matrix of the RHS of $\dot{\mathbf{f}} = \mathbf{g}(t, \mathbf{f})$ differ greatly in magnitude.

A.3.3 Ordinary boundary value problems

Ordinary boundary value problems consist of an ODE and boundary conditions for both ends of the domain. The domain starts at x^{min} and ends at x^{max}. The number and type of boundary conditions to render a problem sensible depends on the order of the ODE. Here, we will concentrate on problems of second order; for problems of higher order see Axelsson and Barker (1984) and Burnett (1987). The *standard form* of a *second order linear boundary value problem* consists of an ODE:

$$u''(x) + a_1(x)u'(x) + a_2(x)u(x) = a_3(x) \tag{A.133}$$

and two boundary conditions:

$$\alpha_1 u(x^{min}) + \beta_1 u'(x^{min}) = \gamma_1 \tag{A.134}$$
$$\alpha_2 u(x^{max}) + \beta_2 u'(x^{max}) = \gamma_2 \tag{A.135}$$

The functions $a_i(x)$ are continuous, while α_i, β_i and γ_i are real constants. Furthermore, it is assumed that α_1 and β_1 are not both zero, and also that α_2 and β_2 are not both zero. Problems of this type are called *Sturm–Liouville systems*. This boundary problem is nonhomogeneous because its ODE and its boundary conditions are nonhomogeneous. A boundary problem is called homogeneous when its ODE and its boundary conditions are homogeneous, i.e.:

$$u''(x) + a_1(x)u'(x) + a_2(x)u(x) = 0 \tag{A.136}$$

and the two boundary conditions:

$$\alpha_1 u(x^{min}) + \beta_1 u'(x^{min}) = 0 \tag{A.137}$$
$$\alpha_2 u(x^{max}) + \beta_2 u'(x^{max}) = 0 \tag{A.138}$$

Often, second order boundary value problems can be solved analytically with the techniques used for initial value problems, with appropriate treatment of the boundary conditions. First, we will address the question of existence and uniqueness. This is a subtle subject that deserves special attention because theorems of similar generality as for initial value problems do not exist. Although (sufficient) conditions for existence and uniqueness for the standard form can be given (see, for example, Bronson, 1993, p. 288f), we will focus on the following problem:

$$u''(x) = f(x, u', u) \tag{A.139}$$
$$u(x^{min}) = \gamma_1 \tag{A.140}$$
$$u(x^{max}) = \gamma_2 \tag{A.141}$$

Theorem 15 (Burden and Faires, 1998, p. 425)

Such a problem has a unique solution if all of the following criteria are met:[1]

- f, $\frac{\partial f}{\partial u}$ and $\frac{\partial f}{\partial u'}$ *are continuous;*
- $\frac{\partial f}{\partial u} > 0$;
- $\frac{\partial f}{\partial u'}$ *is bounded.*   □

Note that the differential equation (A.139) includes *nonlinear* ODEs as well, while the boundary conditions cover only Dirichlet conditions. So the ODE is more general, as in the standard problem, while the boundary conditions are more restrictive. Also note that this set of criteria is only sufficient. In general, the theory of existence and/or uniqueness of a solution becomes complicated quickly. The interested reader is referred to Bailey *et al.* (1968).

Example 45

Can a unique solution of the following problem be assured?

$$u''(x) = -2u'(x) + 3u(x) + 9x \tag{A.143}$$
$$y(0) = 1 \tag{A.144}$$
$$y(1) = \gamma_2 \tag{A.145}$$

All criteria are satisfied:

$$f(u, u', x) = -2u'(x) + 3u(x) + 9x \tag{A.146}$$
$$\frac{\partial f}{\partial u} = 3 \quad > \quad 0 \tag{A.147}$$
$$\frac{\partial f}{\partial u'} = -2 \tag{A.148}$$

                                                                                        □

Example 46

The boundary value problem

$$u''(x) = -2u'(x) + 3u \tag{A.149}$$
$$u(0) = 0 \tag{A.150}$$
$$u(1) = 0 \tag{A.151}$$

has a unique solution because the criteria are met:

$$f(u, u', x) = -2u'(x) + 3u(x) \tag{A.152}$$
$$\frac{\partial f}{\partial u} = 3 \quad > \quad 0 \tag{A.153}$$
$$\frac{\partial f}{\partial u'} = -2 \tag{A.154}$$

---

[1] Because of the composition

$$x \to \begin{pmatrix} x \\ u(x) \\ u'(x) \end{pmatrix} \xrightarrow{f} f\left(x, u(x), u'(x)\right) \tag{A.142}$$

in exact notation these conditions read:
- f, $\frac{\partial f}{\partial y}$ and $\frac{\partial f}{\partial z}$ are continuous;
- $\frac{\partial f}{\partial y} > 0$;
- $\frac{\partial f}{\partial z}$ is bounded.

In order to avoid confusion with PDEs, this notation has been avoided.

The solution is the trivial solution $u(x) = 0$ □

Example 47

Consider $u'' = -u$. Existence of a unique solution cannot be concluded from the above set of criteria because $\frac{\partial f}{\partial u} > 0$ is violated. However, the above criteria do not tell us whether no or several solutions exist. Obviously, $u(x) = c \sin(x)$ satisfies the differential equation, but not arbitrary boundary conditions. In fact, for

$$u(0) = 0 \tag{A.155}$$

$$u(b) = B \tag{A.156}$$

$$b = \pi \tag{A.157}$$

$$B \neq 0 \tag{A.158}$$

no solution exists, while for $B = 0$, every number c gives a solution. For

$$b \neq n\pi \tag{A.159}$$

a unique solution exists with $c \sin(b) = B$. The basic reason for these examples to produce no, one or several solutions is that the solution to $u'' = -u$ is $u(t) = c_1 \cos(t) + c_2 \sin(t)$. The coefficients c_1 and c_2 only stretch the solution but cannot make it meet arbitrary boundary conditions.   □

Example 48

Let us consider a nonlinear example (Sewell, 2000, p. 131):

$$u'' - (u')^2 - u^2 + u + 1 = 0 \tag{A.160}$$

$$u(0) = \frac{1}{2} \tag{A.161}$$

$$u(\pi) = -\frac{1}{2} \tag{A.162}$$

Since

$$\frac{\partial f}{\partial u} = 2u - 1 > 0 \tag{A.163}$$

$$\Longleftrightarrow \qquad u > 0.5 \tag{A.164}$$

the above condition is violated because of the RHS in BC (A.162). One solution is given by

$$u(x) = \sin\left(x + \frac{\pi}{6}\right) \tag{A.165}$$

Another solution is given by

$$u(x) = \sin\left(x + \frac{5\pi}{6}\right) \tag{A.166}$$

□

Theorem 16 (Bronson, 1993, p. 288)

Let u_1 and u_2 be two linearly independent solutions of Equation (A.136). The coefficients α_i, β_i are continuous in x. Nontrivial solutions to the homogeneous boundary value problem Equations (A.136)–(A.138) exist, if and only if the determinant D vanishes:

$$D \equiv \begin{vmatrix} \alpha_1 u_1(x^{\min}) + \beta_1 u_1'(x^{\min}) & \alpha_1 u_2(x^{\min}) + \beta_1 u_2'(x^{\min}) \\ \alpha_2 u_1(x^{\max}) + \beta_2 u_1'(x^{\max}) & \alpha_2 u_2(x^{\max}) + \beta_2 u_2'(x^{\max}) \end{vmatrix} = 0 \tag{A.167}$$

The consequence from the theorem is obvious: the homogeneous boundary value problem always admits the trivial solution $u(x) \equiv 0$. Whenever there is a nontrivial solution, this solution cannot be unique. $\square$

Example 49 (Bronson, 1993, p. 290)
Let us consider the following problem:

$$u'' + 2u' - 3u = 0 \tag{A.168}$$
$$u(0) = 0 \tag{A.169}$$
$$u'(1) = 0 \tag{A.170}$$

From the theory of linear ODEs, we can easily find

$$u_1(x) = \exp(-3x) \tag{A.171}$$
$$u_2(x) = \exp(x) \tag{A.172}$$

so that

$$D \equiv \begin{vmatrix} 1 & 1 \\ -3e^{-3} & e \end{vmatrix} = e + 3e^{-3} \neq 0 \tag{A.173}$$

which means that $u(x) = 0$. This solution can also be computed by finding the coefficients of $u(x) = c_1 u_1(x) + c_2 u_2(x)$:

$$c_1 = c_2 = 0 \tag{A.174}$$

$\square$

Theorem 17 (Bronson, 1993, p. 289)
The nonhomogeneous boundary value problem Equations (A.133)–(A.135) has a unique solution if and only if the associated homogeneous problem Equations (A.136)–(A.138) has only the trivial solution. $\square$

In other words: the nonhomogeneous problem possesses a unique solution if and only if the associated homogeneous problem has a unique solution.

Example 50 (Bronson, 1993, p. 290)
Consider the example:

$$u'' + 2u' - 3u = 0 \tag{A.175}$$
$$u(0) = 0 \tag{A.176}$$
$$u'(1) = 2 \tag{A.177}$$

From the fact that the associated homogeneous problem (Example 49) has a unique solution, it can be concluded that the above problem has a unique solution as well. $\square$

Next, we consider a problem with Dirichlet conditions only:

$$u(x^{\min}) = \gamma_1 \tag{A.178}$$
$$u(x^{\min}) = \gamma_2 \tag{A.179}$$

With the following transformation

$$u(x) = v(x) + \gamma_1 \frac{x - x^{\max}}{x^{\min} - x^{\max}} + \gamma_2 \frac{x - x^{\min}}{x^{\max} - x^{\min}} \tag{A.180}$$

the nonhomogeneous boundary conditions can be eliminated so that we now have:

$$v(x^{\min}) = 0 \tag{A.181}$$

$$v(x^{\max}) = 0 \tag{A.182}$$

A further transformation

$$x = \left(x^{\max} - x^{\min}\right) x_{\text{normed}} + x^{\min} \tag{A.183}$$

$$\iff \quad x_{\text{normed}} = \frac{x - x^{\min}}{x^{\max} - x^{\min}} \tag{A.184}$$

allows one to consider only the interval [0, 1]. Having performed the above transformations, the linear two-point boundary value problem with Dirichlet conditions reads

$$L(v) = v'' - bv' - cv = f \tag{A.185}$$

$$v(0) = 0 \tag{A.186}$$

$$v(1) = 0 \tag{A.187}$$

with b, c and f depending on x_{normed}. This is the setting most numerical research publications on this topic prefer.

For linear ODEs, it is often possible to find solutions with undetermined coefficients. The coefficients then have to be computed from the boundary conditions. This approach does not necessarily lead to a solution of the two-point boundary value problem, as we have seen in Example 27. If such a solution is available, the following theorems give some information about existence and form of the solution. The next theorem follows directly from Theorem 17:

Theorem 18 (Grossmann and Roos, 1994, p. 13)
Consider the problem given by Equations (A.185) to (A.187). If the homogeneous problem (i.e. $f = 0$) possesses only the trivial solution $v = 0$, then the nonhomogeneous problem has a unique solution.   □

Theorem 19 (Grossmann and Roos, 1994, p. 13)
Consider the problem given by Equations (A.185) to (A.187). If $c \geq 0$ then the homogeneous problem (i.e. $f = 0$) possesses only the trivial solution $v = 0$.   □

Most two-point boundary value problems from physics and engineering can be written as:

$$\underbrace{\frac{d}{dx}\left[p(x)\frac{d}{dx}u(x)\right]}_{=pu''+p'u'} + q(x)u(x) + \lambda r(x)u(x) = g(x) \tag{A.188}$$

It is assumed that $p(x)$, $p'(x)$, $q(x)$, $r(x)$ and $f(x)$ are piecewise continuous, that $p(x)$ and $r(x)$ are positive and $q(x)$ negative on $(x^{\min}, x^{\max})$. This special case of Equation (A.133) is called *self-adjoint, symmetric* or *Sturm–Liouville*. Self-adjoint problems are rather the normal case than the exception in natural science, while this is not true for finance or economics, where there are hardly any self-adjoint problems. The solution of Equation (A.188) is strongly influenced by the term $\lambda r u$. With $\lambda \neq 0$, Equation (A.188) is called an *eigenvalue problem*; with $\lambda = 0$ Equation (A.188) is a simple *two-point boundary value problem*. Eigenvalue problems require special numerical treatment and are of little use in current economics and finance, so we will concentrate on $\lambda = 0$. Equation (A.188) can be transformed into

$$\tilde{a}_0(x)u'' + \tilde{a}_1(x)u' + \tilde{a}_2(x)u = \tilde{a}_3(x) \tag{A.189}$$

and vice versa. Equation (A.189) can easily be transformed to the standard of Equation (A.133). Note that, in general, $\tilde{a}_i(x) \neq a_i(x)$. Also, note that the first order term and the second order term in Equation (A.189) satisfy the following condition:

$$\tilde{a}_0' = \tilde{a}_1 \tag{A.190}$$

For linear second order differential equations it is always possible to alter the problem from Equation (A.189) to Equation (A.188). This is a two-step procedure:

1. In order to be able to perform this transformation we first define:

$$\mu = \exp\left(\int \frac{\tilde{a}_1}{\tilde{a}_0}\, dx\right) \tag{A.191}$$

Since $\tilde{a}_0 = p > 0$ this operation is always possible.

2. Equation (A.189) must be multiplied by μ:

$$\mu\tilde{a}_0(x)u'' + \mu\tilde{a}_1(x)u' + \mu\tilde{a}_2(x)u = \mu\tilde{a}_3(x) \tag{A.192}$$

$$\Longleftrightarrow \quad \tilde{a}_0(x)\left[\mu u'(x)\right]' + \mu\tilde{a}_2(x)u = \mu\tilde{a}_3(x) \tag{A.193}$$

$$\Longleftrightarrow \quad \left[\mu u'(x)\right]' + \mu\frac{\tilde{a}_2(x)}{\tilde{a}_0(x)}u = \mu\frac{\tilde{a}_3(x)}{\tilde{a}_0(x)} \tag{A.194}$$

because of the chain rule and $\mu' = \dfrac{\tilde{a}_1(x)}{\tilde{a}_0(x)}\mu$.

Vice versa, the transformation from Equation (A.188) to Equation (A.189) can be performed simply by solving the expression $\frac{d}{dx}\left[p(x)\frac{d}{dx}u(x)\right]$.

Example 51 (Bickford, 1990, Example 2.1)

Consider the following problem:

$$x^2u'' + xu' + u + 1 = 0 \tag{A.195}$$

$$u(1) = 1 \tag{A.196}$$

$$u'(2) + u(2) = 0 \tag{A.197}$$

We have:

$$\tilde{a}_0 = x^2, \quad \tilde{a}_1 = x, \quad \tilde{a}_2 = 1, \quad \tilde{a}_3 = -1 \tag{A.198}$$

with the help of Equation (A.191):

$$\mu(x) = \exp\left[\int \frac{\tilde{a}_1(x)}{\tilde{a}_0(x)}\, dx\right] = \exp\left[\int \frac{x}{x^2}\, dx\right] = \exp\left[\ln(x)\right] = x \tag{A.199}$$

In a second step, this result is inserted into Equation (A.193):

$$x^2(xu')' + xu + x = 0 \tag{A.200}$$

$$\Longleftrightarrow \quad (xu')' + \frac{u}{x} + \frac{1}{x} = 0 \tag{A.201}$$

which is the desired self-adjoint formulation. □

Example 52 (Bickford, 1990, Example 2.2)
Consider the following problem:

$$x^2 u'' - 2xu' + 2u - 1 = 0 \tag{A.202}$$

$$u(1) = 1 \tag{A.203}$$

$$u(2) = 0 \tag{A.204}$$

We have:

$$\tilde{a}_0 = x^2, \quad \tilde{a}_1 = -2x, \quad \tilde{a}_2 = 2, \quad \tilde{a}_3 = 1 \tag{A.205}$$

With the help of Equation (A.191):

$$\mu(x) = \exp\left[\int \frac{\tilde{a}_1(x)}{\tilde{a}_0(x)}\, dx\right] = \exp\left[\int \frac{-2x}{x^2}\, dx\right] = \exp\left[-2\ln(x)\right] = \frac{1}{x^2} \tag{A.206}$$

In a second step, this result is inserted into Equation (A.193):

$$x^2 (\frac{1}{x^2} u')' + 2\frac{1}{x^2} u - \frac{1}{x^2} = 0 \tag{A.207}$$

$$\Longleftrightarrow \quad (\frac{1}{x^2} u')' + \frac{2u}{x^4} - \frac{1}{x^4} = 0 \tag{A.208}$$

which is the desired self-adjoint formulation. □

The tremendous advantage of self-adjointness is that this usually gives tight error bounds for most numerical approaches, while for the general non-self-adjoint problems error bounds are technically involved.

In technical and physical problems, the dependent variable usually is not time, while in economics, it normally is time. Also, in economic applications, the right-hand boundary often is infinity ($t \to \infty$), so that these problems formally resemble initial value problems. It is to be emphasized that, from a numerical point of view, initial value problems need to be treated *completely differently* from boundary value problems. Initial value problems do not need to take any other infomation into account than the values given at the starting point. Boundary value problems possess a starting point and a final point. This is also true for boundary value problems with a boundary at infinity, which require special techniques to integrate the boundary condition at infinity. Analytical and numerical approaches differ: to integrate a boundary condition at infinity into an analytical solution, *transversality conditions* are employed. For an introduction to this topic, see Chiang (1992). The numerical treatment is discussed in Judd (1998), Kunkel and von dem Hagen (2000) and McGrattan (1996). Currently, two approaches are employed in economics; see Section 3.1.2. Either parts of the domain are cut off, or the independent variable is transformed in such a way that the semi-infinite domain is mapped into a compact interval. Both techniques manipulate the domain to make the problem solvable by familiar methods.

A.3.4   Partial differential equations of second order

A partial differential equation is an equation involving one or more partial derivatives of an unknown *multivariate* function. This definition is somewhat misleading because it is not the case that arbitrary equations are studied, but the focus is on those PDEs arising from applications

Table A.3 Types of second order PDE

Type of PDE	Eigenvalues
hyperbolic	$(Z=0$ and $P=1)$ or $(Z=0$ and $P=n-1)$
parabolic	$Z=1$ and $b \cdot v \neq 0$ where v is the eigenvector of A corresponding to the eigenvalue 0
elliptic	$(Z=0$ and $P=n)$ or $(Z=0$ and $P=0)$

in various areas of science. The general linear PDE of second order in n variables has the form

$$\sum_i^n \sum_j^n a_{ij} u_{x_i x_j} + \sum_i^n b_i u_{x_i} + cu = d \qquad (A.209)$$

The coefficients a_{ij}, b_i and c are allowed to be functions of $x_1, \ldots, x_n$. It is always possible to transform Equation (A.209) into the form $a_{ij} = a_{ji}$, so that the matrix $[a_{ij}]_{n \times n} = A$ can be regarded as symmetric without loss of generality. Any real, symmetric $n \times n$ matrix has n real eigenvalues; see Section C.1. This fact can be used to distinguish between different PDEs. Let P be the number of positive eigenvalues and Z the number of zero eigenvalues of A. The type of PDE given by Equation (A.209) can then be determined via Table A.3.

Example 53

Consider the (dynamic) heat equation $u_t = u_{xx}$. Relabeling the variables t to x_1 and x to x_2 we get:

$$0 = a_{11} = a_{12} = a_{21} = b_2 = c = d \qquad (A.210)$$

$$1 = a_{22} = -b_1 \qquad (A.211)$$

$$A = \begin{pmatrix} 0 & 0 \\ 0 & 1 \end{pmatrix} \qquad (A.212)$$

$$\det(A) = 0 \qquad (A.213)$$

so that it can be concluded that the heat equation is parabolic.   □

Example 54

Consider the wave equation $u_{tt} = u_{xx}$. Relabeling the variables t to x_1 and x to x_2 we get:

$$0 = a_{21} = a_{12} = b_1 = b_2 = c = d \qquad (A.214)$$

$$1 = a_{11} = -a_{22} \qquad (A.215)$$

$$A = \begin{pmatrix} 1 & 0 \\ 0 & -1 \end{pmatrix} \qquad (A.216)$$

$$\implies Z = 0 \text{ and } P = 1 \qquad (A.217)$$

so that it can be concluded that the wave equation is hyperbolic.   □

Example 55

Consider the (static) heat equation in three spatial variables $0 = u_{x_1 x_1} + u_{x_2 x_2} + u_{x_3 x_3}$. Then:

$$0 = a_{ij} \; \forall \, i \neq j \qquad (A.218)$$

$$0 = b_i = c = d \qquad (A.219)$$

$$1 = a_{11} = a_{22} = a_{33} \qquad \text{(A.220)}$$

$$\mathbf{A} = \begin{pmatrix} 1 & 0 & 0 \\ 0 & 1 & 0 \\ 0 & 0 & 1 \end{pmatrix} \qquad \text{(A.221)}$$

$$\det(\mathbf{A}) = 1 \qquad \text{(A.222)}$$

$$\Longrightarrow \quad Z = 0 \ \ and \ \ P = 3 = n \qquad \text{(A.223)}$$

so that it can be concluded that the above PDE is elliptic. □

For any nonconstant a_{ij}, the type of Equation (A.209) has to be determined for each part of the domain separately.

Example 56 (Wilmott *et al.*, 1996, p.83)
Many pricing models are of the form

$$ax^2 u_{xx} + bx u_x + cu = u_t \qquad \text{(A.224)}$$

$$u(0, t) = g_1(x, t) \qquad \text{(A.225)}$$

$$u(x_{\max}) = g_2(x, t) \qquad \text{(A.226)}$$

$$u(x, 0) = g_3(x) \qquad \text{(A.227)}$$

Examples for this type of pricing model are the Black–Scholes model and its extensions and variants. This problem is parabolic for $x > 0$. □

A partial differential equation only poses a meaningful problem in combination with its boundary conditions. Evolution problems, i.e. parabolic and hyperbolic problems, also need an initial condition. The initial condition describes the state of the problem under inspection at the beginning. The PDE describes the evolution of the state as time goes by. Boundary conditions pin down the solution at the edge of the (spatial) domain on which the PDE is defined. There are various types of BC, which are discussed in Section 3.1.2. Existence, uniqueness and stability of a solution to any PDE can only be given with the appropriate BCs. To put forward a (physically) meaningful model, all necessary BCs need to be specified.

A.3.5 Parabolic problems

The general version of a linear second order parabolic PDE with one spatial variable x is given by:

$$a(x, t)u_{xx} + b(x, t)u_x + c(x, t)u + d(x, t) = u_t \qquad \text{(A.228)}$$

All linear pricing PDEs with one risk factor are special cases of Equation (A.228). Together with the appropriate BCs and IC, this problem looks rather intimidating to most economists, since partial differential equations are usually outside of their curriculum. Physicists and mathematicians, however, are used to this type of equation because it can be employed to model many phenomena in physics, engineering and biology; for an incomplete list, see page 81 of Wilmott (1998). One of the physical phenomena usually modeled with this type of PDE is *heat flow*, which also seems to be the most intuitive to non-physicists. It is often helpful to think of u in terms of heat, in order to be able to have an intuitive feeling of what the solution to a parabolic or elliptic PDE looks like. This is why we briefly derive the heat equation. This derivation is based on Farlow (1993) and Boyce and DiPrima (1992).

Think of a one-dimensional rod of length x^{max} with the following (idealized) properties:

- It consists of a single homogeneous conducting material (such as copper).
- It is laterally insulated, i.e. heat flows only in the x-direction.
- It is so thin that the temperature at all points of a cross-section is constant.

The basis of the derivation of Equation (A.228) is the concept of *conservation of (heat) energy*. Applying this principle to a segment $[x, x + \Delta x]$ of the rod we get:

$$
\begin{aligned}
&\text{Net change of heat inside } [x, x + \Delta x] \\
&= \text{Net flux of heat across the boundaries} \\
&+ \text{Total heat generated inside } [x, x + \Delta x]
\end{aligned}
\tag{A.229}
$$

At any time t, the total amount of heat inside $[x, x + \Delta x]$ is given by:

$$
\text{Total heat inside } [x, x + \Delta x] = \int_x^{x+\Delta x} c\rho A u(x, t) \, \mathrm{d}s
\tag{A.230}
$$

with:

$$
c = \text{thermal capacity of the rod} \tag{A.231}
$$
$$
\rho = \text{density of the rod} \tag{A.232}
$$
$$
A = \text{cross-section of the rod} \tag{A.233}
$$

The parameters A, c and ρ are assumed to be constants for now. Now, Equation (A.229) can be stated in terms of calculus:

$$
\frac{\mathrm{d}}{\mathrm{d}t} \int_x^{x+\Delta x} c\rho A u(s, t) \, \mathrm{d}s
$$
$$
= kA \left[u_x(x + \Delta x, t) - u_x(x, t) \right] + A \int_x^{x+\Delta x} \underbrace{g(u, s, t)}_{\text{heat source}} \, \mathrm{d}s
\tag{A.234}
$$

with k as thermal conductivity and g the heat source. First, we simplify the LHS. With the help of Theorem 8 we get:

$$
\begin{aligned}
&\frac{\mathrm{d}}{\mathrm{d}t} \int_x^{x+\Delta x} c\rho A u(s, t) \, \mathrm{d}s \\
&= c\rho A \int_x^{x+\Delta x} u_t(s, t) \, \mathrm{d}s \\
&= c\rho A u_t(\xi_1, t)(x + \Delta x - x) \\
&= c\rho A u_t(\xi_1, t)\Delta x \quad \text{with} \quad x < \xi_1 < x + \Delta x
\end{aligned}
\tag{A.235}
$$

Applying Theorem 8 to the RHS of Equation (A.234) leads to:

$$
kA \left[u_x(x + \Delta x, t) - u_x(x, t) \right] + Ag(\xi_2, t)\Delta x \quad \text{with} \quad x < \xi_2 < x + \Delta x
\tag{A.236}
$$

Equating the LHS and RHS:

$$
u_t(\xi_1, t) = \frac{k}{c\rho} \left[\frac{u_x(x + \Delta x, t) - u_x(x, t)}{\Delta x} \right] + \frac{g(\xi_2, t)}{c\rho}
\tag{A.237}
$$

Finally, letting $\Delta x \longrightarrow 0$, we get the desired result:

$$u_t(x, t) = \alpha^2 u_{xx}(x, t) + G(u, x, t) \tag{A.238}$$

with:
$$\alpha^2 = \frac{k}{c\rho} \quad \text{(diffusivity of the rod)} \tag{A.239}$$

$$G(u, x, t) = \frac{g(u, x, t)}{c\rho} \quad \text{(heat source density)} \tag{A.240}$$

We assume that G is of the following form, i.e. it can be separated into two heat sources; one depending on space and time and one depending also on the heat level:

$$G(x, t, u) = F(x, t) + q(x, t)u(x, t) \tag{A.241}$$

Next, we drop the assumption that k is a constant and allow $k = k(x)$ so that the term $\frac{k}{c\rho}u_{xx}$ from Equation (A.238) changes to

$$\frac{1}{c\rho}(ku_x)_x \equiv \frac{1}{c\rho}(k_x u_x + k u_{xx}) \equiv \frac{k_x}{c\rho}u_x + \frac{k}{c\rho}u_{xx} \tag{A.242}$$

Combining all terms, we get:

$$\frac{k}{c\rho}u_{xx} + \frac{k_x}{c\rho}u_x + qu + F(x, t) = u_t \tag{A.243}$$

which is easily recognized to be Equation (A.228) with specific coefficients. The individual terms of Equation (A.228) are usually associated with physical phenomena:

$$\underbrace{a(x, t)u_{xx}}_{\text{diffusion}} + \underbrace{b(x, t)u_x}_{\text{convection}} + \underbrace{c(x, t)u}_{\text{reaction}} + d(x, t) = u_t \tag{A.244}$$

When convection becomes more important than diffusion, a formally parabolic problem starts to behave like a hyperbolic conservation law. The dimensionless *Péclet number*, $Pe = (|b|L)/a$, measures the relative importance of convection compared to diffusion (L being the 'length'). Problems with $Pe \gg 1$ are called convection-dominated. The Péclet number might become as large as 25 or 10^7 when modeling seepage or semiconductors, respectively.

Example 57 (Morton, 1996, p. 6)
What is the Péclet number of the Black–Scholes PDE:

$$\frac{\partial V}{\partial t} + \sigma^2 S^2 \frac{1}{2}\frac{\partial^2 V}{\partial S^2} + rS\frac{\partial V}{\partial S} - rV = 0 \tag{A.245}$$

For typical constant values of $r = 0.1$ and $\sigma = 0.2$, the Péclet number relative to the 'length' scale ΔS has the value $5\frac{\Delta S}{S}$, so that the convective effect is dominant when the asset price S is small and/or volatility is low. □

Parabolic (and hyperbolic) PDEs describe processes that evolve in time. To solve problems of this type, a condition is needed describing the initial state of this system, the initial condition. When time goes backward, a final condition is needed.

Definition 8 (initial value problems)
An initial value problem (or Cauchy problem) is a hyperbolic (hyperbolic conservation law) or parabolic PDE or system of PDEs with spatial variables defined on $\mathbb{R}^n$. □

Mathematical problem	Well-formulated	Well-posed	Weakly well-posed (Hadamard)
			Strongly well-posed
			Strongly well-posed in the generalized sense
			...
		Ill-posed	Mildly Ill-posed
			Ill-posed in any sense
			...
	Ill-formulated	Ill-specified	
		Inverse	

Figure A.7 Classification of mathematical problems (based on Bellomo and Preziosi, 1995, p. 344; Kreiss and Lorenz, 1989)

When an initial value problem has to be solved numerically, usually some kind of technique is employed to transform the infinite domain into a finite one.

Definition 9 (initial boundary value problems)
An initial boundary value problem is a hyperbolic or parabolic PDE or system of PDEs with spatial variables defined on a compact subset of $\mathbb{R}^n$.　　　□

Note that problems on semi-infinite domains also arise.

Example 58
See the option pricing problem from Section 2.2. The spatial domain of this problem is $(0, \infty)$.　　　□

Existence, uniqueness and well-posedness

There are various theorems on the well-posedness of parabolic problems scattered in the literature, all of them only covering some problems from finance. More general results can be found with the help of functional analysis; here we are only interested in classical solutions, i.e. functions which are at least C^2. Even for classical solutions, there are several concepts of well-posedness, which differ in the formalization of the continuous dependence on the auxiliary data. Petrovsky was the first to do this for parabolic problems. In modern parlance, his notion of well-posedness is usually called *weak well-posedness* (Kreiss and Lorenz, 1989, p. 79). All theorems discussed below deal with this kind of well-posedness. Various concepts of well-posedness are given in Figure A.7.

There are many theorems on the well-posedness of partial differential equations. These are usually cast in the parlance of functional analysis, which we want to avoid in this book. Some powerful results can be given for classical solutions. The following result is useful in establishing well-posedness of a wide class of problems: parabolic initial boundary value problems. When pure initial value problems are solved numerically, they are usually transformed to initial boundary value problems by adding appropriate boundary conditions, because almost all numerical techniques are designed for finite domains in the spatial variables.[2] Therefore, we will present theorems dealing with initial boundary value problems only.

---

[2] A notable exception to this point is explicit finite differences, which can be used in such a way that only the initial condition needs to be provided. How this can be done in the context of option pricing is shown in Wilmott *et al.* (1996). Also, numerical techniques for infinite or semi-finite domains exist. The employment of this kind of technique for financial problems was suggested by McIver (1996), but to the knowledge of the author, this has not been done yet. A survey of techniques of how to do this for parabolic problems can be found in Givoli (1992).

Theorem 20 (Petrovski, 1954, p. 271ff)

Suppose the following assumptions are met:

1. The quadratic form

$$\sum_{i,j=1}^{d} a_{ij}(t, \boldsymbol{x}) \rho_i \rho_j \tag{A.246}$$

is positive-definite for all $(t, \boldsymbol{x}) \in \Omega$.
2. The functions a_{ij}, b_i, c and d have continuous derivatives of sufficiently high order.
3. The spatial BCs are of the form

$$\frac{\partial u}{\partial n} + ku = f \tag{A.247}$$

4. Γ_Ω is piecewise of class C^1.
5. f is continuous on $[0, T] \times \Gamma_\Omega$.
6. g is continuously differentiable in Ω and on smooth pieces of Γ_Ω.
7. f and g satisfy the following compatibility condition:

$$\frac{\partial g(x)}{\partial n} + kg(x) = f(0, x) \tag{A.248}$$

Then the initial boundary value problem given by

$$\frac{\partial u}{\partial t} = L(u) + f(t, \boldsymbol{x}) \tag{A.249}$$

$$L(u) := \sum_{i,j=1}^{d} a_{i,j}(t, \boldsymbol{x}) \frac{\partial^2 u}{\partial x_i \partial x_j} + \sum_{i,j=1}^{d} b_l(t, \boldsymbol{x}) \frac{\partial u}{\partial x_i} + c(t, \boldsymbol{x})u + d(t, \boldsymbol{x}) \tag{A.250}$$

$$u(0, \boldsymbol{x}) = g(\boldsymbol{x}) \quad \textit{(Initial condition)} \tag{A.251}$$

$$\frac{\partial u}{\partial n} + ku = f \quad \textit{(Boundary condition)} \tag{A.252}$$

is well-posed. □

The above theorem is applicable to many option pricing problems. The requirement of a continuously differentiable initial condition can be relaxed, since any continuous function can be approximated arbitrarily close by a differentiable function. Consequently, well-posedness can be assured for initial conditions of the type max($\cdot$). However, the large and important family of barrier options is not covered because of the requirement of BCs and IC to be compatible. To be more specific: reverse barrier options without rebate and with continuous monitoring fail to satisfy the condition because the BC is incompatible with the FC.

Weinberger and Protter (1967) show how uniqueness in Theorem 20 can be deduced easily from the maximum principle (Theorem 24). The above theorem can be extended to systems of parabolic PDEs.

Theorem 21 (Bellomo and Preziosi, 1995)

Consider the parabolic system

$$\frac{\partial \mathbf{u}}{\partial t} = \mathbf{A} \frac{\partial^2 \mathbf{u}}{\partial x^2} + \mathbf{f}\left(t, x, \mathbf{u}, \frac{\partial \mathbf{u}}{\partial x}\right) \quad \mathbf{u} \in \mathbb{R}^n \tag{A.253}$$

which is parabolic if **A** *has only eigenvalues with a positive real part. The PDE is defined on a compact set which is complemented by Dirichlet and/or Neumann conditions. Each PDE in this system has to be univariate. This problem is well-posed, provided that* **f** *is continuous with respect to* $(\mathbf{x}, t)$ *and locally Lipschitz continuous with respect to* $(\mathbf{u}, \mathbf{u}_x)$. *With* f *being globally Lipschitz continuous in* $(\mathbf{u}, \mathbf{u}_x)$, *the solution* **u** *is a global in time solution.* $\square$

Note that this theorem covers nonlinear PDEs as well. To be more exact: it covers quasilinear PDEs as long as the matrix **A** is of the form $\mathbf{A} = \mathbf{A}(\mathbf{x}, t)$. All of the above theorems deal with fixed spatial domains only. The following theorem deals with domains that are functions of time.

Theorem 22 (Weinberger and Protter, 1967, p. 177)
Assume all conditions of Theorem 20 are met. The domain Γ *may move with time provided its boundary points move with finite speed. Then uniqueness of a solution can be guaranteed.* $\square$

This theorem is useful to ensure the existence of a solution to certain barrier option pricing problems, where the location of the barriers move as time goes by (Kunitomo and Ikeda, 1992).

Well-formulated problems

If the problem is provided with enough initial, final and boundary conditions to find a solution, then it is said to be *well-formulated*. This solution does not have to be unique. Consider the following example. The problem of finding a solution to $u_t = u_{xx}$ with no boundary or initial conditions is well-formulated because it is possible to find solutions:

- $u(t, x) = c$
- $u(t, x) = t + \frac{x^2}{2}$

It is not well-posed, since its solution is not unique. All of the above still holds if an initial condition is added.

Qualitative behavior of parabolic PDEs

There are numerous theorems describing the qualitative behavior of parabolic PDEs. We will concentrate on some theorems which can be used to check numerical computations.

Theorem 23 (DuChateau and Zachmann, 1986, p. 39; Meis and Marcowitz, 1981)
Provided the conditions of Theorem 20 are satisfied, the solution of the PDE will be smooth in the interior of Ω *for all* $t > 0$. *In order for the solution to be smooth on* Ω *including the boundaries, it is necessary to impose smoothness and compatibility conditions on the boundaries.* $\square$

This theorem has a significant financial interpretation which is often useful as a first test of the correctness of a (usually numerical) solution. Neither the value of a derivative security nor any of its derivatives (Greeks) can exhibit jumps. Jump conditions, as induced by dividends or coupon payments, divide the pricing equation into a series of PDEs. For each of these PDEs, Theorem 23 holds. An example for incompatible boundary and final conditions is reverse barrier options, i.e. barrier options under the risk of breaching the barrier when in the money.

Still, inside the boundary conditions, the value of the option is smooth. Theorem 23 does not hold for nonlinear parabolic PDEs.

Theorem 24 (maximum principle (DuChateau and Zachmann, 1986, p. 37f))

Let Ω denote a bounded region in $\mathbb{R}^n$ whose boundary is a smooth closed surface Γ_Ω. Suppose $u(x_1, \ldots, x_n, t)$ to be continuous for $(x_1, \ldots, x_n)$ in $\bar{\Omega}$ and $0 \le t \le T$; for short, in $\bar{\Omega} \times [0, T]$. Let:

- $M_\Gamma = \max[u(x_1, \ldots, x_n, t) : (x_1, \ldots, x_n)$ *on* Γ_Ω *and* $0 \le t \le T]$.
- $M_0 = \max[u(x_1, \ldots, x_n, t) : (x_1, \ldots, x_n)$ *on* Ω *and* $t = 0]$.
- $M = \max[M_\Gamma, M_0]$.
- $m_\Gamma = \min[u(x_1, \ldots, x_n, t) : (x_1, \ldots, x_n)$ *on* Γ_Ω *and* $0 \le t \le T]$.
- $m_0 = \min[u(x_1, \ldots, x_n, t) : (x_1, \ldots, x_n)$ *on* Ω *and* $t = 0]$.
- $m = \min[m_\Gamma, m_0]$.

If the linear differential operator

$$L[\cdot] = \sum_{i,j=1}^{n} a_{ij}(\boldsymbol{x}, t)\frac{\partial^2[\cdot]}{\partial x_i \partial x_j} + \sum_{i=1}^{n} b_i(\boldsymbol{x}, t)\frac{\partial[\cdot]}{\partial x_i} + c(\boldsymbol{x}, t)[\cdot] \qquad \text{(A.254)}$$

is uniformly elliptic (see Section A.3.6) in Ω for each t in $[0, T]$, then the operator

$$\frac{\partial[\cdot]}{\partial t} - L[\cdot] \qquad \text{(A.255)}$$

is said to be uniformly parabolic in $\Omega \times [0, T]$. Given that $\frac{\partial u}{\partial t} - L[u(x_1, \ldots, x_n, t)]$ is uniformly parabolic in $\Omega \times [0, T]$, with coefficients a_{ij} and b_i continuous in $\Omega \times [0, T]$ and coefficient $c = 0$, then the following hold:

1. *If $u_t - L[u] \le 0$ in $\Omega \times (0, T)$, then $u \le M$ in $\bar{\Omega} \times [0, T]$.*
2. *If $u_t - L[u] \ge 0$ in $\Omega \times (0, T)$, then $u \ge m$ in $\bar{\Omega} \times [0, T]$.*
3. *If $u_t - L[u] = 0$ in $\Omega \times (0, T)$, then $m \le u \le M$ in $\bar{\Omega} \times [0, T]$.*

$\square$

This theorem also can be used as a first step in the assessment of a solution to some problems, such as both Kolmogorov equations. Due to the assumption of $c = 0$, this maximum principle cannot be used for most pricing equations because usually the interest rate does not vanish. Sometimes it is possible to employ the maximum principle to a transformed problem, such as the Black–Scholes problem transformed to the heat equation.

Theorem 25 (DuChateau and Zachmann, 1986, p. 38f)

Let $u(\boldsymbol{x}, t)$ be $C^{2,1}$ in $\bar{\Omega} \times [0, T]$ and $L(u)$ uniformly parabolic as in Theorem 24. Also, let $u(\boldsymbol{x}, t)$ satisfy $u_t = L(u)$. Then, either u is constant in $\bar{\Omega} \times [0, T]$ or else:

1. *At any point $\bar{\boldsymbol{x}}$ on Γ_Ω such that $u(\bar{\boldsymbol{x}}, t) = M$, we have:*

$$\frac{\partial u(\bar{\boldsymbol{x}}, t)}{\partial n} > 0 \qquad \text{(A.256)}$$

$$0 \le t \le T \qquad \text{(A.257)}$$

2. *At any point $\bar{x}$ on Γ_Ω such that $u(\bar{x}, t) = m$, we have:*

$$\frac{\partial u(\bar{x}, t)}{\partial n} < 0 \tag{A.258}$$

$$0 \leq t \leq T \tag{A.259}$$

Note that this theorem is not subject to $c = 0$, so it is directly applicable to many pricing PDEs. □

Example 59 (plain vanilla call)

This example is to demonstrate that the approximation of BCs needs to be done with great care. Assume a plain vanilla call on an equity paying no dividends. The associated Black–Scholes pricing problem is:

$$\frac{\partial V}{\partial t} + \frac{1}{2}\sigma^2 S^2 \frac{\partial^2 V}{\partial S^2} + rS\frac{\partial V}{\partial S} - rV = 0 \tag{A.260}$$

$$V(S, T) = \max(0, S - E) \tag{A.261}$$

$$V(0, t) = 0 \tag{A.262}$$

The semi-infinite spatial domain of $[0; \infty]$ is replaced with the interval $[S_{min}, S_{max}]$ with $0 < S_{min} < S_{max} < \infty$. Also, the perspective of time is changed so that the resulting initial boundary problem is forward parabolic in τ:

$$\frac{\partial V}{\partial \tau} - \frac{1}{2}\sigma^2 S^2 \frac{\partial^2 V}{\partial S^2} + rS\frac{\partial V}{\partial S} - rV = 0 \tag{A.263}$$

$$V(S, T) = \max(0, S - E) \tag{A.264}$$

$$V(S_{min}, \tau) = 0 \tag{A.265}$$

$$V(S_{max}, \tau) = S_{max} - Ee^{-r\tau} \tag{A.266}$$

or

$$\frac{\partial V(S_{max}, \tau)}{\partial S} = 1 \tag{A.267}$$

By economic reasoning, M can be found at $(S_{max}; t)$, i.e. deep in the money with the maximal time value. Here, $\frac{\partial u}{\partial n} = \frac{\partial V}{\partial S} = 1 > 0$ for both BCs. At S_{min} we run into a problem. At $S = 0$, the PDE is no longer parabolic, so that the domain is reduced with $S_{min} > 0$. At $S_{min} > 0$, however, the associated value of V is not known exactly, so that $V(S_{min}, \tau) = 0$ is only a proxy for the correct BC. This, however, violates Theorem 25, which states that the outward-pointing gradient $\frac{\partial V(S_{min}, \tau)}{\partial S}$ takes on a negative value requiring $V(S, \tau) < 0$ for $V(S, \tau) < S_{min}$ (at least locally). This, however, contradicts $V(S, \tau) \geq 0 \; \forall \; (S, \tau)$ which can easily be established by financial reasoning. In essence: the approximation of the BC has to be done with great care! □

Example 60 (reverse barrier option)

Assume a call option with an in the money up-and-out barrier at $S = S_u > E$. For simplicity, we assume continuous monitoring and no rebates. Since the RHS BC $V(S_u, \tau)$ is not compatible with the FC $V(S, T)$ at $S = B$, the assumption of u being C^1 in $\bar{\Omega} \times [0, T]$ is violated. Actually, the gradient vanishes on the knock-out barrier S_u without V being a constant function. □

A.3.6 Elliptic PDEs

Elliptic problems can be interpreted as time-independent problems that are derived from parabolic problems when the transient term has vanished. Then, the initial condition does not matter any more and one has to deal with boundary conditions only. The discussion of well-posedness is subtle, so we will restrict ourselves to a fundamental and quite general result on the uniqueness of linear elliptic PDEs.

Definition 10 (uniform ellipticity)
The elliptic operator $L[u]$

$$L[u] = \sum_{i,k=1}^{n} a_{ik} u_{ik} + \sum_{i=1}^{n} b_i u_i + cu \equiv M[u] + cu \qquad (A.268)$$

with

$$u_i = \frac{\partial u}{\partial x_i} \qquad (A.269)$$

$$u_{ik} = \frac{\partial^2 u}{\partial x_i \partial x_i} \qquad (A.270)$$

is uniformly elliptic (in the sense of Legendre–Hadamard) in Ω if there exists a positive constant λ such that

$$\sum_{i,k=1}^{n} a_{ij}(x_1, \ldots, x_n)\xi_i\xi_k \geq \lambda \sum_{i=1}^{n} \xi_i^2 \qquad (A.271)$$

for all $(\xi_1, \ldots, \xi_n)$ in $\mathbb{R}^n$ and all $(x_1, \ldots, x_n)$ in Ω. □

Definition 11 (divergence form)
The elliptic operator $L[u]$ has either the form

$$L[u] = \sum_{i,k=1}^{n} (a_{ik} u_i)_k + \sum_{i=1}^{n} b_i u_i + cu \qquad (A.272)$$

or

$$L[u] = \sum_{i,k=1}^{n} a_{ik} u_{ik} + \sum_{i=1}^{n} b_i u_i + cu \qquad (A.273)$$

Equation (A.272) refers to the divergence form, *while Equation (A.273) refers to the* nondivergence form. □

Provided that $\mathbf{A}(\mathbf{x})$ is symmetric on Ω, Equation (A.271) implies that $Z = 0$ and $P = n$ (from Table A.3). Consequently, an elliptic PDE is always uniformly elliptic.

In the case where $\mathbf{A}$ happens to be nonsymmetric, it can be replaced by

$$\frac{1}{2}\left(\mathbf{A} + \mathbf{A}^{\mathrm{T}}\right) \qquad (A.274)$$

without changing the differential operator.

Theorem 26 (uniqueness (Courant and Hilbert, 1962, p. 320f))

Let the coefficients $a_{ij} = a_{ji}, b_i, c$ be continuous functions of the variables $x_1, x_2, \ldots, x_n$ in a bounded domain Ω of the n-dimensional space $\mathbb{R}^n$. Then, for $c \leq 0$, there exists at most one solution to $L[u] = 0$ which has continuous derivatives up to second order in Ω, is continuous in $\bar{\Omega} = \Omega \cup \Gamma_\Omega$, and assumes prescribed boundary values on the boundary of Ω. □

A direct consequence of this theorem is that a solution that vanishes on Γ_Ω vanishes identically in G, i.e. homogeneous Dirichlet conditions on the entire boundary lead to the trivial solution.

The qualitative behavior of a solution to an elliptic PDE can be characterized with the help of the following two maximum principles.

Theorem 27 (DuChateau and Zachmann, 1986, p. 20)

Let the following hold:

- *Ω is a bounded region with boundary S.*
- *M and m are the maximum and minimum values of $u(x)$, respectively.*
- *The PDE is uniformly elliptic in Ω with $c = 0$ and a_{ij} and b_i continuous in $\bar{\Omega}$.*

Then the following hold:

1. *If $L(u) \geq 0$ in Ω, then $u(x) < M$ $\forall$ $x \in \Omega$, or else $u(x) = M$ in $\bar{\Omega}$.*
2. *If $L(u) \leq 0$ in Ω, then $u(x) > m$ $\forall$ $x \in \Omega$, or else $u(x) = m$ in $\bar{\Omega}$.*
3. *If $L(u) = 0$ in Ω, then $m < u(x) < M$ $\forall$ $x \in \Omega$, or else $m = u(x) = M$ in $\bar{\Omega}$.*

□

Theorem 28 (DuChateau and Zachmann, 1986, p. 20)

Let the following hold:

- *Ω is a bounded region with boundary S.*
- *The PDE $L[u] = f$ is uniformly elliptic in Ω, with $c \leq 0$ and a_{ij}, b_i, c and f continuous in $\bar{\Omega}$.*

Then the following hold:

1. *If $f \leq 0$ in $\bar{\Omega}$ and $u(x)$ is nonconstant, then any negative minimum of $u(x)$ must occur on S and not in Ω.*
2. *If $f \geq 0$ in $\bar{\Omega}$ and $u(x)$ is nonconstant, then any positive maximum of $u(x)$ must occur on S and not in Ω.* □

A.3.7 Hyperbolic PDEs

Hyperbolic PDEs in the sense of Table A.3 are commonly called *hyperbolic wave equations*. Hyperbolic PDEs following this definition describe wave-like phenomena. They have to be distinguished from hyperbolic conservation laws, which will be discussed shortly in Section A.3.8. Hyperbolic wave equations are, like parabolic PDEs, evolution equations depending on time. The solution of a hyperbolic initial value problem cannot be smoother than the IC. Solutions of hyperbolic PDEs can develop singularities from a smooth IC as time goes by. Hyperbolic PDEs do not satisfy a maximum principle (DuChateau and Zachmann, 1986, p. 37).

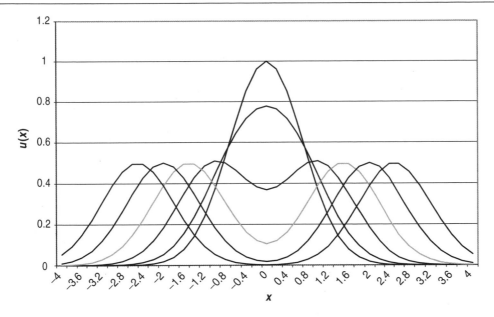

Figure A.8 Wave equation for $t = 0, 0.5, \ldots, 2.5$ with $c = 1$

Example 61 (wave equation (Bellomo and Preziosi, 1995, p. 235))
The second order hyperbolic initial value problem

$$\frac{\partial^2 u}{\partial t^2} = c \frac{\partial^2 u}{\partial x^2} \tag{A.275}$$

$$u(t = 0, x) = \exp(-x^2) \tag{A.276}$$

is solved by

$$u(t, x) = \frac{1}{2} \left\{ \exp\left[-(x - ct)^2\right] + \exp\left[-(x + ct)^2\right] \right\} \tag{A.277}$$

The bell is split in two and both parts travel. One to the left-hand side, one to the right-hand side (see Figure A.8).

□

The statement of BCs for hyperbolic PDEs is similar to parabolic PDEs. Accordingly, well-posedness is often difficult to establish. Since hyperbolic PDEs are rare in finance and unknown in other areas of economics, we will not further dwell on this subject here.

A.3.8 Hyperbolic conservation laws

The prototype of a hyperbolic conservation law in one spatial variable is the transport equation.

Example 62 (transport equation (Bellomo and Preziosi, 1995, p. 233))
The first order hyperbolic initial value problem

$$\frac{\partial u}{\partial t} = -c \frac{\partial u}{\partial x} \tag{A.278}$$

$$u(t = 0, x) = \exp(-x^2) \tag{A.279}$$

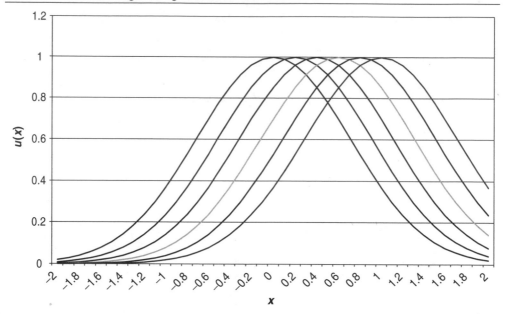

Figure A.9   Transport equation for $t = 0, 0.2, \ldots, 1$ with $c = 1$

is solved by

$$u(t, x) = \exp\left[-(x - ct)^2\right] \tag{A.280}$$

The bell-shaped IC moves to the right as time goes by. While moving to the right, the shape of the curve does not change for different t, as in Figure A.9. □

Similar to the above PDE is the following nonlinear term structure model.

Example 63 (Wilmott *et al.*, 1996, Section 40; Penaud, 2002, Chapter 9)

$$\frac{\partial V}{\partial t} + c\left(\frac{\partial V}{\partial r}\right)\frac{\partial V}{\partial r} - rV = 0 \tag{A.281}$$

$$c(x) = \begin{cases} c^+, & x < 0 \\ c^-, & x > 0 \end{cases} \tag{A.282}$$

$$c^- \leq \frac{dr}{dt} \leq c^+ \tag{A.283}$$

This model can be used to find the worst-case price of an interest rate sensitive portfolio. It has to be solved backwards in time starting from the final condition $V(T, r) = 1$. In terms of time to maturity, this problems reads:

$$-\frac{\partial V}{\partial \tau} + c\left(\frac{\partial V}{\partial r}\right)\frac{\partial V}{\partial r} - rV = 0 \tag{A.284}$$

$$c(x) = \begin{cases} c^+, & x < 0 \\ c^-, & x > 0 \end{cases} \tag{A.285}$$

$$c^- \leq -\frac{dr}{d\tau} \leq c^+ \tag{A.286}$$

□

Hyperbolic conservation laws also arise when computing the BCs to parabolic problems and canceling out the diffusive term; see Randall and Tavella (2000).

A.4 CALCULUS OF VARIATIONS

While *static* optimization deals with the determination of one or several parameters that maximize or minimize a given objective function, *dynamic* optimization aims at finding a *function* which is better in some respect compared to others. Dynamic optimization is usually subdivided into three areas, although there is some overlap:

1. Calculus of variations.
2. Optimal control.
3. Dynamic programming.

All three areas involve the solution of differential equations. In this section, some basic results of the oldest branch of dynamic programming, the calculus of variations, are discussed. The fundamental problem can be stated as:

$$\max / \min \ V(f) = \int_{x^{\min}}^{x^{\max}} G\left[x, f(x), f'(x)\right] dx \tag{A.287}$$

$$f(x^{\min}) = A \tag{A.288}$$

$$f(x^{\max}) = Z \tag{A.289}$$

The quantity to be maximized is V, which is a function of f, which in turn is a function of x. In most economic applications, the independent variable x is time; in most noneconomic settings, it is some physical quantity. The functional relationship of the quantity to be maximized is given by the right-hand side of Equation (A.287). Equations (A.288) and (A.289) are the boundary conditions of the variational problems.

Example 64
A simple variational problem of this type looks like this:

$$V(f) = \int_0^2 \left[12xf + (f')^2\right] dx \tag{A.290}$$

$$f(0) = 0 \tag{A.291}$$

$$f(2) = 8 \tag{A.292}$$

The maximum and the minimum are characterized by the following theorem.   □

Theorem 29 (Euler)
The first order necessary condition for the fundamental problem of calculus of variations Equations (A.287)–(A.289) is:

$$G_f - \frac{d}{dx} G_{f'} = 0 \tag{A.293}$$

$$\Longleftrightarrow \quad G_f - \left(\frac{\partial G_{f'}}{\partial x} + \frac{\partial G_{f'}}{\partial f}\frac{df}{dx} + \frac{\partial G_{f'}}{\partial f'}\frac{df'}{dx}\right) = 0 \tag{A.294}$$

$$\Longleftrightarrow \quad G_{f'f'} f''(x) + G_{ff'} f'(x) + G_{xf'} - G_f = 0 \tag{A.295}$$

such that Equations (A.288) and (A.289) hold, and provided that the minimizer/maximizer is C^2.   □

The Euler equation is an ordinary second order nonlinear differential equation on which the boundary conditions Equations (A.288) and (A.289) are needed to define a solution. It is possible to derive second order conditions and sufficient conditions.

Example 65 (Example 64 continued)

$$G = 12xf + (f')^2 \tag{A.296}$$

$$G_f = 12x \tag{A.297}$$

$$G_{f'} = 2f' \tag{A.298}$$

$$G_{f'f'} = 2 \tag{A.299}$$

$$G_{ff'} = 0 \tag{A.300}$$

$$G_{xf'} = 0 \tag{A.301}$$

By Theorem 29, the Euler equation is

$$f''(x) = 6x \tag{A.302}$$

so that the solution to the Euler equation is:

$$f^*(x) = x^3 + c_1 x + c_2 \tag{A.303}$$

Integrating the boundary conditions leads to $f^*(x) = x^3$.   □

A direct consequence of Theorem 29 is the following theorem, which states the relationship between self-adjoint ordinary boundary value problems and calculus of variations.

Theorem 30 (Burden and Faires, 1998, p. 451)
The function $u(x) \in C^2$ is the unique solution to the boundary value problem

$$\frac{d}{dx}\left(p(x)\frac{d}{dx}u(x)\right) + q(x)u(x) = g(x) \qquad x_{min} \leq x \leq x_{max} \tag{A.304}$$

if and only if $u(x)$ is the unique function in C^2 that minimizes the integral

$$V(f) = \int_{x_{min}}^{x_{max}} \left[q(x)[f(x)]^2 - p(x)[f'(x)]^2 - 2g(x)f(x)\right]dx \tag{A.305}$$

□

With the help of the above theorem and the techniques from Section A.3.3, it is possible to formulate any ordinary linear second order boundary value problem as a problem of calculus of variations. Theorem 29 can be extended in various directions. Here we will only deal with

1. G depending on two univariate functions.
2. G depending also on the second derivative of f.
3. G depending on one bivariate function.

The extensions that we will discuss can be generalized to multiple multivariate functions depending on derivatives of order $n > 2$; many other generalizations are also possible. In order to get a flavor of this discussion, the interested reader is referred to Chiang (1992).

We start with G depending on two univariate functions, so that $G = G(x, f_1, f_2, f_1', f_2')$.

Theorem 31 (Euler)

The first order necessary condition for the problem

$$\max/\min V(f_1, f_2) = \int_{x^{\min}}^{x^{\max}} G\left[x, f_1(x), f_1'(x), f_2(x), f_2'(x)\right] dx \qquad (A.306)$$

$$f_1(x^{\min}) = A_1 \qquad (A.307)$$

$$f_1(x^{\max}) = Z_1 \qquad (A.308)$$

$$f_2(x^{\min}) = A_2 \qquad (A.309)$$

$$f_2(x^{\max}) = Z_2 \qquad (A.310)$$

is a system of Euler equations:

$$G_{f_1'f_1'}f_1''(x) + G_{f_2'f_1'}f_2''(x) + G_{f_1f_1'}f_1'(x) + G_{f_2f_1'}f_2'(x) + G_{xf_1'} - G_{f_1} = 0 \quad (A.311)$$

$$G_{f_1'f_2'}f_1''(x) + G_{f_2'f_2'}f_2''(x) + G_{f_1f_2'}f_1'(x) + G_{f_2f_2'}f_2'(x) + G_{xf_2'} - G_{f_2} = 0 \quad (A.312)$$

Each Euler equation is of the same type as in the univariate case: it is a nonlinear ODE of second order. □

Example 66 (Chiang, 1992, p. 46f)

The task is to find the extrema of:

$$V(f_1, f_2) = \int_0^{x^{\max}} \left[f_1 + f_2 + \left(f_1'\right)^2 + \left(f_2'\right)^2\right] dx \qquad (A.313)$$

First, we compute the derivatives necessary for constructing the Euler equations:

$$G_{f_1'} = 2f_1' \qquad (A.314)$$

$$G_{f_2'} = 2f_2' \qquad (A.315)$$

$$G_{f_1'f_1'} = G_{f_2'f_2'} = 2 \qquad (A.316)$$

$$G_{f_1} = G_{f_2} = 1 \qquad (A.317)$$

All other derivatives vanish, so the Euler equations become:

$$2f_1''(x) - 1 = 0 \qquad (A.318)$$

$$2f_2''(x) - 1 = 0 \qquad (A.319)$$

This is equivalent to:

$$f_1''(x) = f_2''(x) = \frac{1}{2} \qquad (A.320)$$

This system is not coupled, so that both equations can be integrated separately. □

Next, we turn our attention to G depending also on the second derivative of f, so that $G = G\left[x, f, f', f''\right]$:

$$\max/\min V(f) = \int_{x^{\min}}^{x^{\max}} G\left[x, f(x), f'(x), f''(x)\right] dx \qquad (A.321)$$

$$f(x^{\min}) = A_1 \qquad (A.322)$$

$$f(x^{\max}) = Z_1 \qquad (A.323)$$

$$f'(x^{\min}) = A_2 \qquad (A.324)$$

$$f'(x^{\max}) = Z_2 \qquad (A.325)$$

Theorem 32 (Euler–Poisson)

The first order necessary condition for the problem given by Equations (A.321)–(A.325) is:

$$G_f - \frac{\mathrm{d}}{\mathrm{d}x}G_{f'} + \frac{\mathrm{d}^2}{\mathrm{d}x^2}G_{f''} = 0 \tag{A.326}$$

such that Equations (A.288) and (A.289) hold, and provided that the minimizer/maximizer is C^4. $\square$

The Euler–Poisson equation is a nonlinear ODE of fourth order. Sometimes, this ODE happens to be linear, as in the following example:

Example 67 (Chiang, 1992, p. 47)

Consider the variational problem

$$\min / \max V(f) = \int_{x^{\min}}^{x^{\max}} \left\{ x\,[f(x)]^2 + f'(x)f(x) + [f''(x)]^2 \right\} \mathrm{d}x \tag{A.327}$$

$$f(x^{\min}) = A \tag{A.328}$$

$$f(x^{\max}) = Z \tag{A.329}$$

$$f'(x^{\min}) = \alpha \tag{A.330}$$

$$f'(x^{\max}) = \beta \tag{A.331}$$

By differentiating Equation (A.327), the following expressions can be computed:

$$G_f = 2fx + f' \tag{A.332}$$

$$G_{f'} = f \tag{A.333}$$

$$G_{f''} = 2f'' \tag{A.334}$$

Inserting these expressions into the Euler–Poisson equation gives:

$$(2fx + f') - \frac{\mathrm{d}}{\mathrm{d}x}f + \frac{\mathrm{d}^2}{\mathrm{d}x^2}2f'' = 0 \tag{A.335}$$

$$\Longleftrightarrow \qquad 2fx + f' - f' + 2f^{(4)} = 0 \tag{A.336}$$

$$\Longleftrightarrow \qquad\qquad f^{(4)}(x) = f(x)x \tag{A.337}$$

By introducing a new variable $v = f''$, this ODE of fourth order can be reduced to a system of boundary value problems of second order:

$$f''(x) = v(x) \tag{A.338}$$

$$v''(x) = f^{(4)}(x) = -f(x)x \tag{A.339}$$

   $\square$

While all of the material on variational calculus discussed so far is standard in economics, the extension to multivariate functions is not well-known in economics. Some notable exceptions are given in Section 2.6. Last but not least, we will briefly discuss G depending on one bivariate function, so that $G = G\left[x_1, x_2, f(x_1, x_2), f_{x_1}(x_1, x_2), f_{x_2}(x_1, x_2)\right]$. We assume a Dirichlet condition on the boundary of Ω which may possess a more general shape than a rectangle. Again, we state the first order necessary condition:

Theorem 33 (Euler)

The first order necessary condition for the bivariate problem of calculus of variations

$$\max / \min V\left[f(x_1, x_2)\right] = \int_\Omega \int G\left[x_1, x_2, f(x_1, x_2), \frac{\partial f(x_1, x_2)}{\partial x_1}, \frac{\partial f(x_1, x_2)}{\partial x_2}\right] dx_1 dx_2$$

$$\text{(A.340)}$$

is given by the following PDE:

$$G_f - \frac{d}{dx_1} G_{f_{x_1}} - \frac{d}{dx_2} G_{f_{x_2}} = 0 \qquad \text{(A.341)}$$

$\square$

Example 68 (minimum-energy theorem)

The variational problem associated with the Poisson equation

$$\nabla^2 f = f_{x_1 x_1} + f_{x_2 x_2} = g(x_1, x_2) \qquad \text{(A.342)}$$

in a region $\bar{\Omega}$ with $f = 0$ on Γ_Ω is given by

$$V\left[f(x_1, x_2)\right] = \int_\Omega \int \left[f_{x_1}^2 + f_{x_2}^2 + 2fg\right] dx_1 dx_2 \qquad \text{(A.343)}$$

The BCs remain unchanged.

$\square$

In the calculus of variations, the problem usually is to find a differential equation that maximizes or minimizes a given integral. Sometimes, however, one is interested in the variational problem associated with a given differential equation. This is sometimes called the *inverse problem* of calculus of variations (Wan, 1995, Section 8.8). One reason for investigating the variational formulation of a problem at hand is the vast body of techniques available. As an example, the Ritz method is outlined in Section 4.3. There is no general technique for finding the variational problem associated with a differential equation; for some hints see Section 8.8 of Wan (1995) and the references therein. Sometimes, it is not even possible to decide whether a given differential equation possesses an equivalent variational representation. For the Black–Scholes problem, it is possible to show that a variational formulation cannot be given, because it is not possible to twist Theorem 33 in such a way as to include this parabolic PDE. Assume G from Theorem 33 is of the form $G = L(u, v, w) = L(u, u_t, u_x)$. Then, according to Theorem 33, the following holds:

$$\frac{\partial L}{\partial u}(u, u_t, u_x) = \frac{\partial}{\partial t}\left\{\frac{\partial L}{\partial v}(u, u_t, u_x)\right\} + \frac{\partial}{\partial x}\left\{\frac{\partial L}{\partial w}(u, u_t, u_x)\right\} \qquad \text{(A.344)}$$

Parabolic PDEs do not possess terms of the form u_{tt} and u_{tx}. This implies:

$$\frac{\partial L}{\partial v}(u, v, w) = A(u) \qquad \text{(A.345)}$$

Consequently:

$$L(u, v, w) = \underbrace{A(u) \cdot v}_{=:L_0(u,v)} + B(u, w) \qquad \text{(A.346)}$$

Note that $B(u, w)$ does not depend on t. Furthermore:

$$\frac{\partial}{\partial t} \left\{ \frac{\partial L_0}{\partial v}(u, u_t) \right\} - \frac{\partial L_0}{\partial u}(u, u_t) \tag{A.347}$$

$$= \frac{\partial}{\partial t} A(u) - \frac{\partial A}{\partial u}(u) \cdot u_t \tag{A.348}$$

$$= 0 \tag{A.349}$$

This shows that there is no variational problem associated with a parabolic PDE.

The inverse problem of calculus of variations has been studied by various authors. Theorem 30 can be extended to nonlinear ODEs and also certain PDEs. For details see de Matos Pimentão (1986) and the references cited there.

Appendix B
Some Useful Results from Stochastics

B.1 SOME IMPORTANT DISTRIBUTIONS

B.1.1 The univariate normal distribution

The density function

The density function of the univariate normal distribution (shown in Figure B.1) is

$$f(x, \mu, \sigma) = \frac{\exp\left(\frac{-(x-\mu)^2}{2\sigma^2}\right)}{\sqrt{2\pi}\sigma}, \quad x \in \mathbb{R} \tag{B.1}$$

Any normally distributed variable $X \sim N(\mu, \sigma)$ can be transformed to $X_s \sim N(0, 1)$ with the substitution:

$$X_s = \frac{X - \mu}{\sigma} \tag{B.2}$$

Because of this substitution it is sufficient to consider the standardized normal distribution:

$$f(x, 0, 1) = \frac{\exp\left(\frac{-x^2}{2}\right)}{\sqrt{2\pi}}, \quad x \in \mathbb{R} \tag{B.3}$$

The cumulative function

The cumulative function represents the probability $P(x \leq a)$ that some number generated by the standardized normal distribution is lower than a given number a:

$$P(x \leq a) \equiv M(a) = \int_{-\infty}^{a} f(x, 0, 1) \, dx = \frac{1}{\sqrt{2\pi}} \int_{-\infty}^{a} \exp\left(\frac{-x^2}{2}\right) dx \tag{B.4}$$

A closed-form solution for this type of integral is not known, so the integral has to be computed numerically or an analytical approximation has to be employed. There are several analytical approximations available, for which errors are also known (Hull, 2000).

The relationship between the density function and the cumulative function is illustrated in Figure B.2.

B.1.2 The bivariate normal distribution

The density function

The two-dimensional extension of the standardized normal distribution is called the *bivariate normal distribution*. Its density is given by:

$$f(x_1, x_2, \rho) = \frac{1}{2\pi\sqrt{1 - \rho^2}} \exp\left[-\frac{(x_1^2 - 2\rho x_1 x_2 + x_2^2)}{2(1 - \rho^2)}\right], \quad x_i \in \mathbb{R}^2 \tag{B.5}$$

The correlation between the two random variables is denoted by ρ.

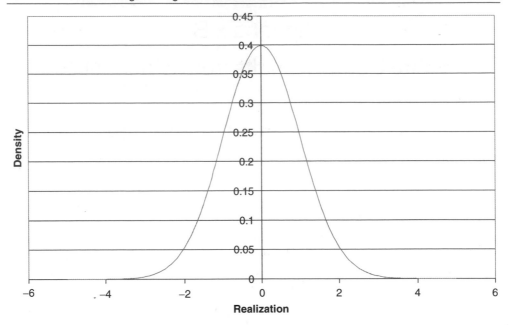

Figure B.1   Density function of the univariate standardized normal distribution

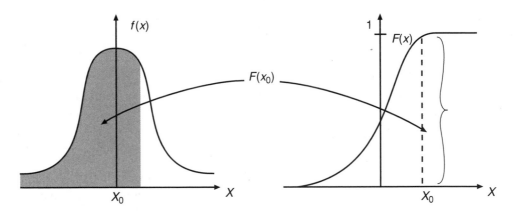

Figure B.2   Relationship between density function and cumulative function

The cumulative function

$$F(x_1, x_2, \rho) = \frac{1}{2\pi \sqrt{1 - \rho^2}} \int_{-\infty}^{a_1} \int_{-\infty}^{a_2} \exp\left[-\frac{(x_1^2 - 2\rho x_1 x_2 + x_2^2)}{2(1 - \rho^2)} \right] \, dx_1 \, dx_2 \qquad \text{(B.6)}$$

This expression cannot be computed analytically. A popular approximation in finance to this problem has been given by Drezner (1978). Since there is a typo in the original paper, see page 260 of Hull (2000) for an adaptation which is accurate to four figures or six figures (Haug, 1997, p. 191f). For a comparative study of advantages and disadvantages of various approximations in the setting of option pricing, see Agča and Chance (2003).

B.1.3   The multivariate normal distribution

The density function

Let the vector-valued random variable $\mathbf{x} = (x_1, \ldots, x_n)^\mathrm{T}$ follow an n-dimensional multivariate normal distribution with mean vector μ and $n \times n$ variance–covariance matrix $\boldsymbol{\Sigma}$, then the multidimensional extension of the normal distribution is given by

$$f(\mathbf{x}, \mu, \boldsymbol{\Sigma}) = \frac{\exp\left[\frac{1}{2}(\mathbf{x} - \mu)^\mathrm{T}\boldsymbol{\Sigma}^{-1}(\mathbf{x} - \mu)\right]}{\sqrt{(2\pi)^n |\boldsymbol{\Sigma}|}} \tag{B.7}$$

The cumulative function

Accordingly, the cumulative function is given by:

$$F(\mathbf{x}, \mu, \boldsymbol{\Sigma}) = \int_{-\infty}^{a_1} \cdots \int_{-\infty}^{a_n} f(\mathbf{x}, \mu, \boldsymbol{\Sigma}) \, \mathrm{d}x_1 \ldots \mathrm{d}x_n \tag{B.8}$$

With increasing n it becomes increasingly difficult to compute this integral (Johnson *et al.*, 2000), since neither accurate approximations nor really fast numerical tools are currently known. Therefore, nowadays simulation estimation methods have become popular (Hajivssiliou, 1993; Hajivssiliou and Ruud, 1993).

B.1.4   The lognormal distribution

Consider a distribution of which the natural logarithm is normally distributed. This distribution is called the *lognormal distribution*. This is somewhat misleading, since it is not the distribution of the logarithm of a normal variable, but of an exponential of such a variable. Its stationary density function (shown in Figure B.3) is:

$$f_{\ln}(x) = \begin{cases} \dfrac{\exp\left(-\frac{[\ln(x)-\mu]^2}{2\sigma^2}\right)}{x\sigma\sqrt{2\pi}}, & x > 0, \ \sigma > 0, \ \mu \in \mathbb{R} \\ 0, & x \leq 0 \end{cases} \tag{B.9}$$

In finance, one is often interested in how a certain stochastic quantity evolves over time. Hence, the dynamic density function is given by:

$$f_{\ln}(x, t) = \frac{\exp\left(-\frac{[\ln(x)-\mu t]^2}{2\sigma^2 t}\right)}{x\sigma\sqrt{2\pi t}}, \quad x > 0 \tag{B.10}$$

B.1.5   The Γ distribution

The density of the Γ distribution (shown in Figure B.4) is given by:

$$f_\Gamma(x) = \frac{\beta^\alpha e^{-\beta x} x^{\alpha-1}}{\Gamma(\alpha)}, \quad x > 0, \ a, b > 0 \tag{B.11}$$

This distribution includes the standard Γ distribution with $\beta = 1$:

$$f_{\Gamma_{\text{Standard}}}(x) = \frac{e^{-x} x^{\alpha-1}}{\Gamma(\alpha)}, \quad x > 0, \ a > 0 \tag{B.12}$$

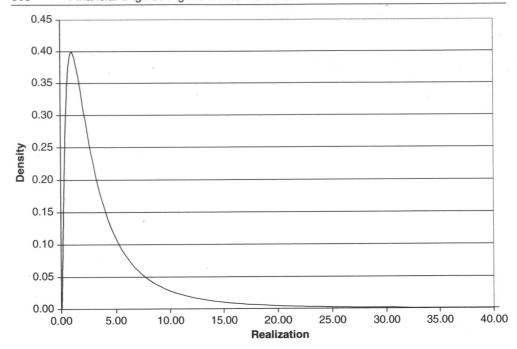

Figure B.3 Density function of the lognormal distribution ($\mu = 0$, $\sigma = 1$)

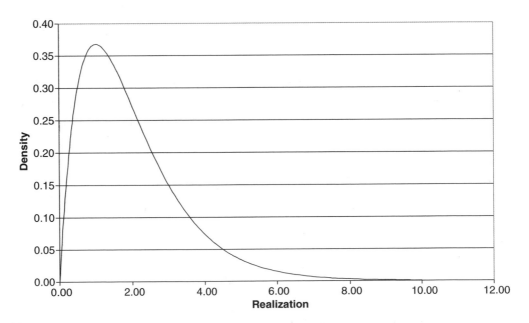

Figure B.4 Density function of the standard Γ distribution ($\alpha = 2$, $\beta = 0.5$)

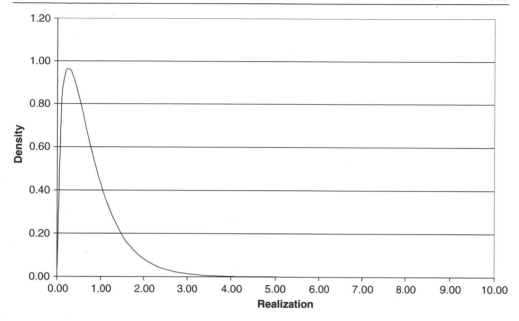

Figure B.5 Density function of the central χ^2 distribution ($\nu = 3$)

and the χ^2 distribution with $\alpha = n/2$ and $\beta = 0.5$. For details on the numerical implementation and approximations of the Γ distribution, see Johnson *et al.* (1994).

B.1.6 The central χ^2 distribution

The probability density function χ_ν^2 (shown in Figure B.5) of a chi-square random variable with ν degrees of freedom is given by:

$$f_{\chi_\nu^2}(x) = \frac{\exp\left(-\frac{x}{2}\right) x^{\frac{\nu}{2}-1}}{2^{\frac{\nu}{2}}\Gamma\left(\frac{\nu}{2}\right)}, \qquad x \geq 0, \quad \nu \in \mathbb{N} \tag{B.13}$$

with $\Gamma(z)$ denoting the (complete) Gamma function given by:

$$\Gamma(z) = \int_0^\infty t^{z-1}\exp(-t)\mathrm{d}t, \quad z > 0 \quad \text{(Euler)} \tag{B.14}$$

$$= \lim_{n\to\infty} \frac{n!\,n^{z-1}}{z(z+1)(z+2)\ldots(z+n-1)}, \quad z \neq 0, -1, -2, \ldots \quad \text{(Gauss)} \tag{B.15}$$

Both definitions are equivalent for $z > 0$, which is satisfied here since $\nu \geq 1$. The Gamma function is just a generalization of the factorial function, which is only defined for integers:

$$z! = \Gamma(z+1) \tag{B.16}$$

The cumulative distribution function is given by:

$$F_{\chi_\nu^2}(x) \equiv \text{Prob}\left[\chi_\nu^2 \leq x\right] = \frac{\Gamma_{\text{incomplete}}\left(\frac{x}{2}, \frac{\nu}{2}\right)}{\Gamma\left(\frac{\nu}{2}\right)} \tag{B.17}$$

with $\Gamma_{\text{incomplete}}(z, \alpha)$ denoting the incomplete Gamma function:

$$\Gamma_{\text{incomplete}}(z, \alpha) = \int_0^z \exp(-t)\ t^{\alpha-1} dt, \qquad \alpha \geq 0 \qquad \text{(B.18)}$$

B.1.7   The noncentral χ^2 distribution

The probability density function of a noncentral chi-square random variable with ν degrees of freedom and the noncentrality parameter λ is given by:

$$f_{\chi^2_{\nu,\lambda}}(x) = \frac{\exp\left(\frac{x-\lambda}{2}\right)}{2^{\nu/2}} \sum_{j=0}^{\infty} \left(\frac{\lambda}{4}\right)^j \frac{x^{\nu/2+j-1}}{j!\ \Gamma\left(\frac{\nu}{2}+j\right)} \qquad \text{(B.19)}$$

The cumulative distribution function is given by:

$$F_{\chi^2_{\nu,\lambda}}(x) \equiv \text{Prob}\left[\chi^2_{\nu,\lambda} \leq x\right] \qquad \text{(B.20)}$$

$$= \begin{cases} \exp\left(\frac{-\lambda}{2}\right) \sum_{j=0}^{\infty} \frac{\left(\frac{\lambda}{2}\right)^j/j!}{2^{\nu/2+j}\Gamma\left(\frac{\nu}{2}+j\right)} \int_0^x y^{\nu/2+j-1} \exp\left(\frac{-y}{2}\right) dy & \text{for} \quad x > 0 \\ 0 & \text{for} \quad x < 0 \end{cases} \qquad \text{(B.21)}$$

$$= \begin{cases} \exp\left(\frac{-\lambda}{2}\right) \sum_{j=0}^{\infty} \frac{\left(\frac{\lambda}{2}\right)^j}{j!} F_{\chi^2_{\nu+2j}} & \text{for} \quad x > 0 \\ 0 & \text{for} \quad x < 0 \end{cases} \qquad \text{(B.22)}$$

B.2   SOME IMPORTANT PROCESSES

B.2.1   Basic concepts

A somewhat informal definition of a stochastic process is *a collection of random variables that are indexed by a parameter such as time or space* $\{Y_t\}_{t\in I}$ (Bhat, 1984, p. 1). To fix these ideas, we will take a look at some examples:

1. Daily temperatures at noon.
2. The diameter of a cable.
3. Weekly number of car accidents at a given place.
4. Number of babies born up to a certain point of time.
5. First order autoregressive process (AR(1)):

$$Y_t = \alpha + \gamma Y_{t-1} + \epsilon_t \qquad \text{(B.23)}$$

$$-1\ <\ \gamma\ <\ 1 \qquad \text{(B.24)}$$

$$\epsilon_t \sim N(0, \sigma) \qquad \text{(B.25)}$$

There are many ways to classify stochastic processes. One is to look at both parameter space and state space (see Table B.1). The value a stochastic process is assuming is called the *state*. All possible states comprise the *state space*. The *parameter space* is the set of possible values for the indexing parameter. In economic applications this is usually *time*, while in technical applications this often is a spatial variable, such as length of a cable, as in the example listed above. Both sets may consist of a number of discrete events or a continuum.

Temperature measured in degrees Celsius or Fahrenheit, as in Example 1, is a continuous quantity. When it is taken daily at noon it is taken at discrete points in time. Therefore, its parameter space is discrete. The same is true for Example 3, however this has a discrete state

Table B.1 Classification of stochastic processes

		State space	
		Discrete	Continuous
Parameter	Discrete	Example 3	Example, 1, 5
space	Continuous	Example 4	Example 2

space, since the number of accidents is an integer by its very nature. The number of babies born, as in Example 4, is also an integer. Since we are interested in the number of babies up to a specific point of time, the parameter space in this example is continuous. Both the diameter of a cable and the corresponding coordinate are continuous quantities, so that in this example, both parameter space and state space are continuous.

Some fundamental processes are *random walks*, *Markov processes* and *martingales*. These processes are not restricted to the discrete parameter space, although this is the space for which we will discuss these processes.

1. *Random walk.* The discrete-time discrete-state random walk begins at a known value X_0 and at times $t = 1, 2, 3, \ldots$ it takes a jump of size 1 either up or down, each with probability 0.5. The jumps are independent of each other so that the process can be described by:

$$Y_t = Y_{t-1} + \epsilon_t \tag{B.26}$$
$$P(\epsilon_t = 1) = P(\epsilon_t = -1) = 0.5 \tag{B.27}$$

For this simple process it is possible to describe the state space. For $Y_0 = 0$ and odd values of t, possible values for Y_t are $(-t, \ldots, -1, 1, \ldots, t)$. For even values of t, possible outcomes belong to $(-t, \ldots, -2, 0, 2, \ldots, t)$. For t steps, the probability of n downward and $t - n$ upward jumps is given by

$$\binom{t}{n} 2^{-t} \tag{B.28}$$

Consequently, the probability that $Y_t = t - 2n$ can be given as:

$$P(Y_t = t - 2n) = \binom{t}{n} 2^{-t} \tag{B.29}$$

This distribution becomes important when we derive the continuous-time counterpart of this process, which is called the *Wiener process* or *Brownian motion*.

2. *Markov processes* are a special class of stochastic process. One of their characteristics is that they unite all information in their last realization. The term y_{n+1} is solely dependent on y_n. Therefore, a Markov process is also called a *process without memory*. To fix these ideas we will discuss two examples:

   – *Gambler (a Markov process).* The wealth of a player H at time t may be V_t^H. Players H and K always gamble with the same amount B. The winner receives the entire bet amount of $2B$, the loser receives nothing. A fair coin decides who wins or loses. For the respective probabilities $p = 0.5$ we have:

$$V_{t+1}^H = V_t^H + B \quad \text{(win)} \tag{B.30}$$
$$V_{t+1}^H = V_t^H - B \quad \text{(loss)} \tag{B.31}$$

The wealth in any period solely depends on the wealth of the previous period. This is called a Markov process.

– *Inn keeper (a non-Markov process).* The inn keeper estimates his turnover U_t^e for a day t according to the following rule:

$$U_t^e = 0.5\,U_{t-1} + 0.5\,U_{t-2} \tag{B.32}$$

U_t being the realized turnover in t. Thus, he requires information that reaches beyond the last period. Consequently, it is not a Markov process.

In general, non-Markov processes are considered to be difficult and Markov processes are – by far – better understood.

3. *Martingales.* Subject to an information set I_t, a *martingale* is defined as

$$E^P(Y_{t+s}|I_t) \;=\; Y_t \tag{B.33}$$

Colloquially, this means, the 'best guess' for a future unknown realization of Y is the last known realization of Y. Accordingly, we can derive the definition of a *supermartingale*. Subject to an information set I_t, a *supermartingale* is defined as

$$E^P(Y_{t+s}|I_t) \;\leq\; Y_t \tag{B.34}$$

Colloquially, this means, the 'best guess' for a future unknown realization of Y is smaller than the last known realization of Y. The definition of a *submartingale* also follows these lines: subject to an information set I_t, a *submartingale* is defined as

$$E^P(Y_{t+s}|I_t) \;\geq\; Y_t \tag{B.35}$$

This means, the 'best guess' for a future unknown realization of Y is larger than the last known realization of Y.

The random walk is an example of a process with the Markov property. Also, the random walk is a martingale. This property, however, cannot be generalized. A martingale does not necessarily have the Markov property, nor does the Markov property imply that the process under inspection is a martingale. This will be demonstrated with some examples, some of which, for simplicity, are *discrete* stochastic processes.

Any unfair gambling (like coin-tossing with an 'unfair' coin) has the Markov property, while being neither a sub- nor a supermartingale. Brownian motion with reflecting barriers is a Markov process without being a martingale (or sub- or supermartingale). For a martingale without the Markov property, consider the following game: two gamblers A and B play a fair coin-tossing game with the following payout: the payout is the number of games plus one that A (respectively B) has already won. The amount of capital held by either player at any time after the first round is a martingale without being Markovian.

B.2.2   Wiener process

If the change ΔX of a continuous stochastic process X in a time interval Δt is related by

$$\Delta X \;=\; \epsilon \sqrt{\Delta t} \quad \text{with} \quad \epsilon \sim N(0,1) \tag{B.36}$$

and if also the values of two Δx for any short intervals Δt are independent, X is called a *Wiener process* or *(pure) Brownian motion*. Figure B.6 shows an example. Taking the limit

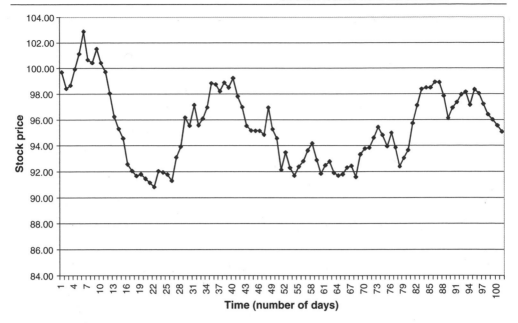

Figure B.6 Stock price as a (pure) Wiener process: the simulation is based on Equation (B.36) with $\Delta t = 1$ day

$\Delta t \to 0$ we get the stochastic differential equation representation

$$dX = \epsilon \sqrt{dt} \qquad (B.37)$$

Most textbooks introduce Brownian motion as the continuous limit of the *random walk*. The Wiener process has the following properties:

- The path of a Brownian motion is continuous everywhere. However, it is nowhere differentiable.
- It possesses the Markov property.
- It is a martingale.
- It is nonstationary, since $\text{Var}[X(t) - X(0)] = T$.
- It does not have any drift, since $E[X(t) - X(0)] = 0$.

B.2.3 Brownian motion with drift

Both drift and volatility are independent of time and Y.

$$dY = \mu\, dt + \sigma\, dX \qquad (B.38)$$

B.2.4 Geometric Brownian motion

The process given by

$$dY = aY\, dt + \sigma Y\, dX \qquad (B.39)$$

$$\frac{dY}{Y} = a\, dt + \sigma\, dX \qquad (B.40)$$

is widely used to model the dynamics of stock prices and of currencies. It was introduced to the economic and finance literature by Samuelson (1965).

B.2.5 Itô process

The Itô process (or Itô diffusion) is defined as

$$dY = A(Y, t) \, dt + B(Y, t) \, dX \qquad (B.41)$$

A is called the *drift*, B^2 the *variance rate* and dX is the Wiener process defined above. This process is a generalization of the Wiener process and geometric Brownian motion. No further generalization will be needed in this text.

B.2.6 Ornstein–Uhlenbeck process

The process[1]

$$dY = a(b - Y) \, dt + \sigma \, dX \qquad (B.42)$$

is a special time-homogeneous Itô process. Again, $a(b - Y)$ is the drift rate and σ^2 the variance rate. The process Y is a Markov process with continuous sample paths and Gaussian increments. It has an average level $b \geq 0$, to which it is forced at rate $a \geq 0$.

B.2.7 A process for commodities

The following SDE has been suggested for the price of gold (Hearst, 2001):

$$dS = v^2 S^{2\gamma - 1} \left[\gamma - \frac{1}{2} - \frac{1}{2a^2} \ln \left(\frac{S}{\bar{S}} \right) \right] dt + v S^{\gamma} dX \qquad (B.43)$$

with the following parameters:

$$v = 16.2 \qquad (B.44)$$
$$\gamma = 0.42 \qquad (B.45)$$
$$\bar{S} = 366 \qquad (B.46)$$
$$a = 0.138 \qquad (B.47)$$

B.3 RESULTS

B.3.1 The transition probability density function

We start by considering the Itô process in terms of a stochastic differential equation

$$dY = A(Y, t) \, dt + B(Y, t) \, dX \qquad (B.48)$$

$A(\cdot)$ and $B(\cdot)$ are known (nonrandom) functions and dX is the increment of a Wiener process. There are (mild) regularity assumptions on $A(\cdot)$ and $B(\cdot)$, assuring the existence and uniqueness of a solution to Equation (B.48), which are usually met in real-world modeling; the interested

---

[1] Different definitions exist. This one is taken from Musiela and Rutkowski (1997) because it fits our purposes best.

reader is referred to Duffie (1996). To simplify the notation we set $Y = y$. The *transition probability density function*, $p(\cdot)$, is defined as:

$$\text{Prob}\left[\underbrace{a \leq y \leq b \text{ at time } t' \mid y}_{=y'} \text{ at time } t\right] = \int_a^b p(y, t, y', t') \, dy' \tag{B.49}$$

This simply translates to the probability that the random variable Y is between a and b at time $t' > t$, when it has started with the value y at t. The *current* values are y and t, while y' and t' are *future* values. If we know where the process is now, we cannot exactly tell where it will be at some time in the future. However, we are able to come up with a probability density function. This problem is called the *forward problem*. Another point of view can be taken: the process has reached a certain point now and we want to know from where it has come. Obviously, we cannot answer this question with certainty. Again, we are able to provide a probability density function to describe a former state of the process. Accordingly, this problem is called the *backward problem*. The open question is how to compute $p(\cdot)$. This can be achieved in different ways for the forward and the backward problems. The appropriate tools are the *Kolmogorov equations*, of which there is a forward and a backward equation.

B.3.2 The backward Kolmogorov equation

Both Kolmogorov equations can be derived in various ways. Less rigorous derivations are based on the binomial or trinomial approximation of the stochastic process y (Cox and Miller, 1965; Dixit and Pindyck, 1994; Wilmott *et al.*, 1996; Wilmott, 1998). Here, we will only state the results, starting with the backward problem for notational convenience. Loosely speaking, the backward problem answers the question from where a stochastic process has come. The direction of time is backwards, i.e. it goes backwards, starting from t'. The probability density function can be determined by solving the PDE:

$$\frac{\partial p(\cdot)}{\partial t} = -\frac{B(y, t)^2}{2} \frac{\partial^2 p(\cdot)}{\partial y^2} - A(y, t) \frac{\partial p(\cdot)}{\partial y} \tag{B.50}$$

This equation is a linear backward parabolic PDE of second order. Whether the spatial domain of the PDE is infinite $(-\infty; \infty)$, semi-infinite $(-\infty; y^{\max}])$ (or $[y^{\min}; \infty)$) or finite $[y^{\min}; y^{\max}]$ depends on the stochastic process at hand. This way the backward Kolmogorov equation is either an initial value problem or an initial boundary value problem. Besides a final condition at $t = t'$, the former type of problem needs a technical condition on $p(\cdot)$ to be met, which guarantees that the solution does not grow without bounds, to assure a unique and stable solution. The final condition is

$$p(y, t', y', t') = \delta(y - y') \tag{B.51}$$

with $\delta(\cdot)$ being the Dirac delta function. Equation (B.51) simply states that

$$p(y, t', y', t') = 0 \quad \text{for} \quad y \neq y' \tag{B.52}$$
$$p(y', t', y', t') \rightarrow \infty \quad \text{for} \quad y = y' \tag{B.53}$$

so that, loosely speaking, all probability mass is concentrated at $y = y'$, from where it spreads out, so that

$$\int_{-\infty}^{\infty} \delta(y - y') \, dy' = 1 \tag{B.54}$$

Boundary conditions and their numerical implementation will be discussed in more detail in the next section.

Example 69 (the pure Wiener process–Brownian motion without drift)
This process is given by

$$dY = dX \tag{B.55}$$

i.e. $A = 0$ and $B = 1$. Consequently, the backward Kolmogorov equation simplifies to:

$$\frac{\partial p}{\partial t} = -\frac{1}{2}\frac{\partial^2 p}{\partial y^2} \tag{B.56}$$

The stochastic process dY is not bounded, so the boundary conditions are:

$$\lim_{y \to -\infty} p(y) = \lim_{y \to \infty} p(y) = 0 \tag{B.57}$$

$\square$

B.3.3   The forward Kolmogorov equation

The forward Kolmogorov equation[2] is employed if we want to know the future distribution of a stochastic process, knowing its current state. Now, time moves as calendar time, i.e. we start at t and move forward to t'. Unfortunately, this makes the notation somewhat unclear. The probability density function can be computed by solving:

$$\frac{\partial p(\cdot)}{\partial t'} = \frac{1}{2}\frac{\partial^2 \left[B(y', t')^2 p(\cdot)\right]}{\partial (y')^2} - \frac{\partial \left[A(y', t') p(\cdot)\right]}{\partial y'} \tag{B.58}$$

This also is a linear parabolic PDE of second order. Because of the direction of time, it is forward parabolic. The initial condition states that all probability mass is concentrated in one point:

$$p(y, t, y', t) = \delta(y' - y) \tag{B.59}$$

The boundary conditions depend on the process under investigation. For example: the stochastic process defined by Equation (B.39) moves freely in $(0; \infty)$. It cannot ever reach $Y = 0$ in the case where it starts with $Y > 0$. In the case where it starts with $Y = 0$, Y remains zero for all times. Consequently, Y can never become negative. For financial applications, processes without barriers and with absorbing barriers are of primary interest. Once the stochastic process has hit an absorbing barrier, it will not leave this state. Therefore, absorbing barriers lead to boundary conditions like

$$p(y = y^{\min}, t, y', t') = 0 \quad \text{and/or} \tag{B.60}$$

$$p(y = y^{\max}, t, y', t') = 0 \tag{B.61}$$

---

[2] This equation is sometimes called the *Fokker–Planck (diffusion) equation*, while the *Kolmogorov equation* is backward by definition. In order to be able to always distinguish both equations at first sight they will be accompanied by their respective adjectives in this book.

depending on whether there are one or two absorbing barriers imposed on the process. If there are no barriers on the process, the spatial domain of $p(\cdot)$ is not bounded. Accordingly, if there is only one barrier, the domain is semi-infinite. Sometimes, it is possible to find analytical solutions to the Kolmogorov PDEs; see, for example, Cox and Miller (1965). These techniques are difficult and sometimes the analytical solutions include infinite series, which have to be approximated numerically, so that it is seems advisable to solve the Kolmogorov PDEs numerically when an analytical solution is not known.

Example 70 (the pure Wiener process–Example 69 revisited)
This process is given by

$$\mathrm{d}Y = \mathrm{d}X \tag{B.62}$$

i.e. $A = 0$ and $B = 1$. Consequently, the forward Kolmogorov equation simplifies to:

$$\frac{\partial p}{\partial t'} = \frac{1}{2}\frac{\partial^2 p}{\partial (y')^2} \tag{B.63}$$

The stochastic process $\mathrm{d}X$ is not bounded, so the boundary conditions are:

$$\lim_{y \to -\infty} p(y') = \lim_{y' \to \infty} p(y') = 0 \tag{B.64}$$

The classical solution to this problem is:

$$p(y, y', t') = \frac{\exp\left(-\frac{(y-y')^2}{2t'}\right)}{\sqrt{2\pi t'}} \tag{B.65}$$

for $Y(0) = y'$. □

Example 71 (Brownian motion with drift)
This process is given by

$$\mathrm{d}Y = \mu\mathrm{d}t + \sigma\mathrm{d}X \tag{B.66}$$

i.e. $A = \mu$ and $B = \sigma$. Consequently, the forward Kolmogorov equation simplifies to:

$$\frac{\partial p}{\partial t'} = \frac{\sigma^2}{2}\frac{\partial^2 p}{\partial (y')^2} - \mu\frac{\partial p}{\partial y'} \tag{B.67}$$

The stochastic process $\mathrm{d}X$ is not bounded, so the boundary conditions are:

$$\lim_{y' \to -\infty} p(y') = \lim_{y' \to \infty} p(y') = 0 \tag{B.68}$$

The classical solution to this problem is:

$$p(y, y', t') = p(y, t') = \frac{\exp\left(-\frac{(y-\mu t)^2}{2\sigma^2 t'}\right)}{\sqrt{2\pi\sigma^2 t'}} \tag{B.69}$$

for $Y(0) = 0$ □

Example 72 (Cox–Ingersoll–Ross process (Cox *et al.*, 1985))
This process is given by

$$\mathrm{d}Y = b(a - Y)\,\mathrm{d}t + \sigma\sqrt{Y}\mathrm{d}X \tag{B.70}$$

i.e. $A = b(a - Y)$ and $B = \sigma\sqrt{Y}$. Consequently, the forward Kolmogorov equation simplifies to:

$$\frac{\partial p(\cdot)}{\partial t'} = \frac{\sigma^2}{2}\frac{\partial^2[y'p(\cdot)]}{\partial(y')^2} - \frac{\partial[b(a - y')p(\cdot)]}{\partial y'} \tag{B.71}$$

The BCs are given by

$$\lim_{y'\to 0} -\frac{\sigma^2}{2}\frac{\partial[y'p(\cdot)]}{\partial y'} + b(a - y')p(\cdot) = 0 \tag{B.72}$$

$$\lim_{y'\to\infty} -\frac{\sigma^2}{2}\frac{\partial[y'p(\cdot)]}{\partial y'} + b(a - y')p(\cdot) = 0 \tag{B.73}$$

while the IC is given by Equation (B.59).   □

For multidimensional stochastic processes

$$dY_1 = A_1(Y_1, \ldots, Y_n, t)dt + \sum_{j=1}^{n} B_{1j}(Y_1, \ldots, Y_n, t)dX_j$$

$$\vdots \tag{B.74}$$

$$dY_n = A_n(Y_t, \ldots, Y_n, t)dt + \sum_{j=1}^{n} B_{nj}(Y_1, \ldots, Y_n, t)dX_j$$

with a correlation matrix $\Sigma = (\rho)_{ij}$, there are multidimensional extensions of the Kolmogorov equations. We will discuss only the forward equation, for the multidimensional backward equation see Cox and Miller (1965). The multidimensional forward Kolmogorov equation is a straightforward extension of Equation (B.58):

$$\frac{\partial p(\cdot)}{\partial t'} = -\frac{1}{2}\sum_{i,j,l,m} \rho_{lm}\frac{\partial^2\left[B_{il}(\cdot)B_{jm}(\cdot)p(\cdot)\right]}{\partial y_i'\partial y_j'} - \sum_{i=1}^{n}\frac{\partial[A_i(\cdot)p(\cdot)]}{\partial y_i'} \tag{B.75}$$

$$= -\sum_{i=1}^{n}\sum_{j=1}^{n}\frac{\partial^2\left[a_{ij}p(\cdot)\right]}{\partial y_i'\partial y_j'} - \sum_{i=1}^{n}\frac{\partial[A_i(\cdot)p(\cdot)]}{\partial y_i'} \tag{B.76}$$

with

$$a_{ij} = \frac{1}{2}\left(\mathbf{B\Sigma B^T}\right)_{ij} \tag{B.77}$$

Example 73 (two uncorrelated Brownian motions)
The most simple multidimensional example is that of two uncorrelated Brownian motions. The forward Kolmogorov equation then simplifies to:

$$\frac{\partial p(\cdot)}{\partial t'} = \frac{1}{2}\left(\frac{\partial^2 p(\cdot)}{(\partial y_1')^2} + \frac{\partial^2 p(\cdot)}{(\partial y_2')^2}\right) \tag{B.78}$$

This is the two-dimensional heat equation, which has been studied for many decades.   □

Example 74 (multidimensional geometric Brownian motion)
In an n-dimensional option pricing setting

$$dS_1 = (r - D_1)S_1 dt + \sigma_1 S_1 dX_1$$

$$\vdots \tag{B.79}$$

$$dS_n = (r - D_n)S_n dt + \sigma_n S_n dX_n$$

with correlations given by

$$\Sigma = \begin{pmatrix} \rho_{11} & \rho_{12} & \cdots \\ \rho_{21} & \rho_{22} & \cdots \\ \vdots & \vdots & \ddots \end{pmatrix} = \begin{pmatrix} 1 & \rho_{12} & \cdots \\ \rho_{12} & 1 & \cdots \\ \vdots & \vdots & \ddots \end{pmatrix} \tag{B.80}$$

the forward Kolmogorov equation simplifies to:

$$\frac{\partial p(\cdot)}{\partial t'} = \sum_{i=1}^{n} \sum_{j=1}^{n} \frac{\partial^2 \left[a_{ij} p(\cdot) \right]}{\partial y_i' \partial y_j'} - \sum_{j=1}^{n} \frac{\partial \left[(r - D_j) y_j' p(\cdot) \right]}{\partial y_j'} \tag{B.81}$$

with

$$a_{ij} = \frac{1}{2} \rho_{ij} \sigma_i \sigma_j y_i' y_j' \tag{B.82}$$

and

$$S_i = y_i' \tag{B.83}$$

Assume $n = 2$ and constant Σ, r and D_i. Then Equation (B.81) further simplifies to:

$$\frac{\partial p(\cdot)}{\partial t'} = \frac{\sigma_1^2}{2} \frac{\partial^2 \left[(y_1')^2 p(\cdot) \right]}{(\partial y_1')^2} + \frac{\sigma_2^2}{2} \frac{\partial^2 \left[(y_2')^2 p(\cdot) \right]}{(\partial y_2')^2} + \sigma_1 \sigma_2 \rho_{12} \frac{\partial^2 [y_1' y_2' p(\cdot)]}{\partial y_1' \partial y_2'}$$

$$- (r - D_1) \frac{\partial \left[y_1' p(\cdot) \right]}{\partial y_1'} - (r - D_2) \frac{\partial \left[y_2' p(\cdot) \right]}{\partial y_2'} \tag{B.84}$$

The initial condition is similar to the one-dimensional problem: all probability is concentrated at one point. The boundary conditions depend on the process under investigation. □

One has to point out, however, that, in general, the forward and backward equations are completely different. As an example, let us take Equation (B.39).

Example 75 (geometric Brownian motion)
The backward equation is simply:

$$-p_t = \frac{\sigma^2}{2} y^2 p_{yy} + (r - D) y p_y \tag{B.85}$$

with $y = Y$ for the spatial variable and subscripts denoting derivatives. The forward equation is

$$p_{t'} = \frac{\sigma^2}{2} \left[(y')^2 p \right]_{y'y'} - (r - D) \left[y' p_{y'} \right] \tag{B.86}$$

$$= \frac{\sigma^2}{2} \left[2p + 4y' p_{y'} + (y')^2 p_{y'y'} \right] - (r - D) \left[p + y' p_{y'} \right] \tag{B.87}$$

Although the PDEs are different, there is a common solution to both problems:

$$p(y, t, y', t') = \frac{1}{\sigma y' \sqrt{2\pi (t' - t)}} \exp \left[\frac{-\left[\ln \left(\frac{y'}{y} \right) - (r - D - \frac{\sigma^2}{2})(t - t') \right]^2}{2\sigma^2 (t' - t)} \right] \quad \text{(B.88)}$$

□

Applications of the Kolmogorov equations in finance are numerous. The backward equation can be used to price contingent claims, not limited to the usual Black–Scholes framework, but with any risk factor given by Equation (B.48). How this works with geometric Brownian motion is shown in the following example.

Example 76 (expectations and Black–Scholes)
The (heuristic) outline of the relationship between the backward Kolmogorov equation and the Black–Scholes pricing PDE (2.30) follows (Wilmott, 2000, Section 10.10). Assuming an asset S following geometric Brownian motion $dS = \mu\, S\, dt + \sigma\, S\, dX$ *and paying no dividends, the backward Kolmogorov equation reads:*

$$-p_t = \frac{\sigma^2}{2} S^2 p_{SS} + \mu S p_S \quad \text{(B.89)}$$

The expected value of a given function $F(S)$ at time T is given by Equation (B.89), together with the FC $p(S, T) = F(S)$. Further assuming that $p(S, T)$ represents the payoff of an option, the present value of this option requires discounting $e^{-r(T-t)} p(S, t)$. We substitute:

$$p(S, t) = e^{r(T-t)} V(S, t) \quad \text{(B.90)}$$

Consequently:

$$\frac{\partial p}{\partial t} = -r\, e^{r(T-t)} V(S, t) + e^{r(T-t)} \frac{\partial V}{\partial t} \quad \text{(B.91)}$$

$$\frac{\partial p}{\partial S} = e^{r(T-t)} \frac{\partial V}{\partial S} \quad \text{(B.92)}$$

$$\frac{\partial^2 p}{\partial S^2} = e^{r(T-t)} \frac{\partial^2 V}{\partial S^2} \quad \text{(B.93)}$$

Inserting the above expressions into Equation (B.89) reads:

$$\frac{\partial V}{\partial t} + \frac{1}{2} \sigma^2 S^2 \frac{\partial^2 V}{\partial S^2} + \mu S \frac{\partial V}{\partial S} - rV = 0 \quad \text{(B.94)}$$

If, in the above equation, μ were equal to r, PDE (B.94) would represent the Black–Scholes equation. Also, if, instead of $dS = \mu\, S\, dt + \sigma\, S\, dX$, we had taken

$$dS = r\, S\, dt + \sigma\, S\, dX \quad \text{(B.95)}$$

we would have arrived at the Black–Scholes equation. Equation (B.95) is called the risk-neutral random walk.   □

A central theorem of modern finance is the following.

Theorem 34 (Wilmott, 2000, p. 148)
The fair value of an option is the present value of the expected payoff at expiry under a risk-neutral random walk for the underlying. Or, in mathematical parlance:

$$V(S, t) = e^{-r(T-t)} \mathbf{E}_Q \{payoff(S) | \mathcal{F}_t\} \quad \text{(B.96)}$$

The subscript Q denotes that the expectations have to be taken in the risk-neutral measure. The conditional $\mathcal{F}_t$ states a further assumption for Equation (B.95), i.e. that a filtration comes with it. Roughly speaking, this means that all information about S comes continuously to the decision maker, who never forgets past data.   □

Further applications of the one-dimensional Kolmogorov equations are the construction of the local volatility function (Andersen and Brotherton-Ratcliffe, 1998; Rebonato, 1999), or the design of the process for stochastic volatility (Oztukel and Wilmott, 1998). In the calibration of term structure models, both Kolmogorov equations are employed, usually implicitly (Rebonato, 1998). The forward equation can be used to determine lower and upper bounds for uncertain parameters which are necessary for worst-case pricing (Topper, 2001b).

B.3.4 Steady-state distributions

When taking the limit $t' \to \infty$, some stochastic processes approach a steady-state distribution:

$$\lim_{t' \to \infty} p(y', t', y, t) = p_\infty(y')$$ (B.97)

Examples for processes possessing a steady-state distribution are the mean-reverting processes

$$dy = \kappa(\bar{y} - y)^n dt + \sigma dX \quad n = 1, 3, \ldots$$ (B.98)

with $\bar{y}$ and κ being the mean-reversion level and adjustment speed, respectively, or the CIR-process

$$dy = \kappa(\bar{y} - y)dt + \sigma\sqrt{y}dX$$ (B.99)

Examples for processes *not* converging to a steady state are Brownian motion and the lognormal random walk. If there is a steady-state distribution p_∞, it can be computed by solving an *ordinary* differential equation, or to be more exact, a two-point boundary value problem:

$$\frac{1}{2}\frac{d^2\left(B_\infty^2 p_\infty\right)}{d(y')^2} - \frac{d\left(A_\infty p_\infty\right)}{dy'} = 0$$ (B.100)

with $\lim_{t' \to \infty} A(\cdot) = A_\infty$ and $\lim_{t' \to \infty} B(\cdot) = B_\infty$. The boundary conditions are the same as those applied to Equation (B.58). Since Equation (B.100) is a linear ODE, it is often possible to find a closed-form solution.

Example 77
Consider Equation (B.98) with $n = 1$, which is the famous Ornstein–Uhlenbeck process. It can be shown (Wilmott et al., 1996) that the solution of Equation (B.100) is then:

$$p_\infty(y') = \sqrt{\frac{\kappa}{\sigma^2 \pi}} \exp\left(\frac{-\kappa}{\sigma^2}\left(y' - \bar{y}\right)^2\right)$$ (B.101)

With the help of an algebraic system it can easily be verified that:

$$\frac{d\left(A_\infty p_\infty\right)}{dy'} = \frac{d}{dy'}\left[\kappa(\bar{y} - y')p_\infty\right]$$ (B.102)

$$= -\kappa p_\infty + \kappa(\bar{y} - y)\frac{dp_\infty}{dy'}$$ (B.103)

$$= \zeta$$ (B.104)

$$\frac{d^2\left(B_\infty^2 p_\infty\right)}{d(y')^2} = \frac{d^2}{d(y')^2}\left[\sigma^2 p_\infty\right] \tag{B.105}$$

$$= \sigma^2 \frac{d^2 p_\infty}{d(y')^2} \tag{B.106}$$

$$= \sigma^2 \frac{2}{\sigma^2}\,\zeta \tag{B.107}$$

with

$$\zeta = \frac{e^{\frac{2\kappa\bar{y}y - \kappa y^2 - \kappa\bar{y}^2}{\sigma^2}}\,\kappa^{3/2}\left(2\kappa y^2 - 4\kappa\bar{y}y' + 2\kappa\bar{y}^2 - \sigma^2\right)}{\sqrt{\pi}\sigma^3} \tag{B.108}$$

so that Equation (B.100) is satisfied. Equation (B.101) is the density of the normal distribution with a mean of $\bar{y}$ and a variance of $\frac{\sigma^2}{2\kappa}$. Taking into account the data $\sigma = 0.02$ and $\kappa = \bar{y} = 0.1$, and again using an algebraic system, it is easy to verify that:

$$\int_{-\infty}^{\infty} p_\infty dy' = 1 \tag{B.109}$$

i.e. stating that the integral over the density function is one.   □

Example 78 (Example 72 continued)
The steady-state distribution is given by:

$$0 = \frac{\sigma^2}{2}\frac{\partial^2[y'p(\cdot)]}{\partial(y')^2} - \frac{\partial[b(a - y')p(\cdot)]}{\partial y'} \tag{B.110}$$

There is an analytical solution to this ODE involving the Γ distribution Equation (B.11) (Lipton, 2002).   □

B.3.5   First exit times

In this section we discuss the tools to answer the following questions:

- What is the expected time of a stochastic process to reach a given state or to leave a given region?
- What is the probability of a stochastic process leaving a given region before some specified point of time t' in the future?

As before, we will only present these models. The reader interested in a derivation of these models is referred to Cox and Miller (1965), Schuss (1980), Wilmott *et al.* (1996) and Wilmott (1998). First exit time is a stochastic quantity, so only its *expected value* can be computed. It is given by an inhomogeneous two-point boundary value problem:

$$\frac{B(\cdot)^2}{2}\frac{d^2u}{dy^2} + A(\cdot)\frac{du}{dy} = -1 \tag{B.111}$$

This equation only holds for coefficients and regions independent of time. The question of whether a steady-state distribution does actually exist is subtle and involves mathematical concepts beyond the scope of this text. The interested reader is encouraged to consult Mandl (1968). Two boundary conditions have to be specified, one for y^{min} and one for y^{max}. When the stochastic process has reached either y^{min} or y^{max}, the expected exit time is zero, since the process has just left the prescribed region. Sometimes, one is only interested in the expected exit time for one side of the region. For example: when holding a down-and-out barrier option, one is only interested in the expected time of the asset hitting this barrier, which is $y^{\min}$. The region is not limited to the right-hand side, so that $y^{\max} = \infty$. The boundary conditions are consequently:

$$u(y^{\min}) = 0 \tag{B.112}$$

$$u(y^{\max}) = 0 \tag{B.113}$$

It is often possible to solve Equation (B.111) analytically.

The multidimensional extension of Equation (B.111) associated with the multidimensional Itô process, as given by Equation (B.74), is called *Dynkin's equation* (Schuss, 1980):

$$\frac{1}{2}\sum_{j=1}^{n}\sum_{k=1}^{n}\sum_{l=1}^{n}B_{jl}(\cdot)B_{lk}(\cdot)\frac{\partial^2 u(\cdot)}{\partial y_j \partial y_k} + \sum_{j=1}^{n}A_j(\cdot)\frac{\partial u(\cdot)}{\partial y_j} = -1 \tag{B.114}$$

This is a linear elliptic PDE. Together with the boundary condition

$$u(y_1, \ldots, y_n) = 0 \tag{B.115}$$

on the entire boundary, it admits a unique and stable solution. Dynkin's equation answers the question: what is the expected exit time for the *first* out of n stochastic processes leaving a prescribed region? A financial application could ask for the expected exit time of the first out of several assets to leave their respective ranges.

Example 79 (2D geometric Brownian motion)
Assume a problem with two risk factors, as given in Example 74. First, we explicitly state Equation (B.114) for $n = 2$:

$$\frac{B_{11}B_{11} + B_{12}B_{21}}{2}\frac{\partial^2 u}{\partial y_1 \partial y_1} + \frac{B_{11}B_{12} + B_{12}B_{22}}{2}\frac{\partial^2 u}{\partial y_1 \partial y_2} +$$
$$\frac{B_{21}B_{11} + B_{22}B_{21}}{2}\frac{\partial^2 u}{\partial y_2 \partial y_1} + \frac{B_{21}B_{12} + B_{22}B_{22}}{2}\frac{\partial^2 u}{\partial y_2 \partial y_2} + \tag{B.116}$$
$$A_1\frac{\partial u}{\partial y_1} + A_2\frac{\partial u}{\partial y_2} = -1$$

Since

$$S_i = y_i \tag{B.117}$$

$$A_i = (r - D_i)S_i \tag{B.118}$$

$$B_{ii} = \sigma_i S_i \tag{B.119}$$

$$B_{ij} = 0 \quad \forall i \neq j \tag{B.120}$$

we get:

$$\frac{1}{2}\left(\sigma_1^2 S_1^2 \frac{\partial^2 u}{\partial S_1^2} + \sigma_2^2 S_2^2 \frac{\partial^2 u}{\partial S_2^2}\right) + (r - D_1)S_1\frac{\partial u}{\partial S_1} + (r - D_2)S_2\frac{\partial u}{\partial S_2} = -1 \tag{B.121}$$

Usually, in option pricing applications, the risk factors are correlated. Assume a correlation coefficient $\rho = \rho_{12} = \rho_{21}$ *between the two risk factors. With the following ansatz:*

$$X_1 = (1 + c^2)^{-1/2}(W_1 + cW_2) \tag{B.122}$$
$$X_2 = (1 + c^2)^{-1/2}(cW_1 + W_2) \tag{B.123}$$

the W_i *being independent Brownian motions. The correlation is given by:*

$$Corr(X_1, X_2) = \frac{2c}{1 + c^2} = \rho \tag{B.124}$$

so that

$$c = \frac{1}{\rho} \pm \left(\frac{1}{\rho^2} - 1\right)^{\frac{1}{2}}, \quad -1 \le \rho \le 1 \tag{B.125}$$

This leads to the following systems of SDEs:

$$dS_1 = (r - D_1)S_1 dt + \sigma_1/\sqrt{1 + c^2}S_1(dW_1 + c\,dW_2) \tag{B.126}$$
$$dS_2 = (r - D_2)S_2 dt + \sigma_2/\sqrt{1 + c^2}S_2(c\,dW_1 + dW_2) \tag{B.127}$$

so that the coefficients of Dynkin's equation are given by:

$$A_1(S, t) = (r - D_1)S_1 \tag{B.128}$$
$$A_2(S, t) = (r - D_2)S_2 \tag{B.129}$$
$$B_{11}(S, t) = (\sigma_1/\sqrt{1 + c^2})S_1 \tag{B.130}$$
$$B_{12}(S, t) = (c\sigma_1\sqrt{1 + c^2})S_1 \tag{B.131}$$
$$B_{21}(S, t) = (c\sigma_2/\sqrt{1 + c^2})S_2 \tag{B.132}$$
$$B_{22}(S, t) = (\sigma_2\sqrt{1 + c^2})S_2 \tag{B.133}$$

$\square$

Example 80 (purely Brownian motion)

In the case of an n-dimensional Brownian motion, Equation (B.114) reduces to:

$$\frac{1}{2}\left(\sum_{i=1}^{n} u_{y_i y_i}\right) = -1 \tag{B.134}$$

In the case where the domain of this differential equation is an n-dimensional sphere of radius R with center point $(c_1, \ldots, c_n)$, *an analytical solution exists:*

$$u(y_1, \ldots, y_n) = \frac{R^2 - \left[(y_1 - c_1)^2 + \ldots + (y_n - c_n)^2\right]}{n} \tag{B.135}$$

Now we address the probability of an Itô process leaving a prescribed region before t'. *This question can be answered by solving*

$$\frac{\partial Q}{\partial t} = -\frac{A(y, t)}{2}\frac{\partial^2 Q}{\partial y^2} - B(y, t)^2\frac{\partial Q}{\partial y} \tag{B.136}$$

This is the same as the backward Kolmogorov equation; however, the boundary conditions are different. The boundary conditions are

$$Q(y^{\min}) = 1 \tag{B.137}$$
$$Q(y^{\max}) = 1 \tag{B.138}$$

meaning that when the process has already reached the boundary, the associated probability is one. At $t' = t$ the process cannot leave the prescribed region anymore, therefore:

$$Q(t') = 0 \qquad (B.139)$$

Although boundary and final conditions are incompatible at (y^{min}, t') and (y^{max}, t'), the problem is well-posed. □

B.3.6 Itô's lemma

Suppose the value of a variable Y follows an Itô diffusion defined as

$$dY = A(Y, t)\, dt + B(Y, t)\, dX \qquad (B.140)$$

Itô's lemma states that a function of Y, say $F(Y)$, is an Itô process as well. The variable $F(Y)$ follows the process

$$dF = \left(A\frac{\partial F}{\partial Y} + \frac{\partial F}{\partial t} + \frac{B^2}{2}\frac{\partial^2 F}{\partial Y^2} \right) dt + B\frac{\partial F}{\partial Y}\, dX \qquad (B.141)$$

$$= \left(\frac{\partial F}{\partial t} + \frac{B^2}{2}\frac{\partial^2 F}{\partial Y^2} \right) dt + \frac{\partial F}{\partial Y}\, dY \qquad (B.142)$$

Thus, Itô's lemma allows us to construct a differential from a function which depends on a stochastic variable. It can be interpreted as a stochastic extension of the Taylor series approximation to a deterministic $\lim_{\Delta \to 0} \Delta F = dF$. Most derivatives textbooks give a derivation of this important result, so we will not repeat it here.

Example 81
A standard Brownian motion is given by X. What is dF of $F(X, t) = X^2$? Since $A = 0, B = 1$, and $dX = dY$:

$$dF = \left(\frac{\partial F}{\partial t} + \frac{1}{2}\frac{\partial^2 F}{\partial Y^2} \right) dt + \frac{\partial F}{\partial Y}\, dX \qquad (B.143)$$

$$= \left(0 + \frac{1}{2}\, 2 \right) dt + 2X\, dX \qquad (B.144)$$

$$= dt + 2X\, dX \qquad (B.145)$$

□

Example 82
The relationship between a stock paying no dividends and a forward is given by:

$$F = S \exp[r(T - t)] \qquad (B.146)$$

Assuming that the stock price dynamics are represented by:

$$dS = \mu S\, dt + \sigma S\, dX \qquad (B.147)$$

what are the dynamics of F? We need several derivatives:

$$\frac{\partial F}{\partial S} = \exp[r(T - t)] \qquad (B.148)$$

$$\frac{\partial^2 F}{\partial S^2} = 0 \qquad (B.149)$$

$$\frac{\partial F}{\partial t} = -r S \exp[r(T - t)] \qquad (B.150)$$

The Itô process for F is given by:

$$dF = \left(e^{r(T-t)}\mu S - rSe^{r(T-t)}\right)dt + e^{r(T-t)}\sigma S\,dz \qquad (B.151)$$

$$= (\mu - r)F\,dt + \sigma F\,dz \qquad (B.152)$$

□

Itô's lemma can be extended in a straightforward manner to functions depending on n stochastic variables which are all governed by Itô diffusions:

$$dY_1 = A_1(Y_1, \ldots, Y_n, t)\,dt + B_1(Y_1, \ldots, Y_n, t)\,dX_1$$

$$\vdots \qquad (B.153)$$

$$dY_n = A_n(Y_1, \ldots, Y_n, t)\,dt + B_n(Y_1, \ldots, Y_n, t)\,dX_n$$

with a correlation matrix Σ as defined in Equation (B.80). Such a function F can be represented by the following Itô diffusion:

$$dF = \left(\sum_{i=1}^{n} A_i \frac{\partial F}{\partial Y_i} + \frac{\partial F}{\partial t} + \frac{1}{2}\sum_{i,j=1}^{n} \rho_{ij} B_i B_j \frac{\partial^2 F}{\partial Y_i \partial Y_j}\right)dt + \sum_{i=1}^{n} B_i \frac{\partial F}{\partial Y_i}dX_i \qquad (B.154)$$

Example 83
The single asset working horse for option pricing is

$$dS = \mu S\,dt + \sigma S\,dX \qquad (B.155)$$

which is easily extended to the multi-asset setting:

$$dS_i = \mu_i S_i\,dt + \sigma_i S_i\,dX_i \quad with \quad \Sigma \qquad (B.156)$$

For a function $V = V(S_1, \ldots, S_n)$ it holds that:

$$dV = \left(\sum_{i=1}^{n} \mu_i S_i \frac{\partial V}{\partial S_i} + \frac{\partial V}{\partial t} + \frac{1}{2}\sum_{i,j=1}^{n} \rho_{ij}\sigma_i\sigma_j S_i S_j \frac{\partial^2 V}{\partial S_i \partial S_j}\right)dt$$

$$+ \sum_{i=1}^{n} \sigma_i S_i \frac{\partial V}{\partial S_i}dX_i \qquad (B.157)$$

$$= \left(\frac{\partial V}{\partial t} + \frac{1}{2}\sum_{i,j=1}^{n} \rho_{ij}\sigma_i\sigma_j S_i S_j \frac{\partial^2 V}{\partial S_i \partial S_j}\right)dt + \sum_{i=1}^{n} \mu_i S_i \frac{\partial V}{\partial S_i}dt$$

$$+ \sigma_i S_i \frac{\partial V}{\partial S_i}dX_i \qquad (B.158)$$

$$= \left(\frac{\partial V}{\partial t} + \frac{1}{2}\sum_{i,j=1}^{n} \sigma_i\sigma_j \rho_{ij} S_i S_j \frac{\partial^2 V}{\partial S_i \partial S_j}\right)dt + \sum_{i=1}^{n} \frac{\partial V}{\partial S_i}\,dS_i \qquad (B.159)$$

□

B.4   NOTES

The lognormal distribution and its associated stochastic process, geometric Brownian motion, is the working horse of today's derivative pricing, on which both the Black–Scholes and the Heath–Jarrow–Morton paradigms are built. The reader is encouraged to go thoroughly through the careful explanations of this distribution's implications for pricing in Bellalah *et al.* (1998), Jarrow and Turnbull (1996) and/or Kwok (1998). A more complete list of stochastic processes used in finance and their properties can be found in Sawyer (1993).

The classical references for all aspects of univariate distributions are Johnson *et al.* (1994) and Johnson *et al.* (1995). Especially useful for financial applications is the discussion of how to approximate cumulative functions for distributions that are not commonly used.

Appendix C
Some Useful Results from
Linear Algebra

C.1 SOME BASIC FACTS

Definition 12 (Singularity)
If a square matrix **A** *possesses no inverse matrix* $\mathbf{A}^{-1}$, *it is called a singular matrix; if a square matrix* **A** *has an inverse* $\mathbf{A}^{-1}$, *it is said to be nonsingular. The determinant of a singular matrix vanishes, i.e. only nonsingular matrices can be inverted.*

Any linear system $\mathbf{Ax} = \mathbf{b}$ *associated with a regular square matrix* **A** *possesses a unique solution.* □

Definition 13 (Diagonal dominance (Hadamard))
An $n \times n$ *matrix* $\mathbf{A} = [a_{ij}]$ *is called diagonal-dominant if*

$$|a_{ii}| > \sum_{j \neq i} |a_{ij}| \ for \ 1 \leq i \leq N \tag{C.1}$$

i.e. the absolute value of the diagonal element a_{ii} *has to be larger than the sum of the absolute values of all other elements in column* i. *Accordingly, an* $n \times n$ *matrix* $\mathbf{A} = [a_{ij}]$ *is called weakly diagonal-dominant if*

$$|a_{ii}| \geq \sum_{j \neq i} |a_{ij}| \ for \ 1 \leq i \leq N \tag{C.2}$$

□

Definition 14 (Sparseness)
A matrix having mostly zero entries is called sparse. □

There is no clear-cut definition saying what the percentage of zeros has to be to make a matrix sparse; but usually it is assumed that a sparse matrix has at least 90% zeros. Many typical applications in FE analysis have stiffness matrices with more than 99% of all entries being zero.

The advantage of sparse matrices is that most modern algorithms make use of the fact that zero entries can be ignored in the process of adding numbers. Besides this, zeros need not be stored.

Definition 15 (Bandedness)
An $n \times n$ *matrix is called a band matrix if integers* p *and* q, *with* $1 < p, q < n$, *exist with the property that* $a_{ij} = 0$ *whenever* $i + p \leq j$ *or* $j + q \leq i$. *The bandwidth is* $w = p + q - 1$ *(Burden and Faires, 1998, p. 279).* □

Example 84 (Burden and Faires, 1998, p. 279)

$$\mathbf{A} = \begin{pmatrix} 7 & 2 & 0 \\ 3 & 5 & -1 \\ 0 & -5 & -6 \end{pmatrix} \tag{C.3}$$

is a band matrix with $p = q = 2$ *and bandwidth* $w = 3$. □

Definition 16 (Eigenvalues)
Every real, symmetric n × n matrix has n real eigenvalues. These eigenvalues are the (possibly repeated) zeros of an nth degree polynomial in λ, $\det([\mathbf{A} - \lambda\mathbf{I}])$. ☐

Example 85
What are the eigenvalues of the matrix $\mathbf{A} = \begin{pmatrix} 3 & 2 \\ 2 & 3 \end{pmatrix}$?

$$\det(\mathbf{A} - \lambda\mathbf{I}) = \begin{vmatrix} 3 - \lambda & 2 \\ 2 & 3 - \lambda \end{vmatrix} = (3 - \lambda)^2 - 4 \overset{!}{=} 0 \tag{C.4}$$

This quadratic equation is solved by $\lambda_1 = 1$ *and* $\lambda_2 = 5$. *These are the only eigenvalues of* $\mathbf{A}$. ☐

Definition 17
A quadratic form q in n variables is given by:

$$q(u_1, \ldots, u_n) = \sum_{i=1}^{n} \sum_{j=1}^{n} d_{ij} u_i u_j \quad (\text{with } d_{ij} = d_{ji}) \tag{C.5}$$

$$= \mathbf{u}^{\mathrm{T}} \mathbf{D} \mathbf{u} \tag{C.6}$$

with $\mathbf{D}$ *being a symmetric n × n matrix, and* $\mathbf{u}$ *denoting a column vector with n entries. Accordingly,* $\mathbf{u}^{\mathrm{T}}$ *is a row vector with n entries.* ☐

Definition 18 (Definiteness)
A quadratic form q is said to be:

- *positive-definite if q is invariably positive* ($q > 0$);
- *positive-semidefinite if q is invariably nonnegative* ($q \geq 0$);
- *negative-semidefinite if q is invariably nonpositive* ($q \leq 0$);
- *negative-definite if q is invariably negative* ($q < 0$). ☐

There are various tests on definiteness and semidefiniteness (Takayama, 1985). These are usually based on the structure of D, since it is difficult to apply Definition 16 to the quadratic form directly. Here, we will state a theorem that is simple to apply.

Theorem 35
A necessary and sufficient condition for the positive definiteness of a quadratic form, as defined in Definition 15, is that all principal minors are positive:

$$|\mathbf{D}_1| \equiv d_{11} \tag{C.7}$$

$$|\mathbf{D}_2| \equiv \begin{vmatrix} d_{11} & d_{12} \\ d_{21} & d_{22} \end{vmatrix} \tag{C.8}$$

$$\vdots$$

$$|\mathbf{D}_n| \equiv \begin{vmatrix} d_{11} & d_{12} & \ldots & d_{1n} \\ d_{21} & d_{22} & \ldots & d_{2n} \\ \ldots & \ldots & \ldots & \ldots \\ d_{n1} & d_{n2} & \ldots & d_{nn} \end{vmatrix} \tag{C.9}$$

The corresponding condition for negative definiteness is that the principal minors alternate in sign as follows:

$$|\mathbf{D}_1| < 0 \tag{C.10}$$

$$|\mathbf{D}_2| > 0 \tag{C.11}$$

$$|\mathbf{D}_3| < 0 \tag{C.12}$$

$$\vdots$$

□

The corresponding conditions for semidefiniteness *cannot* be generated by exchanging '<' with '≤' and '>' with '≥', but are more complicated; see Takayama (1985).

C.2 ERRORS AND NORMS

Before the notion of an ill-conditioned matrix is introduced, measures for errors of scalars, vectors and matrices need to be defined.

Let us start with scalars. Assume x to be the true solution of a given problem. The magnitude or size of x is given by its absolute value $|x|$. Further, assume an approximation for x given by $\bar{x}$. Then we can define the following error measures:

1. Absolute error: $|\bar{x} - x|$.
2. Relative error: $\frac{|\bar{x}-x|}{|x|}$.
3. Percentage error: $\frac{\bar{x}-x}{x}[\%] = \left[\frac{\bar{x}}{x} - 1\right] \cdot 100\,\%$.

If the relative error of $\bar{x}$ is of order 10^{-n}, the approximate solution is accurate to n decimal places.

This definition cannot be generalized directly to vectors, since only scalars can be compared on an ordinal basis. Consequently, some kind of norm is needed which associates a vector with a scalar. The magnitude of a vector is given by its vector norm, denoted by $||\mathbf{x}||_p$

$$||\mathbf{x}||_p = \left(\sum_{i=1}^{n} |x_i|^p \right)^{\frac{1}{p}} \tag{C.13}$$

where p is a positive integer. Usually, p is set to be either 1, 2 or ∞:

- $p = 1$: $||\mathbf{x}||_1 = \sum_{i=1}^{n} |x_i|$. The 1-norm (or absolute norm) is the sum of absolute values of the vector.
- $p = 2$: $||\mathbf{x}||_2 = \left(\sum_{i=1}^{n} |x_i|^2\right)^{0.5}$. The 2-norm is the Euclidian distance of a given point x from the origin. It is also called the Euclidian norm.
- $p = \infty$: $||\mathbf{x}||_\infty = \max_{1 \le i \le n} |x_i|$. The ∞-norm (or maximum norm) is the largest component in absolute value among the elements of the vector.

Example 86

Consider the vector $\mathbf{x} = (-0.8, 0.6)$. *The norms are:*

$$||\mathbf{x}||_1 = |-0.8| + |0.6| = 1.4 \tag{C.14}$$

$$||\mathbf{x}||_2 = \sqrt{(-0.8)^2 + (0.6)^2} = \sqrt{0.64 + 0.36} = 1 \tag{C.15}$$

$$||\mathbf{x}||_\infty = \max[|-0.8|, |0.6|] = 0.8 \tag{C.16}$$

□

For any vector, the following holds:

$$||\mathbf{x}||_\infty \le ||\mathbf{x}||_2 \le ||\mathbf{x}||_1 \tag{C.17}$$

As with scalars, $||\mathbf{x} - \bar{\mathbf{x}}||_p$ refers to the absolute error and $||\mathbf{x} - \bar{\mathbf{x}}||_p / ||\mathbf{x}||_p$ refers to the relative error. As above, if the relative error is of order 10^{-n}, then $\bar{\mathbf{x}}$ is accurate to n decimal digits.

Example 87 (Lapack)

$$\mathbf{x} = \begin{pmatrix} 1 \\ 100 \\ 9 \end{pmatrix}, \quad \bar{\mathbf{x}} = \begin{pmatrix} 1.1 \\ 99 \\ 11 \end{pmatrix} \tag{C.18}$$

$$||\bar{\mathbf{x}} - \mathbf{x}||_1 = 3.1, \quad \frac{||\bar{\mathbf{x}} - \mathbf{x}||_1}{||\mathbf{x}||_1} = 0.0282$$

$$||\bar{\mathbf{x}} - \mathbf{x}||_2 = 2.238, \quad \frac{||\bar{\mathbf{x}} - \mathbf{x}||_2}{||\mathbf{x}||_2} = 0.0223 \tag{C.19}$$

$$||\bar{\mathbf{x}} - \mathbf{x}||_\infty = 2, \quad \frac{||\bar{\mathbf{x}} - \mathbf{x}||_\infty}{||\mathbf{x}||_\infty} = 0.02$$

It follows that $\bar{\mathbf{x}}$ *approximates* $\mathbf{x}$ *to two decimal digits.*

□

Errors in matrices are also measured with norms. The p-norm of a matrix $\mathbf{A}$ is defined as:

$$||\mathbf{A}||_p = \max_{\mathbf{x} \neq \mathbf{0}} \frac{||\mathbf{A}\mathbf{x}||_p}{||\mathbf{x}||_p} \tag{C.20}$$

Commonly used matrix norms are:

$$||\mathbf{A}||_1 = \max_j \sum_{i=1}^n |a_{ij}| \tag{C.21}$$

$$||\mathbf{A}||_\infty = \max_i \sum_{j=1}^n |a_{ij}| \tag{C.22}$$

Example 88

Consider

$$\mathbf{A} = \begin{pmatrix} 2 & -1 & 1 \\ 1 & 0 & 1 \\ 3 & -1 & 4 \end{pmatrix} \tag{C.23}$$

Then:

$$||\mathbf{A}||_1 = \max[6, 2, 6] = 6 \tag{C.24}$$

$$||\mathbf{A}||_\infty = \max[4, 2, 8] = 8 \tag{C.25}$$

□

The matrix 1-norm $||\mathbf{A}||_1$ is the maximum absolute column sum, and the ∞-norm $||\mathbf{A}||_\infty$ is the maximum absolute row sum. Both can be evaluated easily for any matrix $\mathbf{A}$. As before,

$||\bar{\mathbf{A}} - \mathbf{A}||_p$ is the absolute error, $||\bar{\mathbf{A}} - \mathbf{A}||_p/||\mathbf{A}||_p$ is the relative error, and a relative error of order 10^{-n} means that $\bar{\mathbf{A}}$ is accurate to n decimal places.

The *condition number* of a matrix is defined as:

$$\text{cond}_p(\mathbf{A}) \equiv ||\mathbf{A}||_p \cdot ||\mathbf{A}^{-1}||_p \tag{C.26}$$

where $\mathbf{A}$ is square and invertible and p is usually set to either 1 or ∞.

Example 89 (Example 88 continued)
The inverse of a matrix $\mathbf{A}$ *is given by:*

$$\mathbf{A}^{-1} = \begin{pmatrix} 0.5 & 1.5 & -0.5 \\ -0.5 & 2.5 & -0.5 \\ -0.5 & -0.5 & 0.5 \end{pmatrix} \tag{C.27}$$

such that:

$$||\mathbf{A}^{-1}||_1 = \max[1.5, 4.5, 1.5] = 4.5 \tag{C.28}$$

$$||\mathbf{A}^{-1}||_\infty = \max[2.5, 3.5, 1.5] = 3.5 \tag{C.29}$$

$$\text{cond}_1(\mathbf{A}) = 6 \cdot 4.5 = 27 \tag{C.30}$$

$$\text{cond}_\infty(\mathbf{A}) = 8 \cdot 2.5 = 20 \tag{C.31}$$

$\square$

C.3   ILL-CONDITIONING

The linear system $\mathbf{Ax} = \mathbf{b}$ has a unique solution only if $\mathbf{A}$ is square and invertible. If $\bar{\mathbf{x}}$ is an approximation to the solution of this linear system, then for any norm as defined in Section C.2, the following relation holds:

$$\frac{||\mathbf{x} - \bar{\mathbf{x}}||_p}{||\mathbf{x}||_p} \leq \underbrace{||\mathbf{A}||_p \cdot ||\mathbf{A}^{-1}||_p}_{=\text{cond}_p(\mathbf{A})} \underbrace{\frac{||\mathbf{b} - \mathbf{A}\bar{\mathbf{x}}||_p}{||\mathbf{b}||_p}}_{\text{relative residual}} \tag{C.32}$$

provided $\mathbf{b} \neq \mathbf{0}$ and $\mathbf{x} \neq \mathbf{0}$. Consequently, the condition number provides an indication of the connection between the residual vector and the accuracy of the approximation. The error is bounded by the product of the condition number and the relative residual for this approximation. The condition number is usually used as an indicator for the reliability of the solution of a linear system. Its lower bound is one because of the following relationship:

$$1 = ||\mathbf{E}||_p = ||\mathbf{A} \cdot \mathbf{A}^{-1}||_p \leq ||\mathbf{A}||_p \cdot ||\mathbf{A}^{-1}||_p = \text{cond}_p(\mathbf{A}) \tag{C.33}$$

There is no upper bound for the condition number. The matrix $\mathbf{A}$ is called *well-conditioned* if the condition number is close to one. Accordingly, it is called *ill-conditioned* if the condition number is significantly greater than one.

C.4   SOLVING LINEAR ALGEBRAIC SYSTEMS

Gaussian elimination

As should have become clear, an important auxiliary task in working with FE and PDEs is the solution of algebraic systems of the form

$$\mathbf{A} \cdot \mathbf{x} = \mathbf{b} \tag{C.34}$$

By the previous discussion we can expect the matrix $\mathbf{A}$ to be real and nonsingular, in some cases even symmetric. The first two properties make it possible to bring the matrix into an upper triangular form, where the elements below the diagonal are zero. This renders the system of equations to be of the following form:

$$a_{11}x_1 + a_{12}x_2 + \ldots + a_{1n}x_n = b_1$$
$$a_{22}x_2 + \ldots + a_{2n}x_n = b_2$$
$$\vdots \qquad \vdots \qquad\qquad\qquad \text{(C.35)}$$
$$a_{nn}x_n = b_n$$

In general, the coefficients a_{ij} in the above system are different from the ones from the original matrix $\mathbf{A}$. If $a_{ii} \neq 0$ for all $i = 1, \ldots, n$, the system has the unique solution

$$x_n = b_n/a_{nn}$$
$$x_{n-1} = (b_{n-1} - a_{n-1,n}x_n)/a_{n-1,n-1}$$
$$\vdots \qquad \vdots \qquad\qquad\qquad \text{(C.36)}$$
$$x_1 = (b_1 - a_{1n}x_n - \ldots - a_{13}x_3 - a_{12}x_2)/a_{11}$$

This is called back-substitution. However, given a generic matrix, we have to perform a little work to arrive at the upper triangular form of Equation (C.35).

The standard method to bring a nonsingular matrix into an upper triangular form is so-called Gaussian elimination. This method relies on the fact that the set of solutions does not change if we add a multiple of one equation (i.e. one line of (C.34)) to another one. Also, the set of solutions is invariant under interchange of the order of equations. We can use these properties to arrive at a form of Equation (C.34) where all the elements below the diagonal are zero (*eliminated*).

First, we label the above generic form $\mathbf{A}^{(0)}\mathbf{x} = \mathbf{b}^{(0)}$ and start by subtracting $a_{i1}^{(0)}/a_{11}^{(0)}$ times the first equation from equations $i = 2, \ldots, n$. The resulting system is labeled $\mathbf{A}^{(1)}\mathbf{x} = \mathbf{b}^{(1)}$. It has the same set of solutions as $\mathbf{A}^{(0)}\mathbf{x} = \mathbf{b}^{(0)}$, but has zero entries below the diagonal in the first column:

$$\mathbf{A}^{(1)} = \begin{pmatrix} a_{11}^{(1)} & a_{12}^{(1)} & \cdots & a_{1n}^{(1)} \\ 0 & a_{22}^{(1)} & \cdots & a_{2n}^{(1)} \\ \vdots & \vdots & \ddots & \vdots \\ 0 & a_{n2}^{(1)} & \cdots & a_{nn}^{(1)} \end{pmatrix}; \quad \mathbf{b}^{(1)} = \begin{pmatrix} b_1^{(1)} \\ b_2^{(1)} \\ \vdots \\ b_n^{(1)} \end{pmatrix} \qquad \text{(C.37)}$$

We repeat this procedure by subtracting $a_{i2}^{(1)}/a_{22}^{(1)}$ times the second equation from the ith equation, where $i = 3, \ldots, n$ now, and so on until we arrive at the form Equation (C.35) after $n - 1$ steps, which in our notation would correspond to $\mathbf{A}^{(n-1)}\mathbf{x} = \mathbf{b}^{(n-1)}$. If, however, $a_{ii}^{(i-1)} = 0$ in the ith step, this method breaks down. E.g. the system

$$\begin{pmatrix} 0 & 1 \\ 1 & 0 \end{pmatrix}\begin{pmatrix} x_1 \\ x_2 \end{pmatrix} = \begin{pmatrix} 1 \\ 1 \end{pmatrix}$$

obviously has the solution $x_1 = x_2 = 1$, but our unmodified method would not be able to find it. We will deal with that problem in the following section.

Pivoting

An effective and simple solution is to either switch rows or columns or both, since the solution set does not change under these operations. Assuming we have performed i steps as above, our matrix looks like this:

$$
\mathbf{A}^{(i)} =
\begin{pmatrix}
a_{11}^{(i)} & a_{12}^{(i)} & \cdots & a_{1i}^{(i)} & a_{1,i+1}^{(i)} & \cdots & a_{1n}^{(i)} \\
0 & a_{22}^{(i)} & \cdots & a_{2i}^{(i)} & a_{2,i+1}^{(i)} & \cdots & a_{2n}^{(i)} \\
\vdots & \vdots & \ddots & \vdots & \vdots & & \vdots \\
0 & 0 & & a_{ii}^{(i)} & a_{i,i+1}^{(i)} & \cdots & a_{in}^{(i)} \\
0 & 0 & \cdots & 0 & a_{i+1,i+1}^{(i)} & \cdots & a_{i+1,n}^{(i)} \\
\vdots & \vdots & & \vdots & \vdots & \ddots & \vdots \\
0 & 0 & \cdots & 0 & a_{n,i+1}^{(i)} & \cdots & a_{nn}^{(i)}
\end{pmatrix}
\tag{C.38}
$$

Now, let $a_{i+1,i+1}^{(i)} = 0$ and $a_{kk}^{(i)} \neq 0$ for all $k = 1, 2, \ldots, i$. Since we do not want to destroy the partial triangular form of $\mathbf{A}^{(i)}$, we restrict our range of rows and columns on which to operate to the part starting with index $i + 1$. Since we required the original matrix to be nonsingular (i.e. there will be at least n nonzero entries) we will have a choice of entries for the new diagonal element. As a guide, we should choose the element most favorable for our computations. Let us have a look at the basic operation to be performed:

$$
a_{jk}^{(i+1)} = a_{jk}^{(i)} - \frac{a_{j,i+1}^{(i)}}{a_{i+1,i+1}^{(i)}} \cdot a_{i+1,k}^{(i)}, \quad i+2 \leq j, k \leq n
\tag{C.39}
$$

Choosing an element close to zero would result, compared to $a_{jk}^{(i)}$, in a large fraction, and therefore computing the difference would suffer from a decrease in accuracy. If, however, we choose the largest possible element, the fraction multiplying $a_{i+1,k}^{(i)}$ would stay below one and the magnitude of the two terms in the difference Equation (C.39) would stay approximately the same. Therefore, it can also be advantageous to replace the diagonal element $a_{i+1,i+1}^{(i)}$ with another, even if it is not zero. It is clear that we can follow three strategies to choose a new diagonal element. The three strategies applied to $\mathbf{A}^{(i)}$ are

1. *Column search.* Among the elements $a_{j,i+1}^{(i)}$, $j = i+1, i+2, \ldots, n$, choose the element with the highest absolute value as the new diagonal element. If it is in row j_0, change row j_0 with row $i + 1$.
2. *Row search.* Among the elements $a_{i+1,k}^{(i)}$, $k = i+1, i+2, \ldots, n$, choose the element with the highest absolute value as the new diagonal element. If it is in column k_0, change column k_0 with column $i + 1$.
3. *Global search.* Among the elements $a_{jk}^{(i)}$, $j, k \geq i+1$, choose the element with the highest absolute value as the new diagonal element. If it is in row j_0 and column k_0, change row j_0 with row $i + 1$ and column k_0 with column $i + 1$.

The current element used for elimination is then called the *pivot element*.

When dealing with large matrices, a simplification of the above strategies is useful: search as above, but abort if the absolute value of the element exceeds a predefined threshold s. Furthermore, since actually interchanging rows and columns of a matrix can become costly and is cumbersome to do, an effective modification is to construct a vector below and next to the matrix with its entries keeping track of the various permutations of columns and rows. Finally, by using the aforementioned vector, for the first two strategies, $(n-1)+(n-2)+\ldots+1 = n(n+1)/2$ search operations are needed, for the third we need $n^2+(n-1)^2+\ldots+1 = n^3/3+n^2/2+n/6$ steps.

Iterative techniques to solve systems of linear equations

In this section we will be concerned with iterative procedures to solve linear equation systems of the form:

$$\mathbf{A}\mathbf{x} = \mathbf{b} \tag{C.40}$$

In the following we will study three such methods.

Jacobi iterative method

The underlying concept in all three cases is to decompose the matrix $\mathbf{A}$ into its diagonal part $\mathbf{D}$, its upper-triangular part $-\mathbf{U}$ and its lower-triangular part $-\mathbf{L}$, rendering Equation (C.40) into the form

$$(\mathbf{D} - \mathbf{U} - \mathbf{L})\mathbf{x} = \mathbf{b} \tag{C.41}$$

This can readily be transformed into:

$$\mathbf{x} = \mathbf{D}^{-1}(\mathbf{U}+\mathbf{L})\mathbf{x} + \mathbf{D}^{-1}\mathbf{b} \tag{C.42}$$
$$= \mathbf{T}_j\mathbf{x} + \mathbf{c}_j$$

Here $\mathbf{T}_j$ and $\mathbf{c}_j$ have been introduced for later use. Equation (C.42) can be solved iteratively. Explicitly, one finds in the kth iteration for the components $x_i^{(k)}$:

$$x_i^{(k)} = \frac{-\sum\limits_{j=1, j\neq i}^{n} a_{ij}\, x_j^{(k-1)} + b_i}{a_{ii}} \tag{C.43}$$

This is the *Jacobi iterative method*.

Gauss–Seidel iterative method

An obvious improvement to the Jacobi method can be generated by using the already known values of $x_i^{(k)}$ in the kth iteration. This is realized in the *Gauss–Seidel iterative method*, where $x_i^{(k)}$ is given by:

$$x_i^{(k)} = \frac{-\sum\limits_{j=1}^{i-1} a_{ij}\, x_j^{(k)} - \sum\limits_{j=i+1}^{n} a_{ij}\, x_j^{(k-1)} + b_i}{a_{ii}} \tag{C.44}$$

Here, for $j < i$ the (known) kth iteration is used, and for $j > i$, the $(k - 1)$th iteration. In matrix notation this corresponds to:

$$\mathbf{x}^{(k)} = (\mathbf{D} - \mathbf{L})^{-1} \mathbf{U} \mathbf{x}^{(k-1)} + (\mathbf{D} - \mathbf{L})^{-1} \mathbf{b} \tag{C.45}$$
$$= \mathbf{T}_g \mathbf{x}^{(k-1)} + \mathbf{c}_g$$

where $\mathbf{T}_g$ and $\mathbf{c}_g$ are introduced as for the Jacobi method. Let us now consider the convergence properties of both methods. To this end we write out Equation (C.42):

$$\mathbf{x}^{(k)} = \mathbf{T}_j \mathbf{x}^{(k-1)} + \mathbf{c}_j$$
$$= \mathbf{T}_j \left(\mathbf{T}_j \mathbf{x}^{(k-1)} + \mathbf{c}_j \right) + \mathbf{c}_j \tag{C.46}$$
$$\vdots$$
$$= \mathbf{T}_j^k \mathbf{x}^{(0)} + \sum_{l=0}^{k-1} \mathbf{T}_j^l \mathbf{c}_j$$

Since we are interested in the limit $k \to \infty$, convergence of this expression implies that

$$\lim_{k \to \infty} \mathbf{T}_j^k \mathbf{x}^{(0)} = \mathbf{0} \tag{C.47}$$

which means that the eigenvalues of $\mathbf{T}_j$ (and of $\mathbf{T}_g$) have to be smaller than one. In fact, it can be shown that both the Jacobi and the Gauss–Seidel iterative methods converge for any start vector $\mathbf{x}^{(0)}$ as long as the spectral radius $\rho(\mathbf{T}) < 1$. The spectral radius ρ of a matrix $\mathbf{T}$ is given by the largest absolute eigenvalue of $\mathbf{T}$. For practical purposes, this implies that both iterative methods work best for linear systems where the coefficient matrix $\mathbf{A}$ is a diagonally dominated matrix (see Definition 11), since then the entries in matrix $\mathbf{T}$ tend to be sufficiently small to guarantee $\rho(\mathbf{T}) < 1$. Unfortunately, in general, it is not clear whether the Jacobi or the Gauss–Seidel method leads to faster convergence. For the special case of a linear system characterized by a matrix $\mathbf{A}$ with the properties

$$a_{ij} < 0 \quad , \quad i \neq j \quad \text{and} \quad a_{ii} > 0 \tag{C.48}$$

it can be demonstrated that, when one method converges, the other method converges as well, and the Gauss–Seidel converges faster. Similarly, if one method does not converge, the other does not converge either, and the divergence is more pronounced for the Gauss–Seidel method.

Successive over-relaxation method (SOR)

Next, we study how the above methods can be further improved. To that end, let $\bar{\mathbf{x}}$ be an approximate solution of the system $\mathbf{A}\mathbf{x} = \mathbf{b}$. Then one can introduce the residual vector $\mathbf{r}$ defined as

$$\mathbf{r} = \mathbf{b} - \mathbf{A}\bar{\mathbf{x}} \tag{C.49}$$

More precisely, $\mathbf{r}$ is given by (using the Gauss–Seidel approximation scheme of Equation (C.44))

$$\mathbf{r}_i^{(k)} = \left(r_{1i}^{(k)}, r_{2i}^{(k)}, \ldots, r_{ni}^{(k)} \right) \qquad \text{with} \tag{C.50}$$

$$r_{mi}^{(k)} = b_m - \sum_{j=1}^{i-1} a_{mj} x_j^{(k)} - \sum_{j=i+1}^{n} a_{mj} x_j^{(k-1)} - a_{mi} x_i^{(k-1)}$$

By considering the diagonal elements $r_{ii}^{(k)}$ and using the result of Equation (C.44), one can characterize the Gauss–Seidel method in the following way:

$$x_i^{(k)} = x_i^{(k-1)} + \frac{r_{ii}}{a_{ii}} \tag{C.51}$$

As it turns out, a generalization of this result can help to reduce the norm of the vector $\mathbf{r}_i^{(k)}$:

$$x_i^{(k)} = x_i^{(k-1)} + \omega \frac{r_{ii}}{a_{ii}} \tag{C.52}$$

Such generalizations are called relaxation methods. For $0 < \omega < 1$, the procedures are called *under-relaxation methods*, and for $\omega > 1$ they are called *over-relaxation methods*. The latter are used to accelerate the convergence of the Gauss–Seidel method and go under the name of *successive over-relaxation (SOR)*. With a few algebraic manipulations, one can show that the components $x_i^{(k)}$ obtained this way are a weighted sum of the results of the Gauss–Seidel iteration in the kth and in the $(k - 1)$th iteration:

$$x_i^{(k)} = (1 - \omega) x_i^{(k-1)} + \frac{\omega}{a_{ii}} \left[b_i - \sum_{j=1}^{i-1} a_{ij} x_j^{(k)} - \sum_{j=i+1}^{n} a_{ij} x_j^{(k-1)} \right] \tag{C.53}$$

$$= \omega x_i^{(k)} + (1 - \omega) x_i^{(k-1)}$$

In matrix notation, the linear system is expressed as:

$$\mathbf{x}^{(k)} = (\mathbf{D} - \omega \mathbf{L})^{-1} \left[(1 - \omega) \mathbf{D} + \omega \mathbf{U} \right] \mathbf{x}^{(k-1)} + \omega \, (\mathbf{D} - \omega \mathbf{L})^{-1} \, \mathbf{b} \tag{C.54}$$

It remains an open question which value for ω gives the best convergence. For the special case of positive-definite and tridiagonal matrices, it can be demonstrated that the optimal ω is given by:

$$\omega = \frac{2}{1 + \sqrt{1 - \rho(\mathbf{T}_j)^2}} \tag{C.55}$$

This indicates that, at least for these matrices, the SOR method is most effective (as compared to the Gauss–Seidel scheme) when $\rho(\mathbf{T}_j)$ is not too small, i.e. when the diagonal elements of $\mathbf{A}$ are not too large. Otherwise, the SOR method is automatically reduced to the Gauss–Seidel method, since $\omega \to 1$.

We close this section by comparing the convergence of the discussed models. To this end, we look at the linear system generated by

$$\mathbf{A} = \begin{pmatrix} 5 & -3 & 2 \\ -3 & 5 & 0 \\ 2 & 0 & 5 \end{pmatrix} \quad \text{and} \quad \mathbf{b} = \begin{pmatrix} 6 \\ 25 \\ -11 \end{pmatrix} \tag{C.56}$$

The solution of the system is given by $\mathbf{x}^{\mathrm{T}} = (9.45, 13.75, -7.7)$. Since $\mathbf{A}$ is tridiagonal and positive-definite (see Theorem 34), the rule Equation (C.55) applies. The spectral radius of the corresponding matrix $\mathbf{T}_j$ is found to be

$$\rho(\mathbf{T}_j) = 13/25 \quad \Rightarrow \quad \omega = 1.078 \tag{C.57}$$

The results for all three methods, as obtained with the starting vector $\mathbf{x}^{(0)} = (0, 0, 0)$, are given in Table C.1.

Table C.1   Results of the three iteration schemes discussed in the text for various iteration steps

	1	5	10	20
	1.2	7.2192	9.0661	9.4354
Jacobi	5	11.384	13.227	13.730
	2.2	−6.2128	−7.444	−7.6903
	1.2	8.773	9.4242	9.4500
Gauss–Seidel	5.72	13.163	13.728	13.750
	−4.488	−7.465	−7.6911	−7.7000
	1.2936	9.1000	9.4450	9.4500
SOR	6.2267	13.481	13.746	13.750
	−5.0566	−7.6015	−7.6985	−7.7000

C.5   NOTES

All material in this section is elementary and can be found in most textbooks on linear algebra. Easy to read, but far beyond the scope required here, is Robbins (1995).

There are numerous numerical techniques to solve linear systems. For an introduction to this topic, see Burden and Faires (1998). Up to the 1990s, FE codes almost universally used Gaussian elimination and some of its variants. Techniques in use today are discussed in Knabner and Angermann (2000) and Grossmann and Roos (1994).

The economist familiar with input–output models is warned that the definition of *diagonal dominance* used here is different from the one used in the context of linear economic models. The usual definition given here goes back to J. Hadamard. For a discussion of differences and relationships, see Takayama (1985).

Appendix D

A Quick Introduction to PDE2D

PDE2D is the current name of an easy-to-use, interactive, general purpose partial differential equation solver, which has been in development for 30 years (Sewell, 1982, 1985, 1993). Since the very beginning (Sewell, 1976), the program has been able to solve PDEs in 2D polygonal regions using a Galerkin finite element method and triangular elements, and for many years now it has been able to solve problems in general 2D regions with arbitrary curved boundaries, with high accuracy, by using up to 4th degree isoparametric elements. The user supplies an initial triangulation with just enough elements to define the region, and the program automatically refines and grades the triangulation to the user's specifications (adaptive refinement is also available). The result is that, for over 20 years, PDE2D and its predecessors have been able to achieve $O(h^5)$ (h = element diameter) accuracy in solving general systems of linear or nonlinear, time-dependent, steady-state and eigenvalue PDEs in completely general 2D regions – and now, even when there is a curved interior interface, along which material properties are discontinuous. By comparison, finite difference schemes and finite element methods with linear elements, achieve only $O(h^2)$ accuracy.

In 1994, PDE2D was extended to handle three-dimensional PDE systems; however, a very different algorithm was chosen for the 3D problem: a collocation finite element method, with tricubic Hermite basis functions. The collocation method has some important advantages over the Galerkin brand with regard to ease of use. In particular, the user does not have to put his/her equations into the 'divergence' form required by the Galerkin method, so not only the PDEs, but also the boundary conditions, are usually more convenient to formulate. However, this algorithm is easily applied only to problems in rectangular 3D regions, so initially, although the 3D solver, like the 2D solver, was designed to solve very general systems of nonlinear and linear time-dependent, steady-state and eigenvalue problems, only problems in rectangular regions could be handled. After the abilities to handle periodic and 'no' boundary conditions were added, it became possible to solve, with high accuracy, $O(h^4)$, problems in a wide range of simple nonrectangular domains, such as spheres, cylinders, tori, pyramids, ellipsoids and cones, by writing the PDEs in terms of an appropriate system of variables with constant limits. However, rewriting the partial differential equations in the new coordinate system is often extremely unpleasant. Recently, most of the pain has been removed from this process: now the user only has to supply the global coordinate transformation equations and can then write his/her PDEs in their usual Cartesian form. Rewriting the equations in terms of the new coordinates is handled automatically. Thus, it is now very easy to solve general 3D problems in a wide range of simple 3D regions.

PDE2D also solves 1D problems, using a collocation method similar to that used by the 3D solver. For 2D and 3D problems, a variety of direct and iterative linear system solvers are available, including some MPI-based parallel solvers.

PDE2D features an interactive user interface – which makes it exceptionally easy to use for 1D, 2D, 3D or even '0D' problems – and extensive graphical output capabilities. All documentation, including 15 prepared examples, is available online through the interactive driver – in fact, there is no user's manual! The user only has to answer a series of interactive questions about the region, partial differential equations and boundary conditions, and select the solution method and graphical output options. The questions are worded to reflect the number and names of the unknowns, the type of system and other information supplied in answers to previous questions. If a single linear steady-state problem is solved, for example, the user will not be bothered with information or options relevant only to time-dependent or nonlinear problems, or systems of several PDEs. Extensive error checking is done during the interactive session, and also at execution time.

Based on the user's answers, PDE2D creates a FORTRAN program which contains calls to precompiled PDE2D library subroutines designed to solve the PDEs. The FORTRAN program thus created is highly readable – most of the interactive questions are repeated in the comments – so the user can make corrections and modifications directly to the FORTRAN program, and does not have to work through a new interactive session each time minor modifications to the problem or solution method are desired. The PDE2D user has all the flexibility of FORTRAN at his/her disposal, for example, any PDE or boundary condition coefficient can be defined by a user-written function subprogram, or a DO loop can be put around the main program so that the program is run multiple times, with different parameters, in a single run. It is also very easy to add calls to user-supplied subroutines to plot or otherwise postprocess the PDE2D solution.

More complete documentation on the algorithms used by PDE2D, and the range of problems it addresses, can be found in Chapter 6 of Sewell (2000) and at the website www.pde2d.com. The website also contains a list of over 150 journal publications in which PDE2D has been used to generate numerical results.

References

Agča, S. and D.M. Chance (2003), "Speed and accuracy comparison of bivariate normal distribution approximations for option pricing," *Journal of Computational Finance*, **6**(4), 61–96.

Ahn, D.-H., S. Figlewski and B. Gao (1999a), "Pricing discrete barrier options with an adaptive mesh model," *Journal of Derivatives*, **6**, 33–43.

Ahn, H., A. Penaud and P. Wilmott (1999b), "Exotic passport options," *Asia-Pacific Financial Markets*, **6**, 171–182.

Ahn, H., A. Penaud and P. Wilmott (1999c), "Various passport options and their valuation," *Applied Mathematical Finance*, **4**, 375–393.

Ames, W.F. (1965), *Nonlinear Partial Differential Equations in Engineering*, Academic Press, New York.

Ames, W.F. (1992), *Numerical Methods for Partial Differential Equations*, Academic Press, Boston.

Andersen, L. and R. Brotherton-Ratcliffe (1997/1998), "The equity option volatility smile: An implicit finite-difference approach," *Journal of Computational Finance*, **1**(2), 5–38.

Angermann, L. and P. Knabner (2000), *Numerik Partieller Differentialgleichungen*, Springer, Berlin.

Apel, T., Winkler, G. and Wystup, U. (2002), "Valuation of options in Heston's stochastic volatility model using finite element methods," in J. Hakala and U. Wystup (eds) *Foreign Exchange Risk – Models, Instruments and Strategies*, Risk Books.

Ashcroft, R.N. (1996), *Asset Pricing with Spectral Methods*, PhD thesis, Stanford University.

Avellaneda, M. and A. Parás (1994) "Dynamic hedging portfolios for derivative securities in the presence of large transaction costs," *Applied Mathematical Finance*, **1**, 165.

Avellaneda, M.M. and R. Buff (1999), "Combinatorial implications of nonlinear uncertain volatility models: The case of barrier options," *Applied Mathematical Finance*, **6**, 1–18.

Avellaneda, M.M., M. Levy and A. Parás (1994), "Pricing and hedging derivative securities in markets with uncertain volatility," *Applied Mathematical Finance*, **1**, 165–194.

Avellaneda, M.M. and A. Parás (1996), "Managing the volatility risk of portfolios of derivative securities: the Lagrangian uncertain volatility model," *Applied Mathematical Finance*, **3**, 21–56.

Axelsson, O. and V.A. Barker (1984), *Finite Element Solution of Boundary Value Problems*, Academic Press, Orlando.

Bailey, P.B., L.F. Shampine and P.E. Waltman (1968), *Nonlinear Two-Point Boundary Value Problems*, Academic Press, New York.

Bakshi, G. and D. Madan (1999), "A unified treatment of average-rate contingent claims with applications," Technical report, Robert H. Smith School of Business, University of Maryland.

Bakstein, D. and P. Wilmott (2002), "Equity dividend models," in P. Wilmott and H. Rasmussen (eds), *New Directions in Mathematical Finance*, John Wiley & Sons, Ltd, Chichester.

Barone-Adesi, G. and R.E. Whalley (1987), "Efficient analytic approximation of American option values," *Journal of Finance*, **42**(2), 301–320.

Beckers, S. (1980), "The constant elasticity of variance model and its implications for option pricing," *Journal of Finance*, **35**, 661–673.

Beckmann, M.J. (1952), "A continuous model of transportation," *Econometrica*, **20**, 642–673.

Bellalah, M., E. Briys, H.M. Mai and F. de Varenne (1998), *Options, Futures and Exotic Derivatives*, John Wiley & Sons, Ltd, Chichester.

Bellomo, N. and L. Preziosi (1995), *Modelling Mathematical Methods and Scientific Computing*, CRC Press, Boca Raton, Florida.

Berger, E. and D. Klein (1998), "Valuing options on dividend-paying stocks," *Bloomberg Markets*, 7(7), 116–120.

Berti, G. (2000), "Insurance optional," *Risk*, 13(4), 87–90.

Bhat, U.N. (1984), *Elements of Applied Stochastic Processes*, John Wiley & Sons, Inc., New York.

Bickford, W.B. (1990), *A First Course in the Finite Element Method*, Irwin, Boston.

Bjoerk, T. (1998), *Arbitrage Theory in Continuous Time*, Oxford University Press, Oxford.

Black, F. and M. Scholes (1972), "The valuation of option contracts and a test of market efficiency," *Journal of Finance*, 27, 399–417.

Black, F. and M. Scholes (1973), "The pricing of options and corporate liabilities," *Journal of Political Economy*, 81, 637–659.

Borouchaki, H. and P.L. George (1998), *Delaunay Triangulation and Meshing: Application to Finite Elements*, Edition HERMES, Paris.

Bos, L.P., A.F. Ware and B.S. Pavlov (2002), "On a semi-spectral method for pricing an option on a mean-reverting asset," *Quantitative Finance*, 2, 337–345.

Boyce, W.E. and R.C. DiPrima (1992), *Elementary Differential Equations and Boundary Value Problems*, 5th edition, John Wiley & Sons, Inc., New York, Chapter 10.

Boyle, P.P. and H.S. Lau (1994), "Bumping up against the barrier with the binomial method," *Journal of Derivatives*, 1(4), 6–14.

Brennan, M. and E. Schwartz (1977), "Convertible bonds: Valuation and optimal strategies for call and conversion," *Journal of Finance*, 32, 1699–1715.

Brennan, M. and E. Schwartz (1978), "Finite difference methods and jump processes arising in the pricing of contingent claims," *Journal of Financial and Quantitative Analysis*, 9, 461–474.

Bronson, R. (1993), *Differential Equations*, 2nd edition, McGraw-Hill, New York.

Buff, R. (2000), "Worst-case scenarios for American options," *International Journal of Theoretical and Applied Finance*, 3, 25–58.

Buff, R. (2002a), Personal communication.

Buff, R. (2002b), *Uncertain Volatility Models – Theory and Applications*, Springer, Berlin.

Burden, R. and J.D. Faires (1998), *Numerical Methods*, 2nd edition, Brooks/Cole Publishing Company, Pacific Grove.

Burnett, D.S. (1987), *Finite Element Analysis – From Concepts to Applications*, Addison-Wesley, Reading, USA.

Carr, P. (1995), "Two extensions to barrier option valuation," *Applied Mathematical Finance*, 2, 173–209.

Chan, S.-S. (1999), "The valuation of American passport options," Discussion paper, University of Wisconsin, School of Business, Madison.

Chawla, S. (1999), "A minmax problem for parabolic systems with competitive interactions," *Electronic Journal of Differential Equations*, 1999(50), 1–18.

Chiang, A.C. (1992), *Elements of Dynamic Optimization*, McGraw-Hill, New York.

Clarke, N. and K. Parrott (1999), "Multigrid for American option pricing with stochastic volatility," *Applied Mathematical Finance*, 6, 177–195.

Clewlow, L. and C. Strickland (1998), *Implementing Derivatives Models*, John Wiley & Sons, Ltd, Chichester.

Collatz, L. (1966), *The Numerical Treatment of Differential Equations*, Springer, Berlin.

Courant, R. and D. Hilbert (1962), *Methods of Mathematical Physics*, volume II, John Wiley & Sons, Inc., New York.

Cox, D.R. and H.D. Miller (1965), *The Theory of Stochastic Processes*, Chapman & Hall, London.

Cox, J.C., J. Ingersoll and S.A. Ross (1985), "A theory of the term structure of interest rates," *Econometrica*, 53, 385–467.

Cox, J.C. and S.A. Ross (1976), "The valuation of options for alternative stochastic processes," *Journal of Financial Economics*, 3, 145–166.

Crandall, S.H. (1956), *Engineering Analysis*, McGraw-Hill, New York.

Crank, J. (1984), *Free and Moving Boundary Problems*, Clarendon, Oxford.

de Matos Pimentão, L.N. (1986), *Anwendungen der Variationsrechnung auf makro-ökonomische Modelle*, Springer, Berlin, Section 2.5.

Dembo, R. and I. Nelken (1991), "Share the load," *Risk*, **4**(4).

Deutsch, H.-P. (2002), *Derivatives and Internal Models*, 2nd edition, Palgrave, Chippenham, Wiltshire, UK.

d'Halluin, Y., P.A. Forsyth, K.R. Vetzal and G. Labahn (2001), "A numerical pde approach for pricing callable bonds," *Applied Mathematical Finance*, **8**, 49–77.

Dixit, A.K. and R.S. Pindyck (1994), *Investment under Uncertainty*, Princeton University Press, Princeton, USA.

Docker, J., V. Protopopescu and R.T. Santoro (1989), "Combat modelling with partial differential equations," *European Journal of Operational Research*, **38**, 178–183.

Drezner, Z. (1978), "Computation of the bivariate normal integral," *Mathematics of Computation*, **32**, 277–279.

DuChateau, P. and D.W. Zachmann (1986), *Partial Differential Equations*, McGraw-Hill, New York.

Duffie, D. (1996), *Dynamic Asset Pricing Theory*, 2nd edition, Princeton University Press, Princeton, USA.

Duffie, D. and M. Huang (1996), "Swap rates and credit quality," *Journal of Finance*, **LI**(2), 921–949.

Duffie, D. and R. Kan (1996), "A yield factor model of interest rates," *Mathematical Finance*, **6**, 379–406.

Dumas, B., J. Fleming and R.E. Whaley (1998), "Implied volatility functions: Empirical tests," *Journal of Finance*, **53**, 2059–2106.

Eisenberg, L.K. and R.A. Jarrow, (1991), "Option pricing with random volatilities in complete markets," *Review of Quantitative Finance and Accounting*, **4**, 5–17.

Embrecht, P. (2000), "Actuarial versus financial pricing of insurance," *Journal of Risk Finance*, **1**(4), 17–25.

Eriksson, K., D. Estep, P. Hansbo and C. Johnson (1996), *Computational Differential Equations*, Cambridge University Press, Lund, Sweden.

Evans, D.J. (1997), *Group Explicit Methods for the Numerical Solution of Partial Differential Equations*, Gordon and Breach, Amsterdam.

Evans, G.C. (1924), "The dynamics of monopoly," *American Mathematical Monthly*, **2**, 77–83.

Evans, G.C. (1930), *Mathematical Introduction to Economics*, McGraw-Hill, New York.

Farlow, S.J. (1993), *Partial Differential Equations for Scientists and Engineers*, Dover Publications, New York, Chapter 4.

Ferguson, B.S. and G.C. Lim (1998), *Introduction to Dynamic Economic Models*, Manchester University Press, Manchester.

Fleming, W.H. and R.W. Rishel (1975), *Deterministic and Stochastic Optimal Control*, Springer, New York.

Fleming, W.H. and H.M. Soner (1993), *Controlled Markov Processes and Viscosity Solutions*, Springer, New York.

Forsyth, P.A., K.R. Vetzal and R. Zvan (2001), "A finite volume approach for contingent claims valuation," *IMA Journal of Numerical Analysis*, **21**, 703–731.

Forsyth, P.A., K.R. Vetzal and R. Zvan (2002), "Convergence of numerical methods for valuing path-dependent options using interpolation," *Review of Derivatives Research*, **5**, 273–314.

Fu, M., D. Madan and T. Wang (1998–1999), "Pricing continuous Asian options: A comparison of Monte Carlo and Laplace transform inversion methods," *Journal of Computational Finance*, **2**(2), 49–74.

Gemmill, G. (1993), *Option Pricing–An International Perspective*, McGraw-Hill, London.

Gentle, D. (1995), "Basket weaving," in R. Jarrow (ed.) *Over the Rainbow*, Risk Publications, London, pp. 143–146.

Givoli, D. (1992), *Numerical Methods for Problems in Infinite Domains*, Elsevier, Amsterdam.

Grossmann, C. and H.-G. Roos (1994), *Numerik Partieller Differentialgleichungen*, Teubner, Stuttgart.

Groth, C. and G. Müller (1997), *FEM für Praktiker*, 3rd edition, Expert-Verlag, Renningen, Germany.

Gürtler, M. (1998), *Lebesguesche Optionspreistheorie: Ein Modell zur Bewertung pfadabhängiger Derivate*, PhD thesis, University of Bonn, Germany.

Hajivssiliou, V.A. (1993), "Simulation estimation methods for limited dependent variable models," in G.S. Maddala, R. Rao and H.D. Vinod (eds), *The Handbook of Statistics*, volume XI, Elsevier, Amsterdam, pp. 519–543.

Hajivssiliou, V.A. and P.A. Ruud (1993), "Classical estimation methods for LDV models using simulation," in R.F. Engle and D.L. McFadden (eds), *The Handbook of Econometrics*, volume IV, Elsevier, Amsterdam, pp. 2383–2441.

Hakala, J. (2002), "Handling differing expiry and delivery dates," in J. Hakala and U. Wystup (eds), *Foreign Exchange Risk*, Risk Publications, London, pp. 25–26.

Haug, E.G. (1997), *The Complete Guide to Option Pricing Formulas*, McGraw-Hill, New York.

Haug, E.G. (2001), "Closed-form valuation of American barrier options," *International Journal of Theoretical and Applied Finance*, **4**(2), 355–359.

Haug, E.G., J. Haug and A. Lewis (2003), "Back to basics: A new approach to the discrete dividend problem," *Wilmott*, September, pp. 37–47.

Hearst, B. (2001), "Gold: A special commodity," *Wilmott*, **1**(1), 25.

Heath, D. and E. Platen (2002), "A variance reduction technique based on integral representations," *Quantitative Finance*, **2**, 362–369.

Heath, M.T. (2002), *Scientific Computing – An Introductory Survey*, 2nd edition, McGraw-Hill, Boston.

Helton, J.W. and M.R. James (1999), *Extending H∞ Control to Nonlinear Systems*, SIAM, Philadelphia.

Henderson, V., D. Hobson and G. Kentwell (2001), "A new class of commodity hedging strategies: A passport options approach," FORC Preprint 113, Warwick Business School.

Heston, S.L. (1993), "A closed-form solution for options with stochastic volatility with applications to bond and currency options," *Review of Financial Studies*, **6**, 327–343.

Heuser, H. (1995), *Gewöhnliche Differentialgleichungen*, 3rd edition, Teubner, Stuttgart.

Hibbitt, Karlsson & Sorensen (1994), *ABAQUS Theory Manual*, edition 5.4, Hibbitt, Karlsson & Sorensen, Inc.

Hofbauer, J. and K. Sigmund (2003), "Evolutionary game dynamics," *Bulletin (New Series) of the American Mathematical Society*, **40**(4), 479–519.

Hoggard, T., A.E. Whalley and P. Wilmott (1994), "Hedging option portfolios in the presence of transaction costs," *Advances in Futures and Options Research*, **7**, 21–35.

Höhn, W. (1982), "Finite elements for parabolic equations backwards in time," *Numerische Mathematik*, **40**, 207–227.

Hritonenko, N. and Y. Yatsenko (1999), *Mathematical Modeling in Economics, Ecology and the Environment*, Kluwer Academic Publishers, Dordrecht.

Hsieh, P.-F. and Y. Sibuya (1999), *Basic Theory of Ordinary Differential Equations*, Springer, New York.

Huge, B. and D. Lando (1999), "The swap pricing model with two-sided default risk in a rating based model," *European Finance Review*, **3**, 239–268.

Hull, J. (2000), *Options, Futures and Other Derivatives*, 4th edition, Prentice Hall, London.

Hull, J. and A. White (1987), "The pricing of options with stochastic volatility," *Journal of Finance*, **42**, 281–300.

Hull, J. and A. White (1990), "Pricing interest-rate-derivative securities," *The Review of Financial Studies*, **3**, 573–592.

Hunziker, J.P. and P. Koch-Medina (1996), "Two-color rainbow options," in I. Nelken (ed.), *The Handbook of Exotic Options*, Irwin, pp. 143–174.

Huynh, C.B. (1994), "Back to baskets," *Risk*, **5**(7), 59–61.

Hyer, T., A. Lipton-Lifschitz and D. Pugachevsky (1997), "Passport to success," *Risk*, **10**(9), 127–131.

Ikeda, M. and N. Kunitomo (1992), "Pricing options with curved boundaries," *Mathematical Finance*, **2**, 275–298.

Izvorski, I. (1998), "A nonuniform grid method for solving pdes," *Journal of Economic Dynamics and Control*, **22**, 1445–1452.

Jackson, N. and E. Süli (1997), "Adaptive finite element solution of 1D option pricing problems," Report 97/05, Oxford University Computing Laboratory, Oxford.

James, J. and N. Webber (2000), *Interest Rate Models*, John Wiley & Sons, Ltd, Chichester.

Jarrow R., (ed.) (1998), *Volatility – New Estimation Techniques for Pricing Derivatives*, Risk Publications, London.

Jarrow, R. and A. Rudd (1983), *Option Pricing*, Irwin, Homewood, Illinois, USA.

Jarrow, R. and S. Turnbull (1996), *Derivative Securities*, South-Western College Publishers, Cincinatti.

Jiang, B. (1998), *The Least-Squares Finite Element Method*, Springer, Berlin.

Johnson, N.L., S. Kotz and N. Balakrishan (1994), *Continuous Univariate Distributions*, 2nd edition, volume 1, John Wiley & Sons, Inc., New York.

Johnson, N.L., S. Kotz and N. Balakrishan (1995), *Continuous Univariate Distributions*, 2nd edition, volume 2, John Wiley & Sons, Inc., New York.

Johnson, N.L., S. Kotz and N. Balakrishan (2000), *Continuous Multivariate Distributions*, 2nd edition, volume 1, John Wiley & Sons, Inc., New York.

Judd, K.L. (1998), *Numerical Methods in Economics*, MIT Press, Cambridge, USA.

Kaliakin, V.N. (2002), *Introduction to Approximate Solution Techniques, Numerical Modeling, and Finite Element Methods*, Marcel Dekker, Inc., New York, Basel.

Kangro, R. (1997), *Analysis of Artificial Boundary Conditions for Black–Scholes Equations*, PhD thesis, Carnegie Mellon University.

Kantorovich, L.V. and V.I. Krylov (1964), *Approximate Methods of Higher Analysis*, 3rd edition, Interscience Publishers, Inc., New York, Translated by C.D. Benster.

Knabner, P. and L. Angermann (2000), *Numerik partieller Differentialgleichungen*, Springer, Berlin.

Knight, F.H. (1921), *Risk, Uncertainty and Profit*, Riverside Press, Cambridge.

Knudsen, T.S., B. Meister and M. Zervos (1999), "On the relationship of the dynamic programming approach and the contingent claim approach to asset valuation," *Finance and Stochastics*, **3**, 433–449.

Korn, R. and P. Wilmott (1996), "Option prices and subjective beliefs. Bericht zur Stochastik und Verwandten Gebieten 96-5," Department of Mathematics, University of Mainz.

Korn, R. and P. Wilmott (1998), "A general framework for hedging and speculating with options," *International Journal of Theoretical and Applied Finance*, 4(1), 507–522.

Kreiss, H.-O. and J. Lorenz (1989), *Initial-Boundary Value Problems and the Navier–Stokes Equations*, Academic Press, Boston.

Krekel, M., J. de Kock, R. Korn and T.-K. Man (2004), "An analysis of pricing methods for basket options," *Wilmott*, July, pp. 82–89.

Kunitomo, N. and M. Ikeda (1992), "Pricing options with curved boundaries," *Mathematical Finance*, 4(2), 272–298.

Kunkel, P. and O. von dem Hagen (2000), "Numerical solution of infinite-horizon optimal-control problems," *Computational Economics*, **16**, 189–205.

Kwok, Y.K. (1998), *Mathematical Models of Financial Derivatives*, Springer, Singapore.

Lai, Y.-L., A. Hadjidimos, E.N. Houstis and J.R. Rice (1994a), "The analysis of iterative elliptic PDE solvers based on the cubic hermite collocation discretization," Technical Report CSD-TR-94-036, Computer Sciences, Purdue University.

Lai, Y.-L., A. Hadjidimos, E.N. Houstis and J.R. Rice (1994b), "General interior hermite collocation methods for second order elliptic partial differential equations," Technical Report CSD-TR-94-004, Computer Sciences, Purdue University.

Lancaster, F.W. (1914), "Aircraft in warfare," *Engineering*, **98**, 422–423.

Lapidus, L. and G.F. Pinder (1981), *Numerical Solution of Partial Differential Equations in Science and Engineering*, John Wiley & Sons, Inc., New York.

Leschhorn, H. and C. Eibl (2004), "Options–Volatilität unter der Lupe". *Die Bank*, **2**, 104–107.

Linetsky, V. (1999), "Step options". *Mathematical Finance*, **9**, 55–96.

Linetsky, V. (2002), "Exotic spectra," *Risk*, **15**(4), 85–89.

Lions, J.L. (1961), *Optimal Control of Systems Governed by Partial Differential Equations*, Springer, Berlin.

Lipton, A. (2001), *Mathematical Models for Foreign Exchange – A Financial Engineer's Approach*, World Scientific, Singapore.

Lipton, A. (2002), "On the log–log linearity of the size distribution for growth stocks," *Risk*, **15**(7), 70.

Liseikin, V.D. (1999), *Grid Generation Methods*, Springer, Berlin.

Lyons, T.J. (1995), "Uncertain volatility and the riskfree synthesis of derivatives," *Applied Mathematical Finance*, **2**, 117–133.

Macbeth, J.D. and L.J. Merville (1980), "Test of the Black–Scholes and Cox call option valuation models," *Journal of Finance*, **35**, 285–303.

Mandl, P. (1968), *Analytical Treatment of Stochastic Processes*, Springer, Berlin.

Maz'ya, V. and T. Shaposhnikova (1998), *Jacques Hadamard – A Universal Mathematician*, American Mathematical Society, Providence.

McGrattan, E.R. (1996), "Solving the stochastic growth model with a finite element method," *Journal of Economic Dynamics and Control*, **20**, 19–42.

McIver, J.L. (1996), "Overview of modeling techniques," in I. Nelken (ed.), *The Handbook of Exotic Options*, Irwin, Chicago, pp. 273–292.

Meis, T. and U. Marcowitz (1981), *Numerical Solution of Partial Differential Equations*. Springer, Berlin.

Merton, R.C. (1973), "Theory of rational options pricing," *Bell Journal of Economics and Management Science*, **4**, 141–183 (reprinted in Merton, 1998).

Merton, R.C. (1998), *Continuous-Time Finance*, Blackwell, Malden, Massachussetts.

Meyer, G.H. and J. van der Hoek (1979), "The valuation of American options with the method of lines," *Advances in Futures and Options Research*, **9**, 265–285.

Milevsky, M.A. and S.E. Posner (1998), "A closed-form approximation for valuing basket options," *Journal of Derivatives*, **5**(4), 54–61.

Mitchell, A.R. and D.F. Griffiths (1980), *The Finite Difference Method in Partial Differential Equations*, John Wiley & Sons, Ltd, Chichester.

Morton, K.W. (1996), *Numerical Solution of Convection–Diffusion Problems*, Chapman & Hall, London.

Morton, K.W. and D.F. Mayer (1994), *Numerical Solution of Partial Differential Equations*, Cambridge University Press, Cambridge, UK.

Musiela, M. and M. Rutkowski (1997), *Martingale Methods in Financial Modelling*, Springer, Berlin.

Nielsen, B.F., O. Skavhaug and A. Tveito (2000), "Penalty methods for the numerical solution of American multi-asset option problems," Technical report, Department of Informatics, University of Oslo.

Nielsen, B.F., O. Skavhaug and A. Tveito (2002), "Penalty and front-fixing methods for the numerical solution of American option problems," *Journal of Computational Finance*, **5**(4), 69–97.

Oden, J.T. and J.N. Reddy (1982), *An Introduction to the Mathematical Theory of Finite Elements*, John Wiley & Sons, Inc., New York.

Oosterlee, C., A. Schüller and U. Trottenberg (2001), *Multigrid*, Academic Press, London.

Ouachani, N. and Y. Zhang (2004), "Pricing cross-currency convertible bonds with PDE," *Wilmott*, January, pp. 54–61.

Overhaus, M., A. Ferraris, T. Knudsen, R. Milward, L. Nguyen-Ngoc and G. Schindlmayr (2002), *Equity Derivatives: Theory and Applications*, John Wiley & Sons, Inc., New York.

Oztukel, A. and P. Wilmott (1998), "Uncertain parameters, an empirical stochastic volatility model and confidence limits," *International Journal of Theoretical and Applied Finance*, **1**(1), 175–190.

Pelsser, A. (2000), "Pricing double barrier options using analytical inversion of Laplace transforms," *Journal of Finance and Stochastics*, **4**(1), 95–104.

Pelsser, A. and T. Vorst (1994), "The binomial model and the Greeks," *Journal of Derivatives*, **1**, 45–49.

Penaud, A. (2002), "Passport options: A review," in P. Wilmott and H. Rasmussen (eds), *New Directions in Mathematical Finance*, John Wiley & Sons, Ltd, Chichester, pp. 33–53.

Petrovski, G.I. (1954), *Lectures on Partial Differential Equations*, Interscience Publications, New York.

Phillips, A.W. (1954), "Stabilization policy in a closed economy," *Economic Journal*, **64**, 290–323.

Pooley, D.M., P.A. Forsyth, K.R. Vetzal and R.B. Simpson, (2000), "Unstructured meshing for two asset barrier options," *Applied Mathematical Finance*, **7**(33–60).

Pooley, D.M., P.A. Forsyth and K.R. Vetzal, (2003a), "Numerical convergence properties of option pricing PDEs with uncertain parameters," *IMA Journal of Numerical Analysis*, **23**, 241–267.

Pooley, D.M., K.R. Vetzal and P.A. Forsyth (2003b), "Convergence remedies for non-smooth payoffs in option pricing," *Journal of Computational Finance*, **6**(4), 25–38.

Prenter, P.M. (1975), *Splines and Variational Methods*, John Wiley & Sons, Inc., New York.

Press, W.H., S.A. Teukolsky, W.T. Vetterling and B.P. Flannery (1992), *Numerical Recipes in FORTRAN – The Art of Scientific Computing*, 2nd edition, Cambridge University Press, Cambridge.

Protopopescu, V. and R.T. Santoro (1989), "Combat modeling with partial differential equations," *European Journal of Optimisation Research*, **38**, 178–183.

Puu, T. (1997a), *Mathematical Location and Land Use Theory*, Springer, Berlin.

Puu, T. (1997b), *Nonlinear Economic Dynamics*, 4th edition, Springer, Berlin.

Randall, C. and D. Tavella (2000), *Pricing Financial Instruments: The Finite Difference Method*, John Wiley & Sons, Inc., New York.

Rannacher, R. (1984), "Finite element solutions of diffusion problems with irregular data," *Numerische Mathematik*, **43**, 389–410.

Rebonato, R. (1998), *Interest-Rate Option Models*, 2nd edition, John Wiley & Sons, Ltd, Chichester.

Rebonato, R. (1999), *Volatility and Correlation*, 2nd edition, John Wiley & Sons, Ltd, Chichester.

Reid, W.T. (1972), *Riccati Differential Equations*, Academic Press, New York.

Reimer, M. (1997), *Examining Binomial Option Price Approximations*, PhD thesis, University of Bonn, Germany.

Reiner, E. and M. Rubinstein (1991), "Breaking down the barriers," *Risk*, **4**(8), 28–35.

Reiss, O. and U. Wystup (2001), "Computing option price sensitivities using homogeneity and other tricks," *Journal of Derivatives*, **9**(2), 41–53.

Ritchken, P. (1996), *Derivatives Markets*, Harper Collins College Publishers, New York.

Robbins, J.W. (1995), *Matrix Algebra*, A.K. Peters, Wellesley, Massachusetts.

Rogers, L.C.G. and Z. Shi (1995), "The value of an Asian option," *Journal of Applied Probability*, **32**, 1077–1088.

Rogers, L.C.G. and E.J. Stapleton (1998), "Fast accurate binomial pricing," *Finance and Stochastics*, **2**, 3–17.

Rosu, I. and D. Stroock (2004), "On the derivation of the Black–Scholes formula," *Séminaire de Probabilités*, **37**, 399–414.

Rubinstein, M. (1991a), "Breaking down barriers," *Risk*, **4**(8), 28–35.

Rubinstein, M. (1991b), "Somewhere over the rainbow," *Risk*, **4**(11), 63–66.

Rudolf, M. (2000), *Zinsstrukturmodelle*, Physika, Heidelberg.

Samuelson, P.A. (1965), "Rational theory of warrant pricing," *Industrial Management Review*, **6**, 13–32.

Sawyer, K.R. (1993), "Continuous time financial models: Statistical applications of stochastic processes," in G.S. Maddala, C.R. Rao and H.D. Vinod (eds), *Handbook of Statistics*, volume 11, North-Holland, Amsterdam.

Schroder, M. (1989), "Computing the constant elasticity of variance option pricing formula," *Journal of Finance*, **44**, 211–219.

Schuss, Z. (1980), *Theory and Applications of Stochastic Differential Equations*, John Wiley & Sons, Ltd, Chichester.

SciComp (1999), *SciFinance Newsletter*, 4. (www.scicomp.com).

Sewell, G. (1976), "An adaptive computer program for the solution of $\div(p(x, y)\mathrm{grad}u) = f(x, y, u)$ on a polygonal region," in J.R. Whiteman (ed.), *The Mathematics of Finite Elements and Applications II*, Academic Press, pp. 543–553.

Sewell, G. (1982), "TWODEPEP, a small general-purpose finite element program," *Angewandte Informatik*, **4**, 249–253.

Sewell, G. (1985), *Analysis of a Finite Element Method: PDE/PROTRAN*, Springer, Berlin.

Sewell, G. (1993), "PDE2D: Easy-to-use software for general two-dimensional partial differential equations," *Advances in Engineering Software*, **17**(2), 105–112.

Sewell, G. (2005), The Numerical Solution of Ordinary and Partial Differential Equations, second edition, John Wiley & Sons, New York.

Seydel, R. (2002), *Tools for Computational Finance*, Springer, Berlin.

Shadwick, B.A. and W.F. Shadwick (2002), "On the computation of option prices and sensitivities in the Black–Scholes–Merton model," *Quantitative Finance*, **2**, 158–166.

Shevchenko, P.V. (2003), "Addressing the bias in Monte Carlo pricing of multi-asset options with multiple barriers through discrete sampling," *Journal of Computational Finance*, **6**(3).

Shreve, S.E. and J. Večeř (2000a), "Options on a trading account: Vacations calls, vacations puts and passport options," *Journal of Finance and Stochastics*, **4**(3), 255–274.

Shreve, S.E. and J. Večeř, (2000b), "Upgrading your passport," *Risk*, **13**(7), 81–83.

Sluzalec, A. (1992), *Introduction to Nonlinear Thermomechanics*, Springer, London.

Stein, E.M. and J.C. Stein (1991), "Stock price distributions with stochastic volatility: An analytic approach," *Review of Financial Studies*, **4**(4) (reprinted in Jarrow, 1998).

Strang, G. and G. Fix (1973), *An Analysis of the Finite Element Method*, Prentice Hall, Englewood Cliffs, New Jersey.

Strogatz, S.H. (1994), *Nonlinear Dynamics and Chaos: With Applications in Physics, Biology, Chemistry, and Engineering*, Addison-Wesley, Reading, Massachusetts.

Swokowski, E.W. (1979), *Calculus with Analytic Geometry*, 2nd edition, Prindle, Weber & Schmidt, Boston.

Takayama, A. (1985), *Mathematical Economics*, 2nd edition, Cambridge University Press, Cambridge, UK.

Taksar, T. (1997), *Analytical Approximate Solutions for the Pricing of American Options*, PhD thesis, State University of New York, Stony Brook.

Tarkianinen, R. and M. Tuomala (1999), "Optimal nonlinear income taxation with a two-dimensional population: A computational approach," *Computational Economics*, **13**(1), 1–16.

Tavella, D. (2001), "Barrier monitoring," *Risk*, **14**(8), 36.

Tintner, G. (1937), "Monopoly over time," *Econometrica*, **7**, 160–170.

Tomas, M.J. (1996), *Finite Element Analysis and Option Pricing*, PhD thesis, Graduate School of Business, Syracuse University.

Topper, J. (1999), "Finite element modelling of exotic options," Technical report, Department of Economics, University of Hannover.

Topper, J. (2001a), "Eine Finite Elemente Implementierung von Passport Optionen," in W. Gruber (ed.), *Gesamtbanksteuerung: Integration von Markt- und Kreditrisiken*, Schaeffer-Poeschl, Stuttgart.

Topper, J. (2001b), "Uncertain parameters and reverse convertibles," *Risk*, **14**(1), S4–S7.

Turnbull, S.M. and L.M. Wakemann (1991), "A quick algorithm for pricing European average options," *Journal of Financial and Quantitative Analysis*, **26**, 377–389.

Vasicek, O.A. (1977), "An equilibrium characterization of the term structure," *Journal of Financial Economics*, **5**, 177–188.

Večeř, J. (2001), "A new PDE approach for pricing arithmetic average Asian options," *Journal of Computational Finance*, **4**(4), 105–113.

Večeř, J. (2002), "Unified pricing of Asian options," *Risk*, **15**(6), 113–116.

Villeneuve, S. (1999a), "Exercise regions of American options on several assets," *Finance and Stochastics*, **3**, 295–322.

Villeneuve, S. (1999b), *Options Américaines dans un Modèle de Black–Scholes Multidimensionnel*, PhD thesis, Mathématiques Appliquées, L'Université de Marne la Vallée.

Wan, F.Y.M. (1995), *Introduction to the Calculus of Variations and its Applications*, Chapman & Hall, New York.

Weinberger, H.F. and M.H. Protter (1967), *Maximum Principles in Differential Equations*, Prentice Hall, Englewood Cliffs, New Jersey.

White, R.E. (1985), *An Introduction to the Finite Element Method with Applications to Nonlinear Problems*, John Wiley & Sons, Inc., New York.

Wiggins, J.B. (1987), "Options values under stochastic volatility," *Journal of Financial Economics*, **19**, 351–372.

Wilmott, P. (1998), *Derivatives*, John Wiley & Sons, Ltd, Chichester.

Wilmott, P. (2000), *Paul Wilmott on Quantitative Finance*, John Wiley & Sons, Ltd, Chichester.

Wilmott, P., J. Dewynne and S. Howison (1996), *Option Pricing – Mathematical Models and Computation*, Oxford Financial Press, Oxford.

Wolf, J.P. and C. Song (1996), *Finite Element Modelling of Unbounded Media*, John Wiley & Sons, Ltd, Chichester.

Zhang, F. (1998a), "Spectral and pseudospectral approximations in time for parabolic equations," *Journal of Computational Mathematics*, **16**(2), 107–120.

Zhang, J.E. (1999), "Arithmetic Asian options with continuous sampling," Technical report, Department of Economics and Finance, University of Hong Kong.

Zhang, P.G. (1998b), *Exotic Options*, 2nd edition, World Scientific, Singapore.

Zvan, R., P.A. Forsyth and K.R. Vetzal (1997/1998c), "Robust numerical methods for PDE models of Asian options," *Journal of Computational Finance*, **1**(2), 39–78.

Zvan, R., P.A. Forsyth and K.R. Vetzal (1998a), "A general finite element approach for PDE option pricing models," Technical report, Department of Computer Science, University of Waterloo.

Zvan, R., P.A. Forsyth and K.R. Vetzal (1998b), "Penalty methods for American options under stochastic volatility," *Journal of Computational and Applied Mathematics*, **91**, 199–218.

Zvan, R., P.A. Forsyth and K.R. Vetzal (2000), "PDE methods for pricing barrier options," *Journal of Economic Dynamics and Control*, **24**, 1563–1590.

Index

Index compiled by Terry Halliday